Artificial Breeding of Non-Domestic Animals

SYMPOSIA OF THE ZOOLOGICAL SOCIETY OF LONDON
NUMBER 43

Artificial Breeding of Non-Domestic Animals

*(The Proceedings of a Symposium held at
The Zoological Society of London
on 7 and 8 September 1977)*

Edited by

P. F. WATSON

*Department of Physiology, The Royal Veterinary College,
Royal College Street, London, England*

Published for

THE ZOOLOGICAL SOCIETY OF LONDON

BY

ACADEMIC PRESS

1978

ACADEMIC PRESS INC. (LONDON) LTD
24/28 Oval Road, London NW1 7DX

United States Edition published by
ACADEMIC PRESS INC.
111 Fifth Avenue, New York, New York 1003

Library of Congress Catalog Card Number: 74-5683
ISBN: 0-12-613343-3

Printed in Great Britain by
J. W. ARROWSMITH LTD, BRISTOL BS3 2NT

Contributors

AAMDAL, J., *Department of Reproductive Physiology and Pathology, Veterinary College of Norway, Oslo, Norway* (p. 241)

AUSTIN, C. R., *Physiological Laboratory, University of Cambridge, Downing Street, Cambridge CB2 3EG, England* (p. 1)

BOYD, L., *Department of Zoology, Washington State University, Pullman, Washington 99164, USA* (p. 73)

CAIN, J. R., *Poultry Science Department, Texas Agricultural Experiment Station, College Station, Texas 77843, USA* (p. 81)

D'SOUZA, F., *International Disaster Institute, 85 Marylebone High Street, London W1, England* (p. 175)

DUKELOW, W. R., *Endocrine Research Unit, Michigan State University, East Lansing, Michigan 48824, USA* (p. 195)

EVENSEN, B. K., *Department of Animal Science, University of Minnesota, St. Paul, Minnesota 55108, USA* (p. 153)

FOUGNER, J. A., *Department of Reproductive Physiology and Pathology, Veterinary College of Norway, Oslo, Norway* (p. 241)

GEE, G. F., *Patuxent Wildlife Research Center, Laurel, Maryland 20811, USA* (pp. 51, 89)

GOULD, K. G., *Yerkes Regional Primate Research Center, Emory University, Atlanta, Georgia 30322, USA* (p. 249)

GRAHAM, C. E., *Yerkes Regional Primate Research Center, Emory University, Atlanta, Georgia 30322, USA* (p. 249)

GRAHAM, E. F., *Department of Animal Science, University of Minnesota, St Paul, Minnesota 55108, USA* (p. 153)

HENDRICKX, A. G., *California Primate Research Center, University of California, Davis, California 95616, USA* (p. 219)

HESS, D. L., *California Primate Research Center, University of California, Davis, California 95616, USA* (p. 219)

JACZEWSKI, Z., *Institute of Genetics and Animal Breeding, Polish Academy of Sciences, Popielno, 12–222 Wejsuny, Poland* (p. 271)

JONES, D. M., *The Zoological Society of London, Regent's Park, London NW1 4RY, England* (p. 317)

JONES, R. C., *Department of Biological Sciences, University of Newcastle, NSW 2308, Australia* (p. 261)

KRZYWIŃSKI, A., *Institute of Genetics and Animal Breeding, Polish Academy of Sciences, Popielno, 12–222 Wejsuny, Poland* (p. 271)

LAKE, P. E., *ARC Poultry Research Centre, King's Buildings, West Mains Road, Edinburgh EH9 3JS, Scotland* (p. 31)

MARTIN, D. E., *College of Allied Health Sciences, Georgia State University, Atlanta, Georgia 30303, USA* (p. 249)

MARTIN, I. C. A., *Department of Veterinary Physiology, University of Sydney, Sydney, NSW 2006, Australia* (p. 127)

†MURTON, R. K., *Monks Wood Experimental Station, Abbots Ripton, Huntingdon PE17 2LS, England* (p. 7)

NELSON, D. S., *Department of Animal Science, University of Minnesota, St Paul, Minnesota 55108, USA* (p. 153)

NYBERG, K., *Department of Reproductive Physiology and Pathology, Veterinary College of Norway, Oslo, Norway* (p. 241)

PLATZ, C., *Institute of Comparative Medicine, Baylor College of Medicine/Texas A and M University, Houston, Texas 77030, USA* (p. 207)

POLGE, C., *ARC Institute of Animal Physiology, Animal Research Station, 307 Huntingdon Road, Cambridge, England* (p. 303)

PRAHALADA, S., *California Primate Research Center, University of California, Davis, California 95616, USA* (p. 219)

RODGER, J. C., *Department of Veterinary Anatomy, University of Queensland, St Lucia, Queensland 4067, Australia* (p. 289)

SCHMEHL, M. K. L., *Department of Animal Science, University of Minnesota, St Paul, Minnesota 55108, USA* (p. 153)

SEAGER, S. W. J., *Institute of Comparative Medicine, Baylor College of Medicine/Texas A and M University, Houston, Texas 77030, USA* (p. 207)

SEXTON, T. J., *Agricultural Research Service, Avian Physiology Laboratory, Beltsville, Maryland 20705, USA* (p. 89)

SKINNER, J. D., *Mammal Research Institute, University of Pretoria, Pretoria, South Africa* (p. 339)

SMITH, G. F., *Milk Marketing Board, Thames Ditton, Surrey KT7 0EL, England* (p. 329)

TEMPLE, S. A., *Department of Wildlife Ecology, University of Wisconsin, Madison, Wisconsin 53706, USA* (p. 51)

THOMPSON, R. S., *California Primate Research Center, University of California, Davis, California 95616, USA* (p. 219)

WATSON, P. F., *Department of Physiology, The Royal Veterinary College, Royal College Street, London NW1 0TU, England* (p. 97)

WHITE, I. G., *Department of Veterinary Physiology, University of Sydney, NSW 2006, Australia* (p. 289)

WILDT, D., *Institute of Comparative Medicine, Baylor College of Medicine/Texas A and M University, Houston, Texas 77030, USA* (p. 207)

† Deceased 12 June 1978.

Organizer and Chairman

ORGANIZER

P. F. WATSON, on behalf of The Zoological Society of London

CHAIRMEN OF SESSIONS

H. M. DOTT, ARC *Institute of Animal Physiology, Animal Research Station, 307 Huntingdon Road, Cambridge, England*

E. F. GRAHAM, *Department of Animal Science, University of Minnesota, St Paul, Minnesota 55108, USA*

R. D. MARTIN, *Department of Anthropology, University College, Gower Street, London WC1, England*

P. J. S. OLNEY, *Zoological Society of London, Regent's Park, London NW1 4RY, England*

S. W. J. SEAGER, *Institute of Comparative Medicine, Baylor College of Medicine/Texas A and M University, Houston, Texas 77030, USA*

P. F. WATSON, *Department of Physiology, The Royal Veterinary College, Royal College Street, London NW1 0TU, England*

with concluding remarks by

J. D. SKINNER, Mammal Research Institute, University of Pretoria, Pretoria, South Africa

Preface

The relatively poor breeding record of non-domestic animals in captivity demands intensive study; the more so, now that it is becoming increasingly difficult to replace captive stock from wild populations. Indeed, in the face of dwindling natural populations, the keeping of wild animals in captivity may arguably be immoral if they are unable to breed in that environment. The severity of this problem demands that all possible resources to assist breeding be utilized. Amongst these resources, techniques for controlling the breeding process artificially have already proved their value to the domestic livestock industry, and offer considerable advantages in non-domestic animals.

Artificial breeding is not primarily a means of enabling an isolated breeding problem to be overcome. In other words, it must not be viewed simply as an elaborate solution to an isolated problem of infertility when all else has failed. Rather, it should be seen as a means of taking maximum advantage of the limited genetic resources of the captive population. A semen bank could help to ensure the continuity of particular genetic lines and enable specific genetic crosses to be conducted. Isolated animals could be brought into a breeding programme and inter-zoo exchanges would become minor exercises of low cost, with a much reduced risk to the animals involved. Moreover, the problems of the spread of disease could be reduced. In all these exercises, ancillary genetic studies will be of paramount importance. In addition, the ability to preserve semen would permit the collection of semen from wild animals to augment the gene pool of the captive stock without reduction in the size of the wild populations. The prospects of embryo preservation and transfer extend the possibilities still further.

The term "artificial breeding" encompasses many aspects of reproductive function (see Austin, p. 1) but in non-domestic animals several of these aspects are, as yet, untouched. This Symposium was arranged between leading scientists interested in this subject to review the progress made so far and to provide an impetus for further work. Although artificial breeding in non-domestic animals is still in its infancy some notable successes have already been recorded. Owing to the scope of the subject, the Symposium had to be restricted to birds and mammals, although some work is in progress on other animals. In drawing together the various contributions, I have had in mind not only those engaged in researching aspects of the problem in order to develop our understanding, but also those who are considering the application of these techniques in their own circumstances. For this reason, several of the papers contain reviews of the literature and general background information which would benefit such people.

As will become apparent from a persual of the papers, more space is given to aspects of semen collection and preservation and less to the latter stages of artificial breeding; notably absent is reference to any comparative work on insemination techniques in mammals. This emphasis does not reflect the relative importance of these areas of study but merely indicates that much of the early work has been directed at solving problems in the male animal, and, as yet, the peculiar difficulties of studying the reproductive physiology of non-domestic female animals have largely not been overcome. One crucial problem which stands in the way of a wider application of artificial breeding appears to be the detection of oestrus and ovulation. No simple or short-term answers can be expected in this area or in others, and it is therefore vital for zoos and other interested organizations to grasp the nature of the possibilities and to fund research now which will reap benefits in years to come.

I would like to thank Dr H. G. Vevers and his staff at the Zoological Society who have made the organization of this Symposium possible. In particular, I am indebted to Miss Unity McDonnell who offered willing assistance and advice at all stages of the organization of the Symposium and the subsequent editing of the proceedings. Finally I would like to thank my wife for preparing the Subject Index.

Royal Veterinary College   P. F. Watson
September 1978

Acknowledgements

The following organizations generously provided assistance with the travel costs of the overseas invited speakers:
Cryotech
IBM (UK) Ltd
Intervet Laboratories Ltd
Lufthansa (Africa) (Pty) Ltd
Planer Products Ltd
Reckitt and Colman Ltd
The Wellcome Trust.

Contents

The Importance of Photoperiod to Artificial Breeding in Birds

†R. K. MURTON

The Principles and Practice of Semen Collection and Preservation in Birds

P. E. LAKE

† Deceased.

Artificial Insemination for Breeding Non-Domestic Birds

G. F. GEE and S. A. TEMPLE

Artificial Insemination of Falcons

L. L. BOYD

Artificial Insemination: A Practical Method for Genetic Improvement in Ring-Necked Pheasants

J. R. CAIN

A Comparative Study on the Cryogenic Preservation of Semen from Sandhill Crane and the Domestic Fowl

T. J. SEXTON and G. F. GEE

A Review of Techniques of Semen Collection in Mammals

P. F. WATSON

The Principles and Practice of Electroejaculation of Mammals

I. C. A. MARTIN

Semen Preservation in Non-Domestic Mammals

E. F. GRAHAM, M. K. L. SCHMEHL, B. K. EVENSEN and D. S. NELSON

The Detection of Oestrus

F. D'SOUZA

Ovulation Detection and Control Relative to Optimal Time of Mating in Non-Human Primates

W. R. DUKELOW

Artificial Breeding of Non-Primates

S. SEAGER, D. WILDT and C. PLATZ

Artificial Insemination and a Note on Pregnancy Detection in the Non-Human Primate

A. G. HENDRICKX, R. S. THOMPSON, D. L. HESS and S. PRAHALADA

Artificial Insemination in Foxes

J. AAMDAL, J. FOUGNER and K. NYBERG

Successful Artificial Insemination in the Chimpanzee

D. E. MARTIN, C. E. GRAHAM and K. G. GOULD

Studies on Handling Spermatozoa from the African Elephant, *Loxodonta africana*

R. C. JONES

Observations on the Artificial Breeding of Red Deer

A. KRZYWIŃSKI and Z. JACZEWSKI

The Collection, Handling and Some Properties of Marsupial Semen

J. C. RODGER and I. G. WHITE

Embryo Transfer and Embryo Preservation

C. POLGE

Handling and Sedation for Procedures Associated with Artificial Breeding

D. M. JONES

Disease Control in Semen Transfer and Artificial Insemination

G. F. SMITH

Concluding Remarks 339

J. D. SKINNER

Author Index 345

Subject Index 359

Symp. zool. Soc. Lond. (1978) No. 43, 1–6

Introduction—Advantages, Difficulties and Dilemmas of Artificial Breeding

C. R. AUSTIN

Physiological Laboratory, Cambridge, England

Some of the biological success stories of recent years have concerned the breeding of certain rare or endangered birds and mammals in captivity. These are praiseworthy accomplishments, but they should also be viewed in a wider context. The number of species deemed rare is very large—the figure cited by Martin (1975) was 815, and no doubt it is even higher now—and the reproduction of only very few of these has yet been obtained under artificial conditions, despite much painstaking endeavour. It is therefore a matter of urgency that we seek to improve our understanding of the requirements for successful breeding by closer observation under natural or near-natural conditions, and to increase our control over the reproductive processes by the application of the special techniques of artificial breeding. The present Symposium is devoted to the second line of action.

The breeding of animals in captivity (with or without the use of artificial methods) has several goals (Table I). *Conservation* has become a

TABLE I

Goals of artificial breeding

Conservation
Education and amusement
Financial gain
Biological information
Gene reservoir

widespread activity and takes place in a variety of locations, such as wildlife refuges, safari parks and even zoos and circuses. Generally, these institutions are not wholly altruistic and costs are offset by admission charges to a viewing public. Though more frankly commercial in intent, zoos and circuses often play some role in conservation, as well as in *education or amusement*. Almost wholly devoted to *financial gain* is the breeding of animals for their fur, feathers or hide; even here a degree of conservation may enter the scene. Then there is the goal of gathering new *biological information* on the lesser known animals, in part at least to test hypotheses based on observations made with domestic or laboratory stock. This line of approach is also certainly educational and potentially conservationist in effect. Finally, breeding in captivity can have the effect of preserving a *gene reservoir* that could prove invaluable to our future needs—by providing new animals with special propensities for use as laboratory animals (Austin, 1959, 1963) or for domestication (Short, 1976), or by providing new genes to be exploited by hybridization.

The more important special techniques of artificial breeding are set out in Table II—each adds a measure of control and in many instances

TABLE II

Special techniques of artificial breeding

Detection of oestrus
Detection of ovulation
Induction of oestrus and ovulation
Inhibition of oestrus and ovulation
Semen collection
Semen preservation
Artificial insemination
Oocyte and embryo recovery
In vitro maturation and fertilization
Embryo culture (artificial incubation of bird eggs)
Embryo transfer (fostering in birds)
Embryo preservation

---

the application of these procedures has proved of great value in controlling reproduction of both domestic and laboratory animals; similar experiences are beginning to be recorded with non-domestic animals, and the research reports appearing in this book provide clear evidence on the progress that is being made. These methods need only be considered briefly here. The first one, the detection of oestrus, is helpful where, for instance, cohabitation at other times leads to fighting, as in Chinese hamsters among which the male risks death if he ventures too close to a non-receptive female. Detection of ovulation adds further precision, and is especially useful when artificial insemination is to be practised. In some non-domestic animals, such as the mink (Enders, 1939) and the raccoon (Llewellyn & Enders, 1954), ovulation is reported to be reflexly evoked by the act of coitus and could therefore probably be induced by hormone injection. The hormonal induction (or promotion) of both oestrus and ovulation in cyclic species could be helpful but is not easily obtained. "Precipitating" ovulation by injection of gonadotrophin during natural oestrus is likely to have been a major factor in the unique achievement of D. J. Williams (pers. comm.), who has been able to establish pregnancy in five lions, with subsequent live births, by means of artificial insemination.

Sometimes inhibition of oestrus and ovulation may be considered desirable in order to suppress fighting in mixed-sex groups or to support a prescribed breeding policy. Seal *et al.* (1976) reported high contraceptive efficiency, without deleterious side effects, during a period of two years in 21 lions given medroxyprogesterone by intramuscular injection or melengestrol acetate in silicone rubber implants. Following this lead, investigators at Knoxville Municipal Zoo, Tennessee, placed implants of melengestrol acetate subcutaneously in three lions and noted evident suppression of pregnancy for two years; subsequently, (the implants now presumably exhausted) two of the lions became pregnant, producing two normal cubs each (G. L. Smith, pers. comm.). This approach could possibly be adopted to aid breeding of temporarily infertile captive animals for it has been found that oestrus and ovulation can be induced in anoestrous sheep by gonadotrophin injection, if this is preceded by a period of one to two weeks' treatment with progestagen (Anderson, 1969).

Semen collection and artificial insemination are time-honoured practices, to which more recently semen preservation has been added. Clearly of direct value for artificial breeding in birds as in mammals, they form the subject matter for much of the present volume. Suffice it to mention here that, in the work of D. J. Williams recorded above, the lion semen was obtained by electrical stimulation with the aid of a bull rectal probe and that J. D. Roussel (pers. comm.) successfully used a ram rectal probe to provoke ejaculation in nine-banded armadillos, *Dasypus novemcinctus*.

Oocyte and embryo recovery, *in vitro* maturation and fertilization, and embryo culture, transfer and preservation are all highly specialized procedures which are more likely to be relevant to the collection of basic data on early development than to aiding in the breeding of non-domestic animals. Nevertheless there could be circumstances in which they would prove vital—to overcome sterility due to genital tract anomaly in a unique animal (along lines currently under investigation for establishing pregnancy in women with occluded oviducts—Edwards & Steptoe, 1974), to enlist the aid of uterine foster mothers for accelerating reproduction rates (as in sheep and cattle—Lawson *et al.*, 1975), and to store valuable genetic material in the form of embryos of selected species or strains (for further discussion on these procedures see the article by Polge, this volume p. 303). By contrast, in birds, embryo culture, as occurs during cabinet incubation of eggs, and fostering, the caring for young birds irrespective of parenthood by broody hens, are commonplace, and both devices are used with non-domestic species.

The artificial breeding of non-domestic animals involves certain problems and dilemmas which deserve consideration at this point (Table III). First of all, since it is hardly feasible to breed, or attempt to breed,

TABLE III

Dilemmas of artificial breeding

Selection of animals
Depletion of wild populations
Unwelcome visitors
Difficulties of restoration
Imprinting
Genetic drift
Fortuitous drift

under artificial conditions all rare or interesting birds and mammals, selection is necessary and great importance could attach to this step. Thanks to the public concern for conservation, attention is likely to be given primarily to animals on the endangered list—and this is of course commendable—but because financial considerations must play a major part in decision-making the tendency will be to give preference to animals with popular appeal, diurnal extroverted creatures which can readily be observed and appreciated. Because of the demands of medical research a high priority rating will also be given to certain non-human primates. It is to be hoped that, despite these powerful influences, those responsible for the selection will additionally take into account representativeness of natural grouping.

At some point, with a diminishing species, the decision has to be made whether or not it is justifiable further to deplete the small natural population in order to attempt breeding in captivity. The known migratory population of whooping cranes in North America consisted of only 15 individuals at one point before artificial methods were instituted (Erickson, 1975). This may seem the limit of brinksmanship but we should not forget that all laboratory golden hamsters now extant are believed to be descended from the one female and 12 young dug from their burrow in 1930.

Artificial breeding often means removal of the animal to a location lacking the predators and other controlling elements of the natural habitat. Inevitably in the course of time animals escape (or are released) and these, becoming feral, could present a serious nuisance—like the coypu or nutria, *Myocastor coypus*, in the fen country of eastern England. Disease introduced with the animal could well represent a health hazard, as does the so-called B virus, *Herpesvirus simiae*, which appears to be endemic in many Old World primates. (Diseases can also be carried in the transhipment of semen or early embryos.) The animal as well as its parts and parasites have the potential of becoming unwelcome visitors.

Motivation for conserving a rare species commonly includes the intention eventually to re-establish a population in the original habitat. Apart from helping thus to restore a natural balance, this policy relieves the breeding centre of carrying the whole burden of preservation. There are, of course, problems attending this step (Brambell, 1977), but we should not forget that wild populations diminish in number commonly because they have come into conflict with growing human populations—animals are killed because they constitute competition or a danger, or they die out because their habitats are destroyed as agriculture expands. Under these circumstances, the supply of artificially bred animals to the original habitat of the species would be both unwelcome and futile. Accordingly we are compelled to face the prospect that more and more species will need to be maintained in captivity; since moderately large numbers are necessary to preserve each species the work of the breeding and holding centres must become a massive undertaking.

In the face of this rather disturbing prognostication we must also acknowledge that conservation under captive conditions, and especially with the employment of artificial breeding methods, will be bound eventually to change the character of a species. For one thing there is the effect known as imprinting, when a young animal is raised in too close association with its keeper, but this can be largely overcome if the right measures are taken. Much more serious are the genetic effects, namely genetic drift and fortuitous drift.

The size of a breeding colony is important not only to permit allowance for natural losses but also to reduce the effect of genetic drift. In terms of practical significance the two numbers are very far apart—a breeding colony of one male and one or two females might be sufficient to save a species, but to prevent genetic drift becoming obvious in three or four generations the colony would need to be many times that size. Some mitigation of this effect could be obtained if several centres kept breeding colonies and exchanged sires (or semen) at intervals. The establishment of semen banks has indeed already been suggested. By fortuitous drift is meant the result that is produced by the conscious or unconscious selection for characters thought to be "normal" or "desirable" in a species; in this way, distinctive strains are likely to be developed. Random selection of mates is a partial answer—just as exchange of sires (or semen) may reduce the effect of genetic drift. In addition, the use of semen samples obtained from individuals in the residual wild population would be very helpful. But none of these remedies can be wholly successful when populations are small. Inevitably the rescue of an endangered species will result in its modification.

REFERENCES

Anderson, L. L. (1969). Sexual behavior and controlling mechanisms in domestic birds and mammals. In *Reproduction in domestic animals*: 541–568. Cole, H. H. & Cupps, P. T. (eds). New York and London: Academic Press.

Austin, C. R. (1959), Prospective experimental animals for medical research. *J. anim. Techn. Assoc.* **10**: 1–6.

Austin, C. R. (1963). Introducing new animals to the laboratory. *New Scient.* **17**: 117–120.

Brambell, M. R. (1977). Reintroduction. *Int. Zoo Yb.* **17**: 112–186.

Edwards, R. G. & Steptoe, P. C. (1974). Control of human ovulation, fertilization and implantation. *Proc. R. Soc. Med.* **67**: 30–34.

Enders, R. K. (1939). Reproductive phenomena in the mink (*Mustela vison*). *Anat. Rec.* **75** Suppl. 122.

Erickson, R. C. (1975). Captive breeding of whooping cranes at the Patuxent Wildlife Research Center. In *Breeding endangered species in captivity*: 99–114. Martin, R. D. (ed.). London and New York: Academic Press.

Lawson, R. A. S., Rowson, L. E. A., Moor, R. M. & Tervit, H. R. (1975). Experiments on egg transfer in the cow and ewe: dependence of conception rate on the transfer procedure and stage of the oestrous cycle. *J. Reprod. Fert.* **45**: 101–107.

Llewellyn, L. M. & Enders, R. K. (1954). Ovulation in the raccoon. *J. Mammal.* **35**: 550.

Martin, R. D. (1975). Introduction. In *Breeding endangered species in captivity.* Martin, R. D. (ed.). London and New York: Academic Press.

Seal, U. S., Barton, R., Mather, L., Olberding, K., Plotka, E. D. & Gray, C. W. (1976). Hormonal contraception in captive female lions (*Panthera leo*). *J. Zoo Anim. Med.* **7** (4): 12–20.

Short, R. V. (1976). The introduction of new species of animals for the purpose of domestication. *Symp. zool. Soc. Lond.* No. 40: 321–333.

Symp. zool. Soc. Lond. (1978) No. 43, 7–29

The Importance of Photoperiod to Artificial Breeding in Birds

R. K. MURTON†

Monks Wood Experimental Station, Institute of Terrestrial Ecology, Huntingdon, England

SYNOPSIS

Photoperiodism is the process whereby the controlling 24-hour oscillation produced by the cycle of day and night (the *Zeitgeber*) is used by animals (and plants) to entrain a host of endogenous circadian rhythms in body function. So far as is known all bird species are photoperiodic in this sense. Included amongst these physiological rhythms are the endocrine secretions—neurohormones and pituitary hormones—which cause the bird to attain reproductive condition. Birds vary in the extent to which they rely on changes in the daily photoperiod to herald the approach of suitable seasons for migration, reproduction, moult or other functions. The special feature of the "photoperiodic clock" is its capacity to measure daylength or light intensity, and current knowledge about how it works is briefly reviewed. So, too, is the phenomenon of photorefractoriness whereby the subject fails to respond to a photostimulus: refractoriness is very apparent in, but not confined to, most north temperate species which breed in spring and early summer but do not do so in late summer and autumn, when daylengths are still relatively long. When held on non-entraining constant schedules (sometimes under constant light, questionably with light:dark 12:12 and frequently in constant dark) the endogenous rhythms of the bird become expressed as free-running periodicities, sometimes referred to as circannual rhythms.

The importance of photoperiodism to the breeding of birds is frequently evident in captivity, where subjects are often kept at latitudes, and hence under light cycles, very different from the natural ones to which they have evolved adaptive responses. This is illustrated by reference to various species of Anatidae held in the collection of the Wildfowl Trust, Slimbridge. Some high latitude species do not receive sufficient photostimulation to breed under the low amplitude daylength cycles experienced at lower latitudes. Some species from mid-latitudes (e.g. several Australian species) are forced into a premature refractory phase under northern photoperiods and they, too, do not breed readily. In a final section the importance of light and specific behaviour patterns in phasing endocrine rhythms between the sexes is emphasized.

INTRODUCTION

All animals are photoperiodic in the sense that the 24-hour cycle of day and night entrains endogenous circadian rhythms of body function. Animals are extremely complex "black boxes" in which a myriad of physiological rhythms have to be ordered in the correct temporal

---

† Professor R. K. Murton died on 12 June 1978 during the preparation of this volume.

sequence, for example, diurnal changes in blood oxygen, plasma calcium or sugar content. Most of the functions which have been studied in detail exhibit circadian rhythms, which in a non-entraining environment (often constant dark will suffice for day-active organisms) display characteristic free-run periodicities which are usually not quite 24 hours in duration. These circadian rhythms are entrained via one, or perhaps more, master clock(s) to the exact 24-hour periodicity of day and night. (Since the cycle of day and night acts as an entraining oscillator it is called a *Zeitgeber*.) The hormones which regulate gonad development, control secondary sexual characters and in turn govern breeding periodicity are but further examples of physiological rhythms which are entrained photoperiodically. The question is not whether or not birds are photoperiodic, but the extent to which they use photoperiodic inputs as reliable predictors of the season during which reproduction, or any other physiological function, is most likely to be effectively accomplished.

PHOTOPERIODISM IN BIRDS

The Photoperiodic Clock

To understand avian photoperiodism it is not immediately helpful to draw a direct analogy with the photographic process, even though the possibility of induction cannot be dismissed completely. The induction concept imagines that light initiates some biochemical change in a photoreceptor, in comparable manner to the way in which light changes the properties of silver halide granules in the photographic film. Obviously, such a simple concept does not explain why an animal may be responsive to a long duration photoperiod but not to a short one. Some time measurement process must be involved in recognizing the length of a photoperiod and it was early appreciated by Bünning (1936) that some kind of endogenous rhythm of photosensitivity was involved. Bünning's theory essentially assumed a circadian rhythm of cell function comprising two half-cycles lasting approximately 12 hours each. One of these cycles was thought to be light requiring (photophil) and one dark requiring (scotophil). If light impinged on the animal while the photophil phase was active photo-induction could occur. Thus, the photographic film idea is partly relevant with the refinement that the camera shutter is opened by some endogenous rhythm. Bünning's hypothesis was refined by Pittendrigh & Minis (1964) and Pittendrigh (1966) who emphasized that the whole cycle of night and day acted as an entraining oscillator (in the physical sense of the word) to phase the endogenous rhythms of photosensitivity and, if coincidence between light and the "photoperiodically inducible phase" occurred, to induce a response. This early Pittendrigh model is an *external coincidence* one and it has for some time been considered appropriate to understanding avian photoperiodism (examples are given by Lofts, Follett & Murton, 1970; Murton & Westwood, 1975, 1977). The significance of this model, still frequently overlooked by authors, is that the position of the rhythm (phase) is not fixed by dawn, but is, like all circadian rhythms, dependent on the frequency of the whole entraining cycle.

We need to update the original Pittendrigh model because for some while it has been suspected that the clock comprises a coupled oscillator system. Activity cycles, produced by fixing a microswitch to record perch-hopping in birds or wheel-turning in rodents, have been used as a means of measuring circadian rhythms in vertebrates and under some circumstances a bimodal pattern can be observed; the bimodality of bird song has long been recognized. Such patterns suggest that a double but closely coupled oscillator might underlie the rhythms and indeed under some conditions activity cycles may be split into two separate components (Hoffmann, 1971; Gwinner, 1975). The phenomenon of rhythm splitting (the term rhythm splitting is reserved for phenomena such as the loss of synchrony between two different rhythms, involving distinct activities, whereas when a rhythm splits in a manner that results in a phase difference of about 180° a different process is involved and Pavlidis (1973) recommends that the term frequency doubling be used) and other evidence makes it reasonably likely that the photoperiodic clock comprises a pair of coupled oscillators which are normally phase-locked. Activity rhythms often become arrhythmic under certain light regimes—for example, some species show this when held under constant light—as if two oscillators become dissociated so that morning activity extends earlier in one direction and the afternoon activity extends into the evening and joins with the next day's onset. Observations of this kind have led Pittendrigh and his colleague Daan (Pittendrigh & Daan, 1976a,b,c; Daan & Pittendrigh, 1976a,b) to modify the model of the photoperiodic clock to something like the form illustrated in Fig. 1. The two oscillators (morning and evening oscillators) appear to be affected by light in opposite ways so that increases in light intensity result in positive (advance) and negative (delay) phase shifts, respectively (the opposite may occur in nocturnal species). In oscillatory terms, light intensity can be increased by absolute changes in intensity of the illumination (measured in lux or ft candles) or by varying the proportion of light in a 24-hour light:dark cycle (of course experimentally a day can be contrived which is not 24 hours in duration).

With short photoperiods activity begins before dawn but it is sporadic until lights-on. With a very long photoperiod (light:dark—LD—22:2) the coupling between the oscillators breaks and the "evening" oscillator (dotted) jumps to become the "morning" oscillator of the second day. The kind of activity pattern shown has recently been identified in our laboratory with starlings, *Sturnus vulgaris*; indeed the activity cycles are based on starlings kept in our experimental rooms. It is supposed that gonadotrophin secretion is initiated in a similar manner to activity rhythms but this has yet to be proved.

The photoperiodic clock which controls wheel-running activity in rodents appears to be sited in the suprachiasmatic nucleus, which lies in

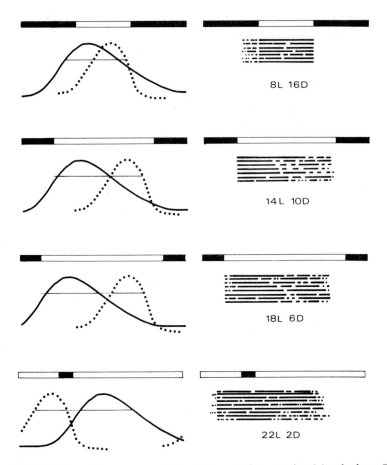

8L 16D

14L 10D

18L 6D

22L 2D

FIG. 1. Illustration of how two coupled oscillators might control activity rhythms. For a detailed model of how such a system works see Pittendrigh & Daan (1976c). The left-hand panels illustrate the behaviour of a pair of coupled light-sensitive oscillators, with one oscillator entrainable to light and the other closely entrained to the first, but having an opposite frequency change in relation to a change in light intensity or duration. As the duration of the daily photoperiod increases, the coupling between the two oscillators weakens and they move apart. With a very long photoperiod, the coupling breaks and the "evening" oscillator (dotted) jumps to become the "morning" oscillator of the second day. Activity, shown in the right-hand panels, occurs when the oscillators are above some "biochemical" threshold, depicted as a horizontal line in the left-hand panels.

the preoptic region of the anterior ventral hypothalamus, although its exact position in birds remains to be defined. From this region neural fibres fan out to enter the tuberal nucleus as a diffuse network (Oksche *et al.*, 1974), and appropriate lesions can prevent light-induced growth of testes in Japanese quail, *Coturnix coturnix* (Davies & Follett, 1974a,b,

1975a,b). In the guinea-pig, neural and neurosecretory pathways from the preoptic region stimulate neurones in the tuberal hypothalamus, where at least some luteinizing hormone releasing factor (LHRF) cell bodies exist, from whence they send processes to the median eminence. The same appears to hold for birds. When quail are transferred to short photoperiods luteinizing hormone (LH) secretion continues for 7–14 days (Nicholls, Scanes & Follett, 1973) but following deafferentation of fibres in the tuberal nucleus LH levels fall rapidly. Thus, some carry-over stimulation of LH on long days resides outside the tuberal nucleus, presumably nearer the site at which photic stimuli are perceived. Here the rhythmic mechanism whereby the daily photoperiod is measured is converted, by an unknown neural transducer, into a continuous stimulus which leads to the release of LHRF and, in turn, LH.

In birds a single releasing factor (LHRF) appears to be responsible for the release of both LH and follicle stimulating hormone (FSH) judging from the work of Follett (1975, 1976) and in contradiction to the claims of Graber, Frankel & Nalbandov (1967) and Stetson (1969). A separate releasing factor exists for thyroid stimulating hormone (TSH) and this can also stimulate prolactin secretion experimentally (Bolton, Chadwick & Scanes, 1973). When photosensitive quail are initially exposed to stimulatory daylengths both FSH and LH secretion are stimulated and this results in gonad growth (Follett, 1976). In mammals, and presumably also in birds, FSH acts on the intratubular Sertoli cells to produce a specific Sertoli cell protein (androgen binding protein—ABP) with a high affinity for binding androgen (Tindall, Schrader & Means, 1974) while LH stimulates growth and development of the interstitial Leydig cells and androgen secretion (Maung, in press). Androgen is drawn into the tubule lumen by ABP and is essential for the maturation of the germ cells (Lofts & Murton, 1973). Androgen released into the blood stream feeds back on hypothalamic centres to inhibit further LH release and in consequence plasma LH titres decline: androgen also sensitises behaviour centres in the brain. In mammals, and presumably also birds, ABP passes in testicular fluid to the epididymides where it is taken up by the epididymal cells. In these cells inhibin is produced which feeds back on the pituitary to inhibit FSH secretion (Franchimont et al., 1975): inhibin has not yet been identified in birds. Thus, as a result of differential feedback, plasma FSH levels may decline, when LH levels remain high (Follett, 1976). The testes may remain enlarged because androgen secretion can maintain spermatogenesis (Lofts, 1962; Lofts & Murton, 1973).

An approximately similar mechanism operates in the female, for stimulatory photoperiods cause FSH and LH secretion with consequent growth of the ovarian follicles and oestrogen production (see Lofts & Murton, 1973; Murton & Westwood, 1977). However, while males achieve full spermatogenesis under the influence of appropriate pho-toregimes, females of most avian species only grow their follicles to a quarter or less of the size that will eventually be ovulated. Final rapid follicular growth occurs during the 6–12 days that elapse between pair formation and ovulation, following specific courtship behaviour from the male; the availability of a territory and nest site are usually integral components of the stimulus he must provide. Ovulation often corresponds with active nest building, as in the canary, *Serinus canaria* (Hutchison, Hinde & Bendon, 1968) and Barbary dove, *Streptopelia risoria* var. (Lehrman, Brody & Wortis, 1961). During this final growth phase of the follicles specific proteins previously stored in the *pectoralis major* muscle are mobilized and appear to be transported to and incorporated in the developing yolk (Jones & Ward, 1976). It follows that any inability on the part of the female to acquire suitable reserves can prevent her responding to other stimuli or cause her to produce a smaller clutch or lay inadequately provisioned eggs with a lowered viability: doubtless this explains how the quality of the eggs of the red grouse, *Lagopus scoticus*, may vary in different seasons and even affect the status of the progeny in the adult social organization (Watson & Moss, 1969). Jones & Ward (1976) consider the nutritional state of the female to be all important and to be the final factor timing the breeding season to the exclusion of effects of the photoperiod. We reject this viewpoint because it is clear that an integrated system is involved. It is inappropriate to pursue this topic further here but it is relevant to note that provision of a correct diet, together with an appropriate photoperiod, are both important to the successful breeding of birds in captivity.

So far we have only considered the effect of photoperiod on gonadotrophin secretion. Other releasing hormones and pituitary hormones, for example, adrenocorticotrophic hormone (ACTH), TSH and prolactin, respond to daylength though not necessarily to the same degree with a comparable photoperiod or at the same rate. In consequence, a complex sequence of endocrine-mediated events in the annual calender of wild birds, including preparation for migration, fat deposition (King, 1970) and moult as well as gonad development are programmed in a species-specific manner in response to seasonal daylength changes.

Types of Breeding Periodicity

Light-induced gonadotrophin secretion results in gonad maturation and the physiological potential for reproduction. This is well demonstrated in the feral pigeon in which plasma titres of LH increase six-fold in response to an increase in photoperiod, from an 8-hour to 18-hour day (Fig. 2a). Under natural photoperiods this simple response to light results in a breeding cycle which is approximately symmetrically aligned on either side of the summer solstice: gonad growth occurs when daylengths reach a stimulatory level in spring and regression follows in the following late summer, when the daily photoperiod falls below this level. Different *Columba* species have slightly different photosensitive

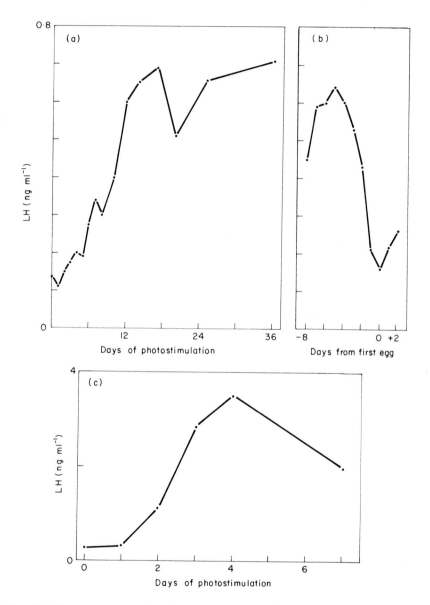

FIG. 2.(a) Increase in mean ($n = 9$) plasma content of luteinizing hormone following transfer of female feral pigeons, *Columba livia* var., from a short (LD 8:16) to long (LD 18:6) photoperiod beginning on Day 0. (b) Changes in plasma content of LH when photostimulated (LD 16:8) feral pigeons were paired. Points are means of eight females and results are plotted relative to the day of laying the first egg. (c) Increase in plasma LH titre in female quail. *Coturnix coturnix japonica*, transferred on Day 0 from short days (LD 8:16) to long days (LD 20:4). Data in (c) from Nicholls *et al.* (1973).

thresholds and so the potential length of the breeding season is shortest in the wood pigeon, *C. palumbus*, slightly longer in the stock dove, *C. oenas*, and longest in the rock dove, *C. livia* (and its domesticated varieties) (Lofts, Murton & Westwood, 1966, 1967a,b). The season of actual breeding depends on ecological conditions so that in London wood pigeons produce most of their eggs and young in spring, but in East Anglia peak breeding is in summer and early autumn (Murton, 1975); a good illustration of how genotypic traits can be modified phenotypically.

Although pigeons respond to photoperiodic stimulation, maximum secretion of gonadotrophin, or more specifically LH, does not occur unless behavioural stimulation is forthcoming from the mate. Figure 2a emphasizes how pairing accompanied by appropriate courtship causes circulating LH levels to be doubled; the subjects were already stimulated by a 16-hour photoperiod. This provides a mechanism which allows the breeding season to be precisely adjusted to ecological conditions, without any unnecessary energy commitment when environmental conditions are poor. In eastern England in June, wood pigeons require the whole day to find food reserves and are unable to devote much time to intra-pair courtship. In consequence, little egg-laying occurs at this time although the birds are physiologically capable (Murton & Isaacson, 1962). This capability for a fine adjustment of actual breeding to times of good food resources may well be important in many tropical species.

In the majority of north-temperate species greater reliance has come to be placed on seasonal changes in daylength for the proximate timing of reproduction. The breeding cycle of the quail, *Coturnix coturnix*, resembles that of the pigeons but LH levels are elevated to a pro-portionately greater extent by photostimulation (eight-fold) than happens in pigeons, that is, intra-pair courtship is much less important in stimulating LH secretion to the point that will induce the final stages of gonad maturation and ovulation (Fig. 2b). Domestic hens, *Gallus domesticus*, have breeding cycles which are extended versions of those seen in quail and their capacity to produce eggs without behavioural stimulation is well attested. Light stimulation is nevertheless important for gonadotrophin secretion, and egg production varies seasonally in response to the photoperiod (Lake, 1971 for references). There is also good evidence that the ovulatory cycle is driven by a circadian rhythm mechanism so that the rate of egg production can be varied by altering the length of the driving oscillator (*Zeitgeber*) (Hutchinson, 1962; Rosales, Biellier & Stephenson, 1968; Murton & Westwood, 1977).

For the majority of northern-hemisphere birds spring and early summer breeding is appropriate, but not egg-laying in late summer and autumn. Hence, a direct response to daylength to produce a symmetrical breeding season as in quail and pigeons would be ecologically non-adaptive. To prevent this a species-characteristic refractoriness to long days is developed so that the reproductive apparatus spontaneously regresses, even though daylengths remain the same or longer than those which initiated a response in the spring. If the subject is maintained on "long day" regimes refractoriness continues and it is only broken by returning the bird to a period of "short day" treatments. In the wild, short days are usually encountered in the autumn so that a species such as the starling, *Sturnus vulgaris*, breaks refractoriness during late September and October, at which time it regains its photosensitivity; that is, by late October or early November starlings can once more be stimulated by long days, although, of course, they do not normally experience these until spring. The actual daylength and duration of exposure necessary to end refractoriness is a relative matter and varies from species to species (see Lofts & Murton, 1968; Storey & Nicholls, 1976; Nicholls & Storey, 1977; Murton & Westwood, 1977, for more details). Lofts & Murton (1968) have emphasized the ecological advantages of photorefractoriness and give examples of the various patterns that can occur in birds but Fig. 3 relies on the starling as a fairly typical illustration.

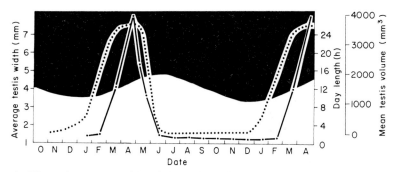

FIG. 3. Change in testes size of free-living male starlings in response to seasonal changes in daylength. Modified from Murton & Westwood (1977) using data for British starlings examined by W. S. Bullough (solid line, testis volume) or free-living populations in the USA studied by R. G. Schwab (dotted line, testis width). Note how the testes regress even though the photoperiod remains long.

The physiological mechanisms underlying photorefractoriness are still not resolved, although there may be an involvement of thyroid hormones. At present, very little is known about cycles of TSH and hence triiodothyronine (T_3) and thyroxin (T_4) secretion in birds and suitable assay methods have only recently been developed. Thyroidectomy of starlings prior to exposure to long photoperiods resulted in the failure of the testes to regress after their initial recrudescence (Wieselthier & van Tienhoven, 1972). If thyroidectomy was delayed until after four weeks of exposure to long days regression did occur, but was followed by a second period of growth. Thyroidectomy performed in autumn did not affect the termination of photorefractoriness. These data, and other evidence from Thapliyal (1969) and Chandola (1972), suggest that with increase in photoperiod both gonadotrophin and thyroid secretion are stimulated and that thyroid hormones reduce the sensitivity of the hypothalamus to photic stimulation so inhibiting gonadotrophin output, since exogenous gonadotrophins can stimulate gonad growth even during the refractory period (Benoit et al., 1950; Lofts, Murton & Thearle, 1973).

In domesticated strains of mallard, Anas platyrhynchos (Pekin ducks), thyroid activity increases during the long days of May and July (Astier, Halberg & Assenmacher, 1970), and LH titres remain high (Jallageas, Assenmacher & Follett, 1974) but LH titres decline in wild strains (Haase, Sharp & Paulke, 1975). Androgen activity decreases partly in consequence of an increased metabolic clearance rate and this effect can be simulated by thyroxine injections during the period of gonad enlargement (Assenmacher, 1974). Furthermore, an experimental thyroxine load prevents the stimulating effects of long days (LD 18:6) on the gonads in winter, apparently by inhibiting androgen activity.

Assenmacher's scheme, therefore, envisages that gonad regression occurs as a result of androgen suppression. Another possibility is that thyroid hormones have a direct effect on the hypothalamus. It is also conceivable that photorefractoriness is a consequence of a changed phasing of endogenous circadian rhythms. We cannot speculate further here on this intriguing question but must be content to know the circumstances under which refractoriness develops.

If a photosensitive starling is exposed to a very long photoperiod, for example LD 18:6, gonad recrudescence is rapidly achieved, but the period of active gametogenesis is relatively short before refractoriness ensues. With a medium photoperiod, say LD 13:11, gonad recrudescence takes longer but the duration of active gametogenesis is lengthened and true refractoriness may not develop (Hamner, 1971; Schwab, 1971). What constitutes a long, medium or short photoperiod varies according to species and the natural daylength cycle to which it has evolved adaptive responses as is discussed in the next section.

At certain critical constant photoperiods (LD 12:12) sequences of gonad recrudescence, active gametogenesis and spontaneous regression occur in the starling with an overall periodicity of approximately 9·5 months. Similar observations apply to the gonad cycle, migratory fattening and moult in the willow warbler, Phylloscopus trochilus, (Gwinner, 1968) and other species (summary in Gwinner, 1975). Gonad growth and development in the collared dove, Streptopelia decaocto, normally a photoperiodic species, can occur in constant dark (Murton & Westwood, 1975). In the wild the wideawake or sooty tern on Ascension Island, like some other tropical seabirds, is known to breed with a periodicity of approximately 9·5 months (Chapin, 1954; Ashmole, 1963). Such observations and other experimental data have led to the idea that seasonal body functions in birds are controlled by circannual rhythms. Circannual rhythms are clearly manifested under certain lighting conditions but it does not follow that they are controlled by circannual oscillators. We hold the view that circannual rhythms are compounded from circadian rhythm mechanisms. In the examples which have been studied, including gonad growth or moult, more than one hormone mechanism is involved so it is possible for two or more endocrine secretions to assume various phase relationships depending on the duration of the daily photoperiod. This is an important topic which we cannot pursue in detail here and readers are recommended to the useful review by Gwinner (1975).

Breeding Cycles in some Captive Waterfowl

The practical consequences of avian photoperiodism as outlined above are conveniently illustrated by reference to captive waterfowl studied in the collection maintained at Slimbridge, Gloucestershire, by the Wildfowl Trust (Murton & Kear, 1973, 1976).

Most of the species form pairs and lay normal clutches of eggs in nests which they construct in the shelter of vegetation or in the special boxes and burrows provided. As in most other birds, the female lays when she receives appropriate courtship from a male at the peak of his reproductive cycle, provided she is not inhibited by a non-stimulatory daylength or inappropriate environmental conditions. Although a female may occasionally lay odd eggs spontaneously, the appearance of a full clutch in a properly constructed nest marks a precise stage in the breeding cycle. Records are kept of the date each year that the first egg is laid by each species while, because most hatching dates are known, the date of last egg-laying can be calculated. For some species records exist covering more than 20 years so that an accurate median date of first egg-laying is available. However, it was established for swans, geese, shelducks and sheldgeese that the variability in laying date from year to year is slight so that even when only a few records are available, reliable data can be extracted.

Some selected records are presented in Figs 4 and 5, where horizontal bars define the egg-laying season at Slimbridge. The bars are plotted against calendar date on the abscissae and against the daylength at which egg-laying begins on the ordinates of the two figures. Murton & Kear (1976) have defined two broad types of egg-laying cycle in Anatidae, a "primitive-type" photoresponse and a "temperate-type" photoresponse (Fig. 4).

In the primitive-type egg-laying extends more or less symmetrically on either side of the solstice, that is, refractoriness to long days does not develop. The Muscovy duck, *Cairina moschata*, has a primitive-type cycle with a fairly low photo-threshold. Egg-laying begins in February when the daily photoperiod reaches 11·7 hours and extends until late September when the daylength falls to 13·5 hours. During this period

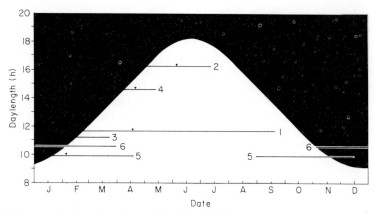

FIG. 4. The egg-laying cycle of some selected waterfowl species held in captivity at 52°N by the Wildfowl Trust. The records are plotted against date on the abscissa and against the daylength at which breeding begins on the ordinate. The daylength cycle, including civil twilight, refers to lat. 52°N. Horizontal bars denote the egg-laying season based on the earliest and latest dates on which eggs were laid, while the arrow marks the median date of first eggs, based on records collected over several years. 1, Muscovy duck; 2, red-billed pintail; 3, Australian shelduck; 4, European shelduck; 5, black swan; 6, Cape Barren goose. Based on Murton & Kear (1976) and Kear & Murton (1976).

Muscovy ducks may produce several clutches and this has clearly facilitated domestication for egg production. The mid-latitude of the natural breeding range is 7°S. In contrast, the red-billed pintail. *Anas erythrorhyncha*, with a breeding range centred on latitude 8°S, apparently has a high photo-threshold and breeding does not begin at Slimbridge until 3 May and only lasts until 12 July. Obviously, the species must respond to shorter photoperiods in its natural range and the pattern displayed at Slimbridge is a consequence of the greater amplitude of photoperiodic change at this northern latitude; however, exact details remain to be elucidated. It would be advantageous to hold species of this kind under an artificial photoperiod of reduced amplitude to encourage breeding over a longer season.

The temperate-type of breeding cycle is illustrated in Fig. 4 by the Australian shelduck, *Tadorna tadornoides*, whose earliest eggs have been laid at Slimbridge on 16 February and the last on 24 March, when refractoriness to long days develops. The Australian shelduck (mid-latitude 35°S), like certain other Australian and low latitude species from the southern hemisphere, is very difficult to breed in captivity, whereas its close relative the New Zealand or paradise shelduck, *T. variegata*, from latitude 41°S, does so relatively easily. Delacour (1939) noted at Clères, France (50°N) that Australian shelduck came into breeding condition in January and February, but usually "nothing happened" and in April they moulted (see Kear & Murton, 1976). The explanation is that the Australian shelduck has evolved from an ancestral line that had photoperiodic responses adapted to suit the southern spring of an Australian continent that once lay much further south, and where refractoriness to summer photoperiods was adaptively advantageous: this is the situation noted in the European shelduck, *T. tadorna* (mid-latitude 45°N) today. The present position of Australia, with a relatively lower amplitude daylength cycle than in the past, necessitates a low photo-threshold so that the Australian shelduck lays during July, August or September, but refractoriness does not rapidly ensue. In Britain, however, the species attempts to breed early and then is rapidly forced into a photorefractory condition. Only by holding the species on a reduced amplitude light cycle could this problem be overcome. The Radjah shelduck, *T. radjah*, is also difficult to breed in captivity in Britain for similar reasons (Kear & Murton, 1976).

The black swan, *Cygnus atrata*, also has a low photo-threshold for initial gonad development to suit present day Australian conditions. But, it also develops a typical temperate-type refractoriness when exposed to temperate photoperiods; since daylengths are shorter in Australia, refractoriness is hardly manifested in the natural range (see Braithwaite & Frith, 1969). The black swan almost certainly evolved in the post-Gondwanaland land mass when Australia lay much further south and breeding in "spring" was appropriate. It has adapted to present day Australia by a drop in the photo-threshold needed to initiate gonadotrophin secretion. When exposed to a more temperate light cycle its ancestry is revealed. In South Island, New Zealand (latitude 44°S) breeding is markedly seasonal in the spring (August to mid-October according to Miers & Williams, 1969). In Britain the breeding pattern is unusual (Fig. 4). First clutches are laid early in the year between 21 January and 28 April—which falsely made some northern aviculturalists think that the species could not adapt to a "right way-up" breeding season in the northern hemisphere (Cunningham, 1958, quoted by Kear & Murton, 1976). The birds then become refractory to the 16-hour photoperiods of late April and remain in this condition until late summer. By then refractoriness is broken and the species can lay eggs again between 30 August and early December, when daylengths again fall below the stimulatory threshold necessary for gonadotrophin secretion. Kear & Murton (1976) mention some other species which have a tendency for a double breeding season in England as a consequence of an imbalance between their photo-thresholds and the *Zeitgeber* at latitude 52°.

The Cape Barren goose, *Cereopsis novaehollandiae*, from latitude 36°S has a very low photo-threshold to allow it to breed during the southern winter (in the Bass Strait some eggs are laid in May, most in June and July and some in August) and it becomes refractory to "summer" photoperiods. In Britain, it lays eggs between 19 November and 31 March (Kear & Murton, 1973, 1976) which can hardly be adaptive under British conditions. Aviculturalists must obviously adjust their rearing programmes to suit such a breeding cycle, unless they prefer to keep their stock on an artificial light regime.

Closely related species, usually members of the same genus, have a similar photoperiodic response mechanism (Murton & Kear, 1973, 1976). As would be expected, physiological differences, like anatomical ones, are least in lines that are phylogenetically close. Usually, congeneric species differ from each other in the photoperiod at which breeding is initiated, that is, variations occur in the photo-threshold necessary to stimulate the same amount of gonadotrophin secretion, or maintain the same rate of release. In contrast, the time at which refractoriness is initiated varies much less markedly between closely related species, as if the mechanisms involved are less amenable to evolutionary change. This is seen to some extent in Fig. 5 which details the breeding

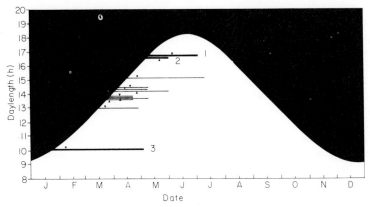

FIG. 5. The egg-laying cycle of different species or races of black geese of the genus *Branta* in the Wildfowl Trust collection. Conventions as for Fig. 4. Heavy bars designate species mentioned in the text the key to these being: 1, red-breasted goose; 2, Pacific brent goose; 3, Hawaiian goose.

season of various geese of the genus *Branta* at Slimbridge. This emphasises two more practical problems in breeding animals in captivity under unnatural photoregimes. First, the red-breasted goose, *Branta ruficollis*, and brent goose, *B. bernicla*, and other sub-arctic breeders, have a very high photo-threshold to prevent them attaining breeding condition until the northern summer arrives in May and June and melts the snow from the tundras they frequent. Daylengths in Britain are sufficiently long in the spring to stimulate gonad recrudescence and egg-laying but not further south than England. Thus, these species will not breed in captivity at latitudes lower than about 50° without the use of artificial lighting.

In contrast, the Hawaiian goose or ne-ne, *Branta sandvicensis*, which evolved from northern *Branta* stock to colonize montane slopes of

Hawaii (20°N), has a reduced photo-threshold. It became extinct in its natural range and has only been saved by breeding in captivity at Slimbridge (Fisher, Simon & Vincent, 1969). At Slimbridge, the Hawaiian goose becomes sexually active in autumn (cf. the black swan) but the short days of winter inhibit further activity until late January, when first eggs are laid. The breeding season then extends until late April (the latest egg so far laid was produced on 4 May). There is a possibility that by now the male and female are not properly in phase (see p. 24) and this may partly explain the high rate (40%) of infertility noted in captivity (Elder, 1958). There is also a risk that artificial breeding will select animals which lay most readily and these need not be the individuals which are most active in the autumn. In other words, animals raised in captivity may not have the endogenous capacity to breed between October and early February when returned to Hawaii. It could be important to control the photoperiod when breeding endangered species in captivity. Hawaiian geese at Slimbridge which were held in a pen which was floodlit until midnight moulted in January and failed to breed (Murton & Kear, 1973).

It was early noted that the median date of egg-laying at Slimbridge of closely related species was related to the mid-latitude of the natural breeding range as shown in Fig. 6 for *Branta*. This relationship becomes obscured if species are not truly close in phylogenetic terms, or if they have moved from their latitude of origin; this last happens in the case of montane species (Murton & Kear, 1973, 1976). Many pheasant (Phasianidae) species are montane derivatives of lowland forms which have been geographically isolated and the time of egg-laying is less clearly related to latitude in such forms (see data in Mallinson & Mallet, 1975). Nevertheless, the fact that the endogenous periodicities of closely related species become expressed in a systematic manner under uniform environmental conditions is of help to the aviculturalist. Information of the kind presented in Fig. 6, while possibly already appreciated empirically, does enable predictions to be made. More records of this kind ought to be collected referring to other taxa, for zoo collections have a wealth of material of potential scientific interest.

CO-ORDINATION OF BEHAVIOUR CYCLES

The hormones which regulate reproduction are initially stimulated and phased by responses to the daily photoperiod. These secretions encourage appropriate behaviour patterns which, in some species, in turn stimulate further increases or changes in endocrine secretion. It is frequently possible, especially under conditions of captivity, for the proper co-ordination of hormone secretion in the male and female to be dislocated so preventing the correct interaction of the internal and external events that regulate a breeding cycle. This is illustrated in Fig. 7

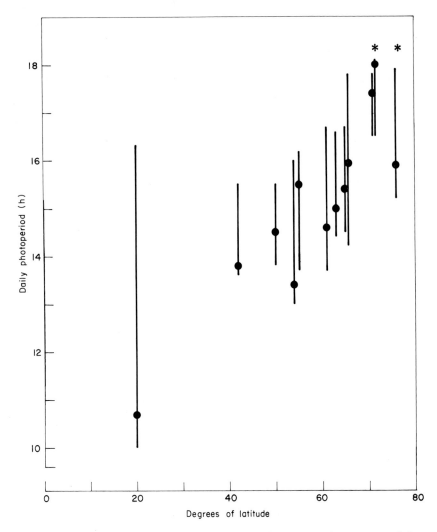

FIG. 6. Relationship between egg-laying season of different species or races of *Branta* geese in the Wildfowl Trust collection and the mid-latitude of their natural breeding range. Vertical bars denote the egg-laying season in terms of the shortest and longest daily photoperiod under which eggs have ever been laid, while solid dots give the median photoperiod for first eggs, based on records collected over a series of years. For two species, marked with asterisks, the bars have a slightly different configuration from those depicted in Fig. 5. In these species egg-laying extends over the summer solstice and so the bars are longer when plotted by date rather than by photoperiod.

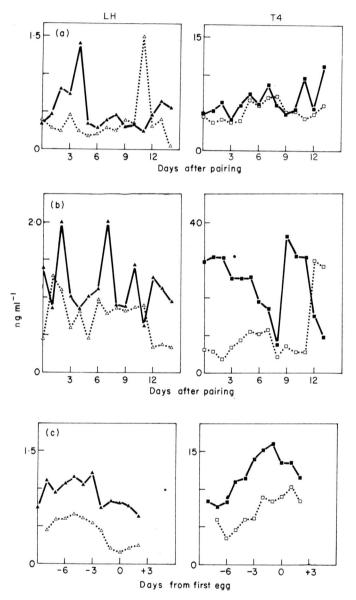

FIG. 7. Changes in plasma luteinizing hormone (LH) and thyroxine (T_4) in male (solid lines) and female (dotted lines) feral pigeons on pairing. (a) Pair produced no eggs. (b) Pair failed to produce eggs after normal time period but went on to lay late. (c) Mean hormone levels (nine pairs of birds) during the normal cycle. The first egg is laid on Day 0. For further details see text.

where the typical pattern of thyroxin and LH secretion during the pre-incubation cycle of the feral pigeon is illustrated: the abscissa gives the number of days elapsing between pairing and egg-laying.

During a normal cycle plasma LH titres increase within a day of pairing and remain elevated until a day or two before oviposition. T_4 levels increase a little later in the cycle and presumably at least partly function to increase the metabolic rate during a period of intense activity, when a nest must be built and the female must mobilize her body reserves to form the eggs. In the first example of an abnormal cycle, the male responded with LH secretion when first paired but the female did not. On the eleventh day the female was able to produce LH but by now the male could not and as a result no eggs were laid at all. T_4 secretion remained at the level noted in unpaired birds. In the second example, the pair missed their first attempt to synchronize their cycles for the female responded on Day 1 whereas the male delayed until Day 2, by which time the female was no longer stimulated. The pair entered a second cycle four days later which this time led to egg-laying, albeit 14 days after the birds were first introduced. At first there was no response in T_4 secretion, but with the second attempt, T_4 secretion increased as would happen during a normal cycle.

We wonder whether a disruption in the correct phasing of the reproductive cycles of the male and female may not occur to a greater extent than in the examples quoted above (cf. Hawaiian geese mentioned on p. 21) and prevent the successful breeding of animals in captivity. In the case of pigeons we know that different colour morphs have different photoresponses which affect their capability to pair successfully and there could be difficulties in attempts to breed animals which, though of the same species, originate from extreme parts of the range and so have slightly different photoresponses. Given that a reasonable approximation of the correct environmental milieu can be provided, the major obstacle that is likely to inhibit reproduction is an inappropriate photoregime. Even recourse to artificial insemination is likely to be unsuccessful if the light schedule under which subjects are kept inhibits the secretion of pituitary hormones.

CONCLUSIONS

The two most important ways in which the photoperiod adversely affects the reproduction of captive animals are by being insufficiently long (daylength cycles having too low an amplitude) to simulate species originating from high latitudes and which have evolved a high photothreshold, or by forcing a premature state of photorefractoriness on species which originate from low latitudes but which have an evolutionary history during which they once evolved refractory mechanisms. Species originating from low latitudes which lack such a mechanism seem to present no difficulties, from a photoperiodic viewpoint, when held captive at high latitudes. It would be sensible for zoos to collect more data of the kind presented in this paper and to give more thought to the manipulation of the daylength cycle. There is much scope for experiment.

REFERENCES

Ashmole, N. P. (1963). The biology of the Wideawake or Sooty tern *Sterna fuscata* on Ascension Island. *Ibis* **103**b: 297–364.

Assenmacher, I. (1974). External and internal components of the mechanism controlling reproductive cycles in drakes. In *Circannual clocks: annual biological rhythms*: 197–251. Pengelley, E. T. (ed.). New York and London: Academic Press.

Astier, H., Halberg, F. & Assenmacher, I. (1970). Rythmes circanniens de l'activité thyroidienne chez le Canard Pékin. *J. Physiol., Paris* **62**: 219–230.

Benoit, J., Mandel, P., Walter, F. X. & Assenmacher, I. (1950). Sensibilité testiculaire à l'action gonadotrope de l'hypophyse chez le canard domestique au cours de la régression testiculaire saisonnière. *C. r. Séanc. Soc. Biol.* **144**: 1400–1403.

Bolton, N., Chadwick, A. & Scanes, C. G. (1973). The effect of thyrotrophin releasing factor on the secretion of thyroid stimulating hormone and prolactin from the chicken anterior pituitary gland. *J. Physiol., Lond.* **238**: 78–79P.

Braithwaite, L. W. & Frith, H. J. (1969). Waterfowl in an inland swamp in New South Wales. III. Breeding. *C.S.I.R.O. Wildl. Res.* **14**: 65–109.

Bünning, E. (1936). Die endogene Tagesrhythmik als Grundlage der photoperiodischen Reaktion. *Ber. dt. bot. Ges.* **54**: 590–607.

Chandola, A. (1972). *Thyroid in reproduction. Reproductive physiology of* Lonchura punctulata *in relation to iodine metabolism and hypothyroidism.* Ph.D. Thesis no. 7774, Banaras Hindu University, Banaras (India).

Chapin, J. P. (1954). The calendar of Wideawake Fair. *Auk* **71**: 1–15.

Daan, S. & Pittendrigh, C. S. (1976a). A functional analysis of circadian pacemakers in nocturnal rodents. II. The variability of phase response curves. *J. comp. Physiol.* **106**: 253–266.

Daan, S. & Pittendrigh, C. S. (1976b). A functional analysis of circadian pacemakers in nocturnal rodents. III. Heavy water and constant light: homeostasis of frequency? *J. comp. Physiol.* **106**: 267–290.

Davies, D. T. & Follett, B. K. (1974a). The effect of intraventricular administration of 6-hydroxydopamine on photo-induced testicular growth in Japanese quail. *J. Endocr.* **60**: 277–283.

Davies, D. T. & Follett, B. K. (1974b). Changes in plasma luteinizing hormone in the Japanese quail after electrical stimulation of the hypothalamus. *J. Endocr.* **63**: 31–32.

Davies, D. T. & Follett, B. K. (1975a). The neuroendocrine control of gonadotrophin release in the Japanese quail. I. The role of the tuberal hypothalamus. *Proc. R. Soc.* (B) **191**: 285–301.

Davies, D. T. & Follett, B. K. (1975b). The neuroendocrine control of gonadotrophin release in the Japanese quail. II. The role of the anterior hypothalamus. *Proc. R. Soc.* (B) **191**: 303–315.

Delacour, J. (1939). The first nesting in captivity of the Australian shelduck.
Avic. Mag. **45**: 159–160.

Elder, W. H. (1958). *A report on the Nene.* Internal report to the Board of
Agriculture and Forestry, USA.

Fisher, J., Simon, N. & Vincent, J. (1969). *The red book: wildlife in danger.*
London: Collins.

Follett, B. K. (1975). Follicle-stimulating hormone in the Japanese quail: variation in plasma levels during photoperiodically induced testicular growth
and maturation. *J. Endocr.* **67**: 19P–20P.

Follett, B. K. (1976). Plasma follicle-stimulating hormone during photoperiodically induced sexual maturation in male Japanese quail. *J. Endocr.* **69**:
117–126.

Franchimont, P., Chari, S., Hagelstein, M. T. & Duraiswami, S. (1975). Existence
of a follicle-stimulating hormone inhibiting factor 'inhibin' in bull seminal
plasma. *Nature, Lond.* **257**: 402–404.

Graber, J. W., Frankel, A. I. & Nalbandov, A. V. (1967). Hypothalamic center
influencing the release of LH in the cockerel. *Gen. comp. Endocr.* **9**: 187–192.

Gwinner, E. (1968). Circannuale Periodik als Grundlage des jahreszeitlichen
Funktionswandels bei Zugvögeln. Untersuchungen am Fitis (*Phylloscopus
trochilus*) und am Waldlaubsänger (*P. sibilatrix*). *J. Orn., Lpz.* **109**: 70–95.

Gwinner, E. (1975). Circadian and circannual rhythms in birds. In *Avian biology*
5: 221–284. Farner, D. S. & King, J. R. (eds). New York and London:
Academic Press.

Haase, E., Sharp, P. J. & Paulke, E. (1975). Seasonal changes in plasma LH levels
in domestic ducks. *J. Reprod. Fert.* **44**: 591–594.

Hamner, W. M. (1971). On seeking an alternative to the endogenous reproductive rhythm hypothesis in birds. In *Biochronometry*: 448–461. Menaker, M.
(ed.). Washington D.C.: National Academy of Sciences.

Hoffmann, K. (1971). Splitting of the circadian rhythm as a function of light
intensity. In *Biochronometry*: 134–151. Menaker, M. (ed.). Washington D.C.:
National Academy of Sciences.

Hutchinson, J. C. D. (1962). The annual rhythm of egg production in fowls.
Wld's Poult. Congr. **12**: 124–129.

Hutchison, R. E., Hinde, R. A. & Bendon, B. (1968). Oviduct development and
its relation to other aspects of reproduction in domestic canaries. *J. Zool.,
Lond.* **155**: 87–102.

Jallageas, M., Assenmacher, I. & Follett, B. K. (1974). Testosterone secretion
and plasma luteinizing hormone concentration during a sexual cycle in
the Pekin duck, and after thyroxine treatment. *Gen. comp. Endocr.* **23**:
472–475.

Jones, P. J. & Ward, P. (1976). The level of reserve protein as the proximate
factor controlling the timing of breeding and clutch-size in the red-billed
quelea *Quelea quelea. Ibis* **118**: 547–574.

Kear, J. & Murton, R. K. (1973). The systematic status of the Cape Barren goose
as judged by its photo-responses. *Wildfowl* **24**: 141–143.

Kear, J. & Murton, R. K. (1976). The origins of Australian waterfowl as
indicated by their photoresponses. *Int. orn. Congr.* **16**: 83–97.

King, J. R. (1970). Photoregulation of food intake and fat metabolism in relation to avian sexual cycles. In *La photorégulation de la reproduction chez les oiseaux et les mammifères*: 365–385. Benoit, J. & Assenmacher, I. (eds). Colloques int. Cent. natn. Rech. scient. No. 172.

Lake, P. E. (1971). The male in reproduction. In *Physiology and biochemistry of the domestic fowl* **3**: 1411–1447. Bell, D. J. & Freeman, B. M. (eds). London and New York: Academic Press.

Lehrman, D. S., Brody, P. N. & Wortis, R. P. (1961). The presence of the mate and of nesting material as stimuli for the development of incubation behavior and for gonadotropin secretion in the ring dove (*Streptopelia risoria*). *Endocrinology* **68**: 507–516.

Lofts, B. (1962). The effects of exogenous androgen on the testicular cycle of the weaver-finch *Quelea quelea*. *Gen. comp. Endocr.* **2**: 394–406.

Lofts, B., Follett, B. K. & Murton, R. K. (1970). Temporal changes in the pituitary-gonadal axis. *Mem. Soc. Endocr.* **18**: 545–575.

Lofts, B. & Murton, R. K. (1968). Photoperiodic and physiological adaptations regulating avian breeding cycles and their ecological significance. *J. Zool., Lond.* **155**: 327–394.

Lofts, B. & Murton, R. K. (1973). Reproduction in birds. In *Avian biology* **3**: 1–107. Farner, D. S. & King, J. R. (eds). New York and London: Academic Press.

Lofts, B., Murton, R. K. & Thearle, R. J. P. (1973). The effects of testosterone propionate and gonadotrophins on the bill pigmentation and testes of the house sparrow *Passer domesticus*. *Gen. comp. Endocr.* **21**: 202–209.

Lofts, B., Murton, R. K. & Westwood, N. J. (1966). Gonad cycles and the evolution of breeding seasons in British Columbidae. *J. Zool., Lond.* **150**: 249–272.

Lofts, B., Murton, R. K. & Westwood, N. J. (1967a). Photoresponses of the woodpigeon *Columba palumbus* in relation to the breeding season. *Ibis* **109**: 338–351.

Lofts, B., Murton, R. K. & Westwood, N. J. (1967b). Interspecific differences in photosensitivity between three closely related species of pigeons. *J. Zool., Lond.* **151**: 17–25.

Mallinson, J. J. C. & Mallet, J. J. (1975). Egg-laying patterns and incubation periods of some bird species at the Jersey Zoological Park. *Rep. Jersey Wildl. Preserv. Trust* **12**: 6–10.

Maung, Z. W. (in press). Effect of LH and cAMP on steroidogenesis in interstitial cells isolated from the testis of the Japanese quail. *Gen. comp. Endocr.*

Miers, K. H. & Williams, M. (1969). Nesting of the black swan at Lake Ellesmere, New Zealand. *Rep. Wildfowl Trust* **20**: 23–32.

Murton, R. K. (1975). Ecological adaptation in avian reproductive physiology. *Symp. zool. Soc. Lond.* No. 35: 149–175.

Murton, R. K. & Isaacson, A. J. (1962). The functional basis of some behaviour in the woodpigeon *Columba palumbus*. *Ibis* **104**: 503–521.

Murton, R. K. & Kear, J. (1973). The nature and evolution of the photoperiodic control of reproduction in certain wildfowl (Anatidae). *J. Reprod. Fert.* Suppl. **19**: 67–84.

Murton, R. K. & Kear, J. (1976). The role of daylength in regulating the breeding seasons and distribution of wildfowl. In *Light as an ecological factor* **2**: 337–360. Bainbridge, R. (ed.). Oxford: Blackwell Scientific Publications.

Murton, R. K. & Westwood, N. J. (1975). Integration of gonadotrophin and steroid secretion, spermatogenesis and behaviour in the reproductive cycle of male pigeon species. In *Neural and endocrine aspects of behaviour in birds*: 51–89. Wright, P., Caryl, P. G. & Vowles, D. M. (eds). Amsterdam: Elsevier.

Murton, R. K. & Westwood, N. J. (1977). *Avian breeding cycles.* Oxford: Clarendon.

Nicholls, T. J., Scanes, C. G. & Follett, B. K. (1973). Plasma and pituitary luteinizing hormone in Japanese quail during photoperiodically induced gonadal growth and regression. *Gen. comp. Endocr.* **21**: 84–98.

Nicholls, T. J. & Storey, C. R. (1977). The effect of duration of the daily photoperiod on recovery of photosensitivity in photorefractory canaries (*Serinus canarius*). *Gen. comp. Endocr.* **31**: 72–74.

Oksche, A., Kirschstein, H., Hartwig, H. G., Oehmke, H. J. & Farner, D. S. (1974). Secretory parvocellular neurons in the rostral hypothalamus and in the tuberal complex of *Passer domesticus. Cell Tissue Res.* **149**: 363–370.

Pavlidis, T. (1973). *Biological oscillators: their mathematical analysis.* New York and London: Academic Press.

Pittendrigh, C. S. (1966). The circadian oscillation in *Drosophila pseudoobscura* pupae: a model for the photoperiodic clock. *Z. Pflanzenphysiol.* **54**: 275–307.

Pittendrigh, C. S. & Daan, S. (1976a). A functional analysis of circadian pacemakers in nocturnal rodents. I. The stability and lability of spontaneous frequency. *J. comp. Physiol.* **106**: 223–252.

Pittendrigh, C. S. & Daan, S. (1976b). A functional analysis of circadian pacemakers in nocturnal rodents. IV. Entrainment: pacemaker as clock. *J. comp. Physiol.* **106**: 291–331.

Pittendrigh, C. S. & Daan, S. (1976c). A functional analysis of circadian pacemakers in nocturnal rodents. V. Pacemaker structure: a clock for all seasons. *J. comp. Physiol.* **106**: 333–355.

Pittendrigh, C. S. & Minis, D. H. (1964). The entrainment of circadian oscillations by light and their role as photoperiodic clocks. *Am. Nat.* **98**: 261–294.

Rosales, A. A., Biellier, H. V. & Stephenson, A. B. (1968). Effect of light cycles on oviposition and egg production. *Poult. Sci.* **47**: 586–591.

Schwab, R. G. (1971). Circannian testicular periodicity in the European starling in the absence of photoperiodic change. In *Biochronometry*: 428–445. Menaker, M. (ed.). Washington D.C.: National Academy of Sciences.

Stetson, M. H. (1969). Hypothalamic regulation of FSH and LH secretion in male and female Japanese quail. *Am. Zool.* **7**: 1078–1079.

Storey, C. R. & Nicholls, T. J. (1976). Some effects of manipulation of daily photoperiod on the onset of a photorefractory state in canaries (*Serinus canarius*). *Gen. comp. Endocr.* **30**: 204–208.

Thapliyal, J. P. (1969). Thyroid in avian reproduction. *Gen. comp. Endocr.* Suppl. **2**: 11–122.

Tindall, D. J., Schrader, W. T. & Means, A. R. (1974). The production of androgen binding protein by Sertoli cells. In *Hormone binding and target cell activation in the testis* (Current topics in molecular endocrinology, I): 167–175. Dufau, M. L. & Means, A. R. (eds). New York and London: Plenum Press.

Watson, A. & Moss, R. (1969). Dominance, spacing behaviour and aggression in relation to population limitation in vertebrates. *Symp. Br. ecol. Soc.* No. 10: 167–220.

Wieselthier, A. S. & van Tienhoven, A. (1972). The effect of thyroidectomy on testicular size and on the photorefractory period in the starling, *Sturnus vulgaris. J. exp. Zool.* **179**: 331–338.

Symp. zool. Soc. Lond. (1978) No. 43, 31–49

The Principles and Practice of Semen Collection and Preservation in Birds

P. E. LAKE

ARC Poultry Research Centre, Edinburgh, Scotland.

SYNOPSIS

Artificial semen collection and preservation *in vitro* has been practised with domestic birds for more than 40 years, particularly in relation to the breeding of turkeys and domestic fowl commercially by artificial insemination (AI). Undiluted semen will deteriorate within 30 to 45 min, or sooner, depending upon initial quality. It must be diluted if the fertilizing ability of spermatozoa is to be preserved for a longer period *in vitro* before insemination and much of our knowledge of the techniques and problems of achieving maximum fertility with stored semen has been derived from studies on the turkey, *Meleagris gallopavo*, and the fowl, *Gallus domesticus*.

Interest is beginning to be shown in the possibility of applying these techniques as aids in the breeding of other bird species. This account discusses some advantages that have already accrued, and some which could be gained, from using AI with fresh or stored semen in breeding several types of birds. Some methods that have already been employed to obtain semen from birds, other than domestic species, are described pointing out the variations in anatomical features of the reproductive organs which govern the technique employed in each case to obtain good quality semen. This is one vital factor influencing the fertility success with stored semen.

Methods of diluting and preserving semen for long and short periods *in vitro* are discussed drawing on experiences from domestic birds. Comments are made on the general features of semen dilution and storage procedures which influence the viability of spermatozoa before insemination. Initial attempts at diluting and storing the semen of a few other birds are also reviewed.

Lastly, during the storage of semen there is a progressive increase in the death rate of spermatozoa and, since the achievement of maximum fertility is dependent upon the number of good quality spermatozoa actually inserted into the female and their being able to survive in the oviduct for many days, factors associated with oviduct physiology and insemination technique that influence the survival of spermatozoa in the oviduct are briefly outlined.

INTRODUCTION

The practice of semen collection and *in vitro* preservation in birds has been confined mainly to domestic species, particularly the turkey and the domestic fowl, where artificial insemination (AI) is commonly practised in commercial breeding (Lake & Stewart, 1978a; Lake & Nixey, 1979).

Semen has also been collected from a few types of non-domestic birds for AI purposes, but there has been little or no attempt to preserve their semen *in vitro*, although the possibility is beginning to be explored.

Our knowledge of the methods and problems associated with collecting and preserving semen *in vitro* has been gained largely from studies on domestic birds, but it is feasible that the general principles will serve to indicate the general course of development of such techniques should there be any need in the future to apply them to aid in the breeding of a wide variety of birds.

There are several reasons for a general interest in methods of semen collection and preservation in birds for short or long periods so that artificial breeding can be practised. Unprecedented declines in the population of some birds, e.g. peregrine falcons (Hickey & Roelle, 1969), other birds of prey and the Copper pheasant (Higuchi, 1972) have already directed attention towards seeking reliable techniques for captive propagation and AI. However, one great problem governing the success of such ventures is the creation of the correct environmental conditions, including housing and diet, to encourage the birds to attain sexual maturity, develop the desire for mating and maintain breeding condition. Mendelssohn & Marder (1970) discussed the many facets of the problem of breeding birds of prey in captivity, Pendergast & Boag (1971) dealt specifically with spruce grouse and Yamashina (1976) with methods of rearing and breeding the Japanese Copper pheasant for successful artificial breeding. Owen (1941) and Dowling (1976) discussed the importance of providing correct housing conditions for pigeons to attain optimum breeding condition, i.e. for males to produce semen and females to produce eggs. Boyd, Boyd & Dobler (1977) did likewise for the prairie falcon, *Falco mexicanus.*

Crossing birds of different genera becomes possible by using semen collection techniques and AI. For example, domestic fowl have been crossed with quail (McFarquhar & Lake, 1964; Ogasawara & Huang, 1963) and the problems of crossing the fowl with the pheasant have been discussed (Shaklee & Knox, 1954; Watanabe *et al.*, 1963). However, the fertilization rates and survival of offspring are problems of this practice which have been discussed generally by Lorenz, Ogasawara & Asmundson (1964), McFarquhar & Lake (1964), Sarvella & Marks (1970) and Makos & Smyth (1970).

Pigeon breeders have developed an interest in AI for several reasons; it will, for example, enable breeding with sexes of different sizes, allow wider use of individual proven males in a basically monogamous species, enable the production of outcrosses, allow the use of a valuable injured male and also allow the transfer of semen to areas in a loft without exposing a male to disease problems where they exist (Owen, 1941; Raises, 1968; Dowling, 1976; Szumowski, Theret & Denis, 1976).

Species of rare exotic birds are often kept in pairs and in small numbers. In these cases, it could be of interest to use AI so that proven fertile males giving large volumes of good quality semen can be bred to many females.

It may be acceptable to veterinary authorities in some circumstances to transport semen of birds rather than live birds in which cases semen storage methods and AI would be valuable assets. In this connection, particularly with domestic birds and pigeons, genes from valuable stock may be saved, if, for example, there is a severe disease problem affecting a breeding site; it would be safer to bring out semen than eggs or surviving birds.

Artificial insemination offers a means of breeding certain non-domestic birds as potential laboratory animals, e.g. blue grouse are difficult to breed in captivity and a semen collection technique, similar to that used for the domestic fowl, has enabled breeding under these conditions (Stirling & Roberts, 1967).

The domestic goose, though not so monogamous as the wild type, may show preference for certain females. Artificial breeding in this case offers a means of inseminating a larger number of females thereby increasing the progeny and reducing the cost of production of goslings. This economic factor, as illustrated by the goose, is important in the breeding of several other types of domestic poultry (Lake & Stewart, 1978a; Lake & Nixey, 1979).

Several advantages thus accrue from being able to carry out artificial breeding. However, in general, caution must be taken to avoid the development or spread of hereditary faults which could easily occur as a result of inbreeding and the concentration of adverse genes in a population. With suitable precautions in domestic and some other birds, in addition to enabling the preservation and propagation of species, AI can be used to increase the rate of genetic improvement in a population by making possible the progeny testing of males and the use of proven sires. In general, controlled breeding between selected dams and sires is made possible.

The use of methods of semen preservation makes it possible to increase the number of hens inseminated by a particular male. Prolonged storage of semen at sub-zero temperature enables the creation of banks of semen from selected desirable sires to be used long after their death. It would provide insurance against the extinction of a species due to habitat deterioration by pollutants or by other causes. Short-term and prolonged storage of semen facilitates the transportation, between and within countries, of semen of desirable sires. Techniques of semen dilution and storage have other obvious economic advantages in the poultry breeding industry (Lake & Stewart, 1978a).

Finally, methods of collecting semen and AI are valuable tools enabling the investigation of some fertility problems by examining quality of semen and testing individual females directly.

SEMEN COLLECTION

Most of our knowledge on the practice of semen collection has been gained from extensive studies on various species of domestic birds, e.g. the fowl, the turkey, the guinea-fowl, ducks and geese. Detailed descriptions of the techniques of semen collection used with these birds are given by Lake & Stewart (1978a). The information may serve to demonstrate how adaptations can be made in attempts to collect semen from a wide variety of other birds. Non-domestic birds are temperamental in yielding semen and the amount obtained is often small and very variable (Table I).

The Form of the Male Reproductive Organs Pertinent to Semen Collection Procedures

The form of the avian male reproductive tract in relation to the collection of semen has been extensively studied in the domestic fowl (see review by Lake, 1971) and it is shown diagrammatically in Fig. 1.

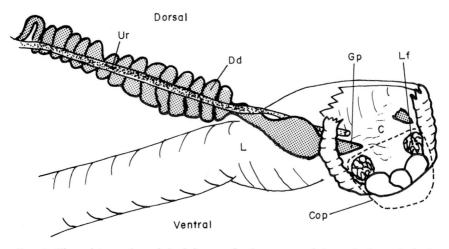

FIG. 1. The pelvic portion of the left reproductive organs of the male domestic fowl, including the distal part of the ductus deferens and the erectile phallic structures in the cloaca. Dd, Ductus deferens; Gp, ejaculatory duct (papilla) protruding into the cloaca; C, cloacal chamber cut away dorsally; Cop, erectile copulatory appendages in the ventral cloaca including lymph folds (Lf) from which watery fluid exudes; Ur, ureter; L, large intestine (from Lake & Stewart, 1978a).

However, whilst it is similar to other galliform birds (turkey, guinea-fowl, partridge, quail, grouse), features of its form are modified in other birds. The intra-abdominal parts of the genital organs, i.e. the testes and the ducti deferentes, are similar in most birds but the form of the erectile copulatory appendages in the cloaca varies between species. For example, in the palmipeds (ducks, geese) a penis-like structure unwinds as it erects from the cloaca during sexual excitation (Fujihara & Nishiyama, 1976a; Fujihara, Nishiyama & Nakashima, 1976; Nishiyama, Nakashima & Fujihara, 1976). The passerine birds and possibly some parrots (see

Rand, 1936; Wolfson, 1958) have densely coiled and swollen distal regions of the ducti deferentes lying beneath the skin dorsolateral to the cloaca which cause a protuberance through the cloacal wall when the birds are in full breeding condition (Salt, 1954—vesper sparrow, *Pooecetes gramineus*; Wolfson, 1954—several passerine species).

The reproductive tract of the bird in general differs from that of most of the mammals in two main respects. First, the testes are internal and, secondly, the accessory reproductive organs, namely seminal vesicles, prostate gland, Cowper's glands and the urethral glands of Littré, are absent. Semen is thus chiefly composed of spermatozoa and fluid secretions from the seminiferous tubules and the lining epithelial cells of the entire reproductive tract. Its chemical composition reflects the absence of the secretions of the above accessory glands (see Lake, 1971) because some of the substances commonly produced by mammalian accessory glands, e.g. fructose, citric acid, ergothioneine and specific prostatic acid phosphatase (Mann, 1964) are absent from fowl semen.

In general, the absence of large volumes of seminal fluid, typical of the semen of several domestic mammals, accounts for the high density of spermatozoa and the high viscosity of the semen of most birds.

However, in domestic fowl, ducks and geese under many circumstances a large volume of a watery fluid originating from erectile glandular tissue in the cloaca enters into the semen, in which case it reduces the concentration of spermatozoa. The significance of this fluid as a seminal component has been discussed by Lake (1971) for the fowl and by Nishiyama *et al.* (1976) for the duck. It can adversely affect the keeping quality of spermatozoa under certain conditions *in vitro* (Lake, 1971; Fujihara & Nishiyama, 1976b) and also fertility after immediate insemination if it is present in a large volume causing a reduction in the density of spermatozoa in the collected semen sample (Van Wambeke, 1976). Semen is easily contaminated with urine and/or faeces during semen collection because of the presence of a cloaca in bird anatomy. These materials also affect spermatozoa adversely and must be avoided, especially if semen is to be preserved *in vitro*.

Methods of Semen Collection

Galliformes

A general rule in artificial breeding is that males should be separated physically from females for maximum amounts of semen to be obtained.

The erection of the copulatory organ and the ejaculation reflex can be elicited in the domestic fowl, turkey, guinea-fowl, pheasant, quail and grouse by gently massaging (stroking with the palm of the hand) the back of the bird, and sometimes the abdomen, towards the tail (Lake & Stewart, 1978a—domestic fowl, turkey, guinea-fowl; Marks & Lepore, 1965—quail; Watanabe *et al.*, 1963; Maru *et al.*, 1966; Spiller, Grahame

& Wise, 1977—pheasant). The hand stroke ends with the thumb and forefinger making a slight upwards movement at the base of the tail. When the copulatory appendages in the cloaca become erect, semen is rapidly expressed from the swollen ejaculatory papillae (Fig. 1) by gently pressing inwards with the thumb and forefinger on either side of the cloaca. The method of holding the bird and the number of personnel required for the operation of semen collection varies according to the species of bird. The procedure is generally performed by two persons, one restraining the male and the other massaging. Either person expresses out the semen and catches it in an open collecting vessel or by suction into a cannula or closed vessel depending upon the size of the bird and the nature of the ejaculation process. For example, sometimes semen squirts readily from the ejaculatory papillae in the cloaca of the fowl, in which case it is best collected in an open vessel, or it wells up on the erectile appendages (turkey, quail, guinea-fowl), in which case it is best aspirated into a closed vessel (see Lake & Stewart, 1978a for details).

The best ejaculatory response is obtained from these birds if they are handled gently. Too much excitement caused by rough handling may inhibit erection and the ejaculatory reflex and sometimes increases the chances of defaecation and urination. Frequent handling (training) will often lead to improved responses to massage. The method of restraining and the response of galliform birds during semen collection depends upon their tameness and size. For example, blue grouse are held with their head and shoulders under an arm of the person holding them, whilst another person massages the bird around the base of the tail (Stirling & Roberts, 1967). Quail are generally held between the hands of one person whilst massage is applied by another (Marks & Lepore, 1965). In the quail foamy material, produced by cloacal glands, is first expelled by applying light pressure with the fingers before expelling the semen (Ogasawara & Huang, 1963). It has no harmful effect on spermatozoa should it contaminate the semen (Sarvella & Marks, 1970; Kobayashi, Okamoto & Matsuo, 1972) but hinders the collection method. The best response from pheasants has been obtained by using gentle methods of restraint (Spiller et al., 1977).

On average, the volume of semen obtained from the quail and the pheasant is small and it is advisable to collect it by an aspiration method into a small-diameter tube (Ogasawara & Huang, 1963; Spiller et al., 1977).

Columbiformes

Semen is collected from these birds in a similar manner to that of the quail and grouse (Raises, 1968; Szumowski et al., 1976). However, a single person may attempt the whole operation (Owen, 1941). Again, the volume of semen is small and it is best collected by aspiration into a micropipette previously filled with a small volume of diluent.

Falconiformes

The co-operative technique of semen collection is generally employed to collect semen from these birds and is dealt with by Gee & Temple (this volume, p. 51). It differs from the traditional massage method described for galliform and other birds because it requires the male birds to be imprinted to human handlers (Berry, 1972—American goshawk, *Accipiter gentilis*; Temple, 1972—red-tailed hawk, *Buteo jamaicensis*; Grier, 1973—golden eagle, *Aquila chrysaetos*; Boyd *et al.*, 1977—prairie falcon, *Falco mexicanus*). Semen from these males is generally deposited on a covered hand of the person carrying out the elicitation of copulation and ejaculation and is pipetted quickly into a container tube.

Often only a small proportion of attempted copulations result in the production of semen (Berry, 1972). Also, a high proportion of ejaculates from some falconiformes have been reported to be unusable because the volume was not great and it is difficult to retrieve (Grier, 1973). In these cases instant dilution of the available semen would be an advantage for handling.

A non-co-operative technique of semen collection by massage has been used successfully with the American kestrel, *Falco sparverius* (Bird, Laguë & Buckland, 1976; Bird & Laguë, 1977).

Anseriformes

These birds have a penis-like copulatory appendage which unwinds from the cloaca in response to massage and pressure applied with the fingers. The penis is made to enter a large tube or other type of collecting vessel (see Lake & Stewart, 1978a) into which the semen is ejaculated; it is then pipetted into a tube. As previously mentioned, the semen of these birds is easily contaminated with a watery fluid that exudes from erectile cloacal tissue; care must be taken to avoid this contamination otherwise the density of spermatozoa is markedly reduced (Nishiyama *et al.*, 1976).

Induction of ejaculation by electrical stimulation has been performed with ducks and is dealt with by Gee & Temple (this volume, p. 51).

Passeriformes

Semen from galliform birds, water-fowl, pigeons and Falconiformes has been collected and used for artificial insemination. That collected from a few species of passeriform birds has not yet been used for this purpose but chiefly for an examination of the breeding condition of the male. The semen of the Brewer blackbird, *Euphagus cyanocephalus*, has been obtained by conditioning the bird to the presence of humans and eliciting ejaculation by training it to copulate with appropriate stuffed specimen birds of its own species (Howell & Bartholomew, 1952).

TABLE I

Semen volume obtained artificially from different birds, and the conditions under which diluents have been used

Species	Mean semen vol (ml) (range)	Diluent	Dilution ratio	Insemination dose (ml)	Holding temperature (°C)	Holding time	Reference
Duck	0·23 (0·1–1·0)	Egg yolk-citrate	1/4	0·3	Moderate ambient	15 to 30 min	Watanabe (1961)
Domestic fowl	0·2 (0·08–0·5)	Fructose-glutamate	1/1	0·04	Moderate ambient	30 min	Lake & Stewart (1978a)
Domestic fowl	NG	Egg albumen-milk-glutamate	1/1	0·1	2 to 3	24 hour	Van Wambeke (1972)
Turkey	0·2 (0·08–0·33)	Fructose-glutamate	1/2	0·06	15	5 hour	Lake & Stewart (1978a)
Guinea-fowl	0·075 (0·05–0·15)	Valine-glutamate	1/1	0·06	Moderate ambient	20 min	M. J. Petitjean (pers. comm.)—see Lake & Stewart (1978a)
Pheasant	0·02 (0·005–0·1)	Ringer's	1/3	0·05	Moderate ambient	15 min	Maru et al. (1966)
Pheasant	NG	Egg yolk-citrate	1/2 or 1/3	0·4	Moderate ambient	15 min to 2·5 hour	Watanabe et al. (1963)

Species						
Pigeon	NG (0·01–0·02)	Ringer's	1/2	0·03	Moderate ambient	15 min	Owen (1941)
Pigeon	0·016 (0·002–0·04)	Ringer's	1/3 or 1/4	0·05 to 0·1	Moderate ambient 20	15 min	Szumowski et al. (1976)
Quail	0·01 (NG)	Modified Ringer's, Locke's or Tyrode's	1/1	0·02	Moderate ambient	15 min	Lepore & Marks (1966)
Golden eagle	NG (0·05–0·2)	Nil	—	0·05	Moderate ambient	30 min	Grier (1973)
Red-tailed hawk	0·1 (NG)	Nil	—	0·05?	Moderate ambient	30 min	Temple (1972)
American goshawk	0·025 (drop to 0·05)	Nil	—	0·05	Moderate ambient	15 min	Berry (1972)

NG, Not given.

When sexually active, several species of passeriform birds have a protuberance in the cloaca consisting of the internal, dilated distal parts of the ducti deferentes (Salt, 1954; Wolfson, 1952, 1958). Semen can be collected by holding the male in the hand and gently squeezing the semen from the bulbous ducti deferentes and extruded papillae in the cloaca with the thumb and forefinger, care being taken not to squeeze out faeces and/or urine. A drop of semen collected in this way can be deposited on a microscope slide. Normally, a small quantity of diluent is added to prevent the semen drying out as it is very viscous and contains a very dense population of spermatozoa (Wolfson, 1952). It is feasible that the diluted semen could be inseminated if required.

SEMEN DILUTION AND PRESERVATION

Experience with domestic birds has shown that sufficient viable spermatozoa to give maximum fertility can be maintained in undiluted, fresh semen for only between 30 and 45 min at common ambient temperatures (between 5 and 20°C). From the moment of collection, spermatozoa begin to die, the progressive death rate depending upon handling conditions *in vitro*. This is an important basic factor to remember, especially when practising semen dilution and preservation, because additional stress is experienced by spermatozoa and losses may be greater. Diluents and semen storage techniques are designed to minimize the rate of loss of viability of spermatozoa *in vitro*.

The main objects of using a diluent are either to enable many females to be inseminated immediately with fresh semen, in which case semen is diluted and used within 30 to 45 min at ordinary ambient temperatures or to preserve semen for a longer period between collection and insemination, in which case it is kept at a low temperature.

Diluents are an asset to enable ease of handling of the minute ejaculate volumes obtained from some birds, e.g. passeriforms, quail, pigeons, goshawks, some eagles and pheasants, especially if the density of spermatozoa is great thus rendering the semen very viscous (see pp. 38 and 40). However, over-dilution must be avoided since, as mentioned below, the numbers of spermatozoa inseminated are important for achieving maximum fertility. This is particularly relevant if it is desired to split the diluted semen between several females.

To achieve maximum fertility with stored semen in birds more spermatozoa must be revived than is the case with, for example, the bull, because spermatozoa are required to survive in the oviduct to fertilize eggs ovulated over many days after insemination.

Details of the initial procedures for handling semen to minimize damage to spermatozoa are given for domestic birds by Lake & Stewart (1978a) in Appendix B. Certain points, however, are applicable to the handling of semen of all birds, particularly the importance of collecting uncontaminated semen containing good quality spermatozoa, the avoidance of too great a temperature differential between the semen as it is ejaculated and the containers or fluids into which the semen is immediately placed, the avoidance of contamination with dust, detergent, other chemicals or water in containers, and the prevention of exposure to direct sunlight.

Short-Term Storage of Semen

In general several common physiological saline solutions suffice for diluting semen that is to be inseminated within about 15 min of collection. A few special diluents have been developed (Table I), however, for keeping the semen of some domestic birds for up to 5 hours above 0°C. The composition of a diluent and methods of storage are very critical if it is desired to store semen beyond this time.

Some general criteria (Mann, 1964) have to be satisfied for diluents to be suitable for sustaining spermatozoa *in vitro*.

Buffering action

During metabolism acids are produced which damage cells. A diluent must be able to neutralize these noxious effects and usually contains substances such as glutamate, phosphates of sodium and potassium or biological buffer salts, e.g. TRIS, MES, BES, TES, for the purpose. A diluent must be controlled at a constant pH throughout the storage period to enable enzymes to function properly to maintain the viability of the spermatozoa. The pH is generally held at some point between 6·8 and 7·4.

Chemical composition and tonicity

The kind and concentration of certain chemical substances is important in a diluent for maintaining the integrity of spermatozoa. Cell membranes, water content and enzymes are some of the features of spermatozoa dependent upon the tonicity of the suspending medium and the presence of certain chemicals for their optimum functioning. Slightly hypertonic solutions are often best for fowl spermatozoa *in vitro* (Van Wambeke, 1977). Calcium, ammonium and chloride ions present above a certain concentration in solution can have adverse effects on fowl spermatozoa (see review by Lake, 1966).

Nutrient content

Spermatozoa require certain sugars, amino acids and other compounds as energy sources for their survival.

Antibacterial agents

If avian semen is used soon after collection or stored for only a brief period before insemination then an increase in the microbial population would be negligible. It is questionable, therefore, whether antibiotics or other antibacterial agents are essential in diluents under these circumstances. However, it is sometimes thought that the presence of such agents may guard against the spread of some diseases in the oviduct of birds from the accidental use of dirty semen and/or oviduct damage occurring during insemination. Antibacterial compounds may usefully be incorporated in diluents when semen is stored for up to 24 hours and there is a possibility of toxins being produced which may harm spermatozoa. However, the use of antibacterials should *never encourage the use of dirty semen.* Certain antibiotics are known to chelate toxic elements (Wilcox, 1959) which may be present in diluted semen. Their presence in these circumstances would render the solution harmless to the spermatozoa.

Keeping quality

Diluents by virtue of their composition are media ideal for the growth of bacteria or moulds. They will not keep indefinitely and should be stored in a refrigerator if not used immediately after manufacture. Freeze-dried preparations or sterile solutions will keep best.

Extensive work has been carried out towards perfecting an ideal diluent for storing the semen of various species of domestic birds above 0°C, e.g. the fowl, the turkey, the duck, the goose and the guinea-fowl (Ogasawara & Ernst, 1970; Van Wambeke, 1972; Sexton, 1976a; Lake & Stewart, 1978a). No work has been done specifically to develop diluents for other types of birds but adaptations of poultry semen diluents and other media have been used in an *ad hoc* manner where required (Watanabe *et al.*, 1963; Higuchi, 1972; Yamashina, 1976—pheasant; Owen, 1941; Szumowski *et al.*, 1976— pigeon; Lepore & Marks, 1966— quail). In these cases they are helpful only for immediate dilution and insemination because 'diluents for long-term storage of semen have more specific properties.

Information about diluents and their usage for various birds is given in Table I.

Long-Term Storage at Sub-Zero Temperature

As was mentioned above the properties of diluents and the physical conditions under which diluted semen is kept vary depending upon whether semen is to be diluted and inseminated within about 15 to 30 min, within 24 hours or kept for an indefinite period before insemination. Semen has to be diluted and frozen if it is to be stored indefinitely. Some recognized advantages of such a procedure in the breeding and conservation of birds are outlined on p. 33.

Methods have been devised for freezing the spermatozoa of the domestic fowl (Mitchell *et al.*, 1975; Sexton, 1976b; Watanabe & Terada, 1976; Lake & Stewart, 1978b) and turkey (Macpherson, Chatterjee &

Friars, 1969) and have been most successful with the former. However, further work is necessary to improve the survival rate of spermatozoa with current methods.

The criterion of success for the avian physiologist in attempting to freeze-store semen is to reproduce the prolonged *fertile period* (Lake, 1975) in the female whereby a succession of fertile eggs are produced over a period of several days or weeks, depending upon the species, after a single insemination. This is unlike the case with the mammal, e.g. the cow or sow, when only a single egg, or several eggs ovulated over a period of about six hours are required to be fertilized.

A summary of the fertility results revealed by the current methods for the domestic fowl shows that compared with an average fertility level of about 92% with fresh semen, frozen semen has given a 30 to 75% range of egg fertilization rate during the first week after insemination. Watanabe & Terada (1976) obtained high fertility but semen was inseminated directly into the uterus (shell gland) instead of intravaginally which is normal practice. Lake & Stewart (1978b) showed that it was possible to obtain 80% fertile eggs from Days 2 to 6 inclusive post-insemination if semen was stored and revived under optimum conditions in their experiment.

In order to achieve the maximum level of fertility in birds, particularly the fowl or turkey, many more spermatozoa need to be revived in a fully viable state after freezing than is the case with the mammal, i.e. more than a 60 to 70% revival rate is necessary. The task is difficult and, although encouraging fertility has been obtained with fowl semen stored in liquid nitrogen for a few years, improvements have still to be made.

Three aspects of the method are known to be vital to success; others may yet be revealed. These are the composition of the diluent, the nature of the cryoprotective agent and the freezing procedure. All need attention for future improvements in technique. The greatest level of fertility in the fowl has so far been achieved with glycerol as a cryoprotective agent (present at a final concentration of 7 to 8% v/v in diluted semen). However, unlike the situation in some mammals, its concentration must be reduced to less than 2% after thawing and before insemination into the vagina otherwise nil or very low fertility is obtained. This necessitates a step in the thawing procedure to remove glycerol and it would be more convenient for field practice if this step could be eliminated. Further research has been directed towards seeking an alternative to glycerol and Sexton (1976b) has recently described a method using dimethyl sulphoxide (DMSO). It is too early to evaluate its efficacy compared with the results that have so far been obtained with glycerol. Glycerol appears to be more harmful to turkey spermatozoa (Marquez & Ogasawara, 1977) in which case fertility with frozen semen has only been obtained with intra-magnal insemination. Brown & Graham (1971) studied the effect of various cryoprotective agents on turkey spermatozoa *in vitro*. They suggested that ethylene glycol may be a more favourable substance than any of the following: glycerol, DMSO, 1,3- or 1,2-propanediol. However, no fertility tests after freezing were performed in this study.

Little work has been reported on direct comparisons of different freezing techniques and diluents for freezing avian spermatozoa, and no detailed specific studies have been reported with non-domestic birds. However, Sexton & Gee (this volume p. 89) report an attempt to modify a method developed for fowl semen for use with semen of the sandhill crane, *Grus canadensis*.

Clearly, the preservation of semen at sub-zero temperature for indefinite periods offers many advantages in the area of bird conservation and it is to be hoped that rapid progress will be made in the improvement of techniques in the not too distant future.

FACTORS AFFECTING FERTILITY FOLLOWING INSEMINATION WITH STORED SEMEN

A basic feature of reproduction in the female bird is that spermatozoa survive in the oviduct after a single insemination to fertilize a number of eggs laid daily in succession. This is known as the *fertile period* for the species (Lake, 1975). It is known that at least 80 to 100 million spermatozoa must be inseminated into a fowl or turkey female to achieve the maximum length of fertile period (Lake & Stewart, 1978a; Sexton, 1977). In this respect, Table II gives examples of the implications of density of spermatozoa in the semen of two domestic birds in determining the dosage of diluted semen for achieving maximum fertility. The

TABLE II

Facts relating average density of spermatozoa in semen, dilution rate and insemination dosage for maximum fertility from single inseminations of domestic fowl and turkeys

	Turkey	Fowl
Average density of spermatozoa in a good semen sample ($10^6 \, ml^{-1}$ semen)	12 000	4000
Dose of semen for maximum fertility per insemination (ml)	0·009	0·025
Dilution rate and insemination dose (ml) per female for semen kept up to 45 min *in vitro*	3-fold; 0·03	2-fold; 0·04
Dilution rate and insemination dose (ml) per female for semen kept up to 5 hours *in vitro*[a]	3- or 4-fold; 0·06	2-fold; 0·06

[a] For special conditions for successful storage, see Lake & Stewart (1978a).

same principles would have to be considered in initial attempts to dilute and store semen of other birds.

The ideal diluent would enable all spermatozoa to survive *in vitro* for an indefinite period. This has not yet been achieved with any animal and with current techniques there is a progressive loss of spermatozoal viability with increasing duration of storage. In birds it is liable to affect the length of the fertile period and reduce overall fertility after a single insemination. Hence, great care must be exercised in inseminating a female bird with stored semen in order to achieve the maximum possible fertility with the remaining viable spermatozoa.

In this respect it is worthwhile recognizing several other factors about insemination procedure and oviduct physiology which influence fertility and assume greater importance with stored semen where numbers of spermatozoa are at a premium. They have been observed with domestic birds but to some extent could be pertinent to other birds. In conclusion, these are briefly described below. Detailed information on the subject has been reviewed by Lake (1969).

Species, Breed, Strain and Crosses

Different species, breeds, strains and their crosses show different lengths of fertile period when inseminated with semen of the same or different kind.

Intensity of Lay

In a given female, fewer infertile eggs and a longer duration of fertility are obtained generally with good intensively-laying females, probably owing to a better environment for the sustenance of spermatozoa in the oviduct of such females.

Age of Hen and/or Advancing Season

A decline in fertility and hatching ability of eggs occurs with increasing age of female and towards the end of a breeding season, particularly in birds with a long breeding season. The precise underlying cause(s) is unknown. More frequent inseminations may alleviate conditions at these times and maintain high fertility especially if a decline in the density of spermatozoa in semen is also partly responsible.

Position of Egg in Oviduct

The presence of a hard-shelled egg in the uterus (shell gland) at insemination often leads to an inadequate insertion of semen, a consequent loss of some spermatozoa and a reduction in the length of the fertile period.

Placement of Semen in the Oviduct

During artificial insemination, the vagina should be everted well and semen placed deep into the vagina near to the sites of the sperm storage glands (Lake, 1975; Ogasawara & Fuqua, 1972) to lessen the chances of regurgitation and poor fertility or complete infertility occurring after an insemination. The male does not place semen so far into the oviduct naturally, but accompanying oviducal contractions probably occur thus aiding in the aspiration of spermatozoa into the storage gland area.

Frequency of Insemination

The frequency of insemination is important to maintain maximum fertility in a flock throughout a particular prolonged breeding period. If semen is greatly diluted, which in effect means reducing spermatozoal number per unit volume, then an increased frequency of insemination may be necessary to achieve a high level of fertility.

Inherent Quality of Spermatozoa

If the quality of a proportion of the spermatozoa in a semen sample is poor, in effect the number of viable spermatozoa is reduced, thus a larger dose of this type of semen, or more frequent insemination in a prolonged breeding season, will be necessary. This remedy assumes that the overall poor quality of the spermatozoa is not carrying undesirable traits into the offspring.

Quality of Semen due to Management of Males and Collection Procedures

Increasing the rest period between ejaculations (El Jack & Lake, 1966) and the collection of contaminated semen effectively reduce the number of viable spermatozoa in a semen sample. The former has important implications for overall fertility in a flock of domestic birds where it is required to sustain maximum fertility over a long breeding season and there is a tendency to keep surplus males.

Uncertain Factors—Disease, Immunity to Spermatozoa, Nutrition, Pollutants

Several unknown factors associated with disease, build-up of immunity to spermatozoa, nutrition or pollutants can operate to affect adversely the oviduct environment or hatching ability. In these cases the period of survival of spermatozoa in the female or the quality of spermatozoa may be affected. Much work is required in these fields to establish the true extent of the effects (see Lake, 1969, 1975).

REFERENCES

Berry, R. B. (1972). Reproduction by artificial insemination in captive American goshawks. *J. Wildl. Mgmt* **36**: 1283–1288.

Bird, D. M. & Laguë, P. C. (1977). Semen production of the American kestrel. *Can. J. Zool.* **55**: 1351–1358.

Bird, D. M., Laguë, P. C. & Buckland, R. B. (1976). Artificial insemination vs. natural mating in captive American kestrels. *Can. J. Zool.* **54**: 1183–1191.

Boyd, L. L., Boyd, N. S. & Dobler, F. C. (1977). Reproduction of prairie falcons by artificial insemination. *J. Wildl. Mgmt* **41**: 266–271.

Brown, K. I. & Graham, E. F. (1971). Effect of some cryophylactic agents on turkey spermatozoa. *Poult. Sci.* **50**: 832–835.

Dowling, J. J. (1976). Artificial insemination. *Racing Pigeon Pictorial*, October **7**: 300–301.

El Jack, M. H. & Lake, P. E. (1966). The effect of resting roosters from ejaculation on the quality of spermatozoa in semen. *J. Reprod. Fert.* **11**: 489–491.

Fujihara, N. & Nishiyama, H. (1976a). Studies on the accessory reproductive organs in the drake. 3. Ejecting mechanism of the fluid from the ejaculatory groove region. *Poult. Sci.* **55**: 1026–1030.

Fujihara, N. & Nishiyama, H. (1976b). Studies on the accessory reproductive organs in the drake. 5. Effects of the fluid from the ejaculatory groove region on the spermatozoa of the drake. *Poult. Sci.* **55**: 2415–2420.

Fujihara, N., Nishiyama, H. & Nakashima, N. (1976). Studies on the accessory reproductive organs in the drake. 2. Macroscopic and microscopic observations on the cloaca of the drake with special reference to the ejaculatory groove region. *Poult. Sci.* **55**: 927–935.

Grier, J. W. (1973). Techniques and results of artificial insemination with Golden eagles. *Raptor Res.* **7**: 1–12.

Hickey, J. J. & Roelle, J. E. (1969). Conference summary and conclusions. In *Peregrine falcon populations: their biology and decline*: 553–567. Hickey, J. J. (ed.). Madison and Milwaukee: The University of Wisconsin Press.

Higuchi, H. (1972). Artificial insemination of the Copper pheasant. *Int. Zoo Yb.* **12**: 151.

Howell, T. R. & Bartholomew, G. A. (1952). Experiments on the mating behaviour of the Brewer blackbird. *Condor* **54**: 140–151.

Kobayashi, S., Okamoto, S. & Matsuo, T. (1972). The influence of the foamy material on the fertilising capacity of Japanese quail (*Coturnix coturnix japonica*) semen. *Agric. Bull., Saga Univ.* **32**: 89–95.

Lake, P. E. (1966). Physiology and biochemistry of poultry semen. *Adv. Reprod. Physiol.* **1**: 93–123. McLaren, A. (ed.). London: Logos Press.

Lake, P. E. (1969). Factors affecting fertility. In *The fertility and hatchability of the hen's egg*: 3–29. (Br. Egg Market. Bd Symp. No. 5). Carter, T. C. & Freeman, B. M. (eds). Edinburgh: Oliver & Boyd.

Lake, P. E. (1971). The male in reproduction. In *Physiology and biochemistry of the domestic fowl* **3**: 1411–1447. Bell, D. J. & Freeman, B. M. (eds). London and New York: Academic Press.

Lake, P. E. (1975). Gamete production and the fertile period with particular reference to domesticated birds. *Symp. zool. Soc. Lond.* No. 35: 225–244.

Lake, P. E. & Nixey, C. (1979). Artificial insemination. In *Turkey production. Bull. Min. Agric. Fish. Fd.*

Lake, P. E. & Stewart, J. M. (1978a). Artificial insemination in poultry. *Bull. Min. Agric. Fish. Fd* No. 213.

Lake, P. E. & Stewart, J. M. (1978b). Preservation of fowl semen in liquid nitrogen—an improved method. *Br. Poult. Sci.* **19**: 187–194.

Lepore, P. D. & Marks, H. L. (1966). Intravaginal insemination of Japanese quail: factors influencing the basic technique. *Poult. Sci.* **45**: 888–891.

Lorenz, F. W., Ogasawara, F. X. & Asmundson, V. S. [1964]. Fertility and gametogenesis in hybrids of galliform birds. *V Int. Congr. Anim. Reprod. Artif. Insem.* [Trento] **7**: 320–325.

McFarquhar, A. M. & Lake, P. E. (1964). Artificial insemination in quail and the production of chicken-quail hybrids. *J. Reprod. Fert.* **8**: 261–263.

Macpherson, J. W., Chatterjee, S. & Friars, G. W. (1969). Frozen turkey semen. *Can. J. comp. Med.* **33**: 37–38.

Makos, J. J. & Smyth, J. R. (1970). A study of fertility following intergeneric crosses among certain gallinaceous birds. *Poult. Sci.* **49**: 23–29.

Mann, T. (1964). *Biochemistry of semen and of the male reproductive tract.* London: Methuen & Co Ltd.

Marks, H. L. & Lepore, P. D. (1965). A procedure for artificial insemination of Japanese quail. *Poult. Sci.* **44**: 1001–1003.

Marquez, B. J. & Ogasawara, F. X. (1977). Effects of glycerol on turkey sperm cell viability and fertilizing capacity. *Poult. Sci.* **56**: 725–731.

Maru, N., Ichinoe, K., Ishijima, Y., Sakuma, Y. & Sasaki, M. (1966). Studies on the artificial insemination in pheasants. *Jap. Poult. Sci.* **3**: 83–87.

Mendelssohn, H. & Marder, U. (1970). Problems of reproduction in birds of prey in captivity. *Int. Zoo Yb.* **10**: 6–11.

Mitchell, R. L., Buckland, R. B., Forgrave, L. & Baker, R. D. (1975). A simple controlled rate freezing apparatus as applied to freezing poultry semen. *Poult. Sci.* **54**: 1796 (Abstr.).

Nishiyama, H., Nakashima, N. & Fujihara, N. (1976). Studies on the accessory reproductive organs in the drake. 1. Addition to semen of the fluid from the ejaculatory groove region. *Poult. Sci.* **55**: 234–242.

Ogasawara, F. X. & Huang, R. (1963). A modified method of artificial insemination in the production of chicken-quail hybrids. *Poult. Sci.* **42**: 1386–1392.

Ogasawara, F. X. & Ernst, R. A. (1970). Effects of three semen extenders on reproduction of turkeys. *Calif. Agric.* 1970 August: 15.

Ogasawara, F. X. & Fuqua, C. L. (1972). The vital importance of the utero-vaginal sperm-host glands for the turkey hen. *Poult. Sci.* **51**: 1035–1039.

Owen, R. D. (1941). Artificial insemination of pigeons and doves. *Poult. Sci.* **20**: 428–431.

Pendergast, B. A. & Boag, D. A. (1971). Maintenance and breeding of spruce grouse in captivity. *J. Wildl. Mgmt* **35**: 177–179.

Rand, A. L. (1936). The distribution and habits of Madagascar birds. *Bull. Am. Mus. nat. Hist.* **72**: 143–499.

Raises, M. B. (1968). Artificial insemination of pigeons. *Austr. Vet. J.* **44**: 486.

Salt, W. R. (1954). The structure of the cloacal protuberance of the Vesper sparrow (*Pooecetes gramineus*) and certain other passerine birds. *Auk* **71**: 64–73.

Sarvella, P. & Marks, H. (1970). Attempts to obtain quail × chicken hybrids. *Poult. Sci.* **49**: 1176–1178.

Sexton, T. J. (1976a). Studies on the dilution of turkey semen. *Br. Poult. Sci.* **17**: 179–184.

Sexton, T. J. (1976a). Studies on the fertility of frozen fowl semen. *VIII Int. Congr. Anim. Reprod. Artif. Insem.* [Krakow] **4**: 1079–1082.

Sexton, T. J. (1977). Relationship between number of sperm inseminated and fertility of turkey hens at various stages of production. *Poult. Sci.* **56**: 1054–1056.

Shaklee, W. E. & Knox, C. W. (1954). Hybridization of the pheasant and fowl. *J. Hered.* **45**: 183–190.

Spiller, N., Grahame, I. & Wise, D. R. (1977). Experiments on the artificial insemination of pheasants. *Wld Pheasant Assoc. J.* No. 2: 89–96.

Stirling, I. & Roberts, C. W. (1967). Artificial insemination of blue grouse. *Can. J. Zool.* **45**: 45–47.

Szumowski, P., Theret, M. & Denis, B. [1976]. Semen and artificial insemination of pigeons. *VIII Int. Congr. Anim. Reprod. Artif. Insem.* [Krakow] **4**: 1086–1089.

Temple, S. A. (1972). Artificial insemination with imprinted birds of prey. *Nature, Lond.* **237**: 287–288.

Van Wambeke, F. (1972). Fertility and hatchability results with fowl spermatozoa stored in fresh and freeze-dried diluent. *Br. Poult. Sci.* **13**: 179–183.

Van Wambeke, F. [1976]. The effect of two different methods of semen collection on fertility and hatchability results obtained with stored fowl semen. *V Eur. Poultry Conf.* [Malta] **2**: 1230–1240.

Van Wambeke, F. (1977). The effect of tonicity of storage media for fowl semen on the occurrence of neck bending of spermatozoa, fertility and hatchability. *Br. Poult. Sci.* **18**: 163–168.

Watanabe, M. (1961). Experimental studies on the artificial insemination of domestic ducks with special reference to the production of Mule ducks. *J. Fac. Fish. Anim. Husb., Hiroshima Univ.* **3**: 439–478.

Watanabe, M. & Terada, T. [1976]. A new diluent for deep freezing preservation of fowl spermatozoa. *VIII Int. Congr. Anim. Reprod. Artif. Insem.* [Krakow] **4**: 1096–1099.

Watanabe, M., Yamane, J., Tsukunaga, S. & Takahashi, T. (1963). An artificial hybrid between White Leghorn cock and Japanese Green pheasant hen produced by insemination. *J. Fac. Fish. Anim. Husb., Hiroshima Univ.* **5**: 129–137.

Wilcox, F. H. (1959). Studies of the effect of oxytetracycline on chicken spermatozoa. *Am. J. vet. Res.* **20**: 957–960.

Wolfson, A. (1952). The cloacal protuberance. *Bird-Banding* **23**: 159–165.

Wolfson, A. (1954). Notes on the cloacal protuberance, seminal vesicles, and a possible copulatory organ in male passerine birds. *Bull. Chicago Acad. Sci.* **10**: 1–23.

Wolfson, A. (1958). The ejaculate and the nature of coition in some passerine birds. *Ibis* **102**: 124–125.

Yamashina, Y. (1976). Notes on the Japanese Copper pheasant. *Phasianus soemmerringii. Wld Pheasant Assoc. J.* No. 1: 23–42.

Symp. zool. Soc. Lond. (1978) No. 43, 51–72

Artificial Insemination for Breeding Non-Domestic Birds

G. F. GEE and S. A. TEMPLE

Patuxent Wildlife Research Center, Laurel, Maryland, USA
University of Wisconsin, Madison, Wisconsin, USA

SYNOPSIS

Captive breeding of non-domestic birds has increased dramatically in this century, and production of young often exceeds that of the same number of birds in their native habitat. However, when infertility is a problem, artificial insemination can be a useful method to improve production. Artificial insemination programs with non-domestic birds are relatively recent, but several notable successes have been documented, especially with cranes and raptors.

Three methods of artificial insemination are described—cooperative, massage, and electroejaculation. Cooperative artificial insemination requires training of birds imprinted on man and is used extensively in some raptor programs. The massage technique generally is used when there are larger numbers of birds to inseminate since it requires less training of the birds than with the cooperative method, and a larger number of attempted semen collections are successful. Although the best samples are obtained from birds conditioned to capture and handling procedures associated with the massage method, samples can be obtained from wild birds. Semen collection and insemination for the crane serves to illustrate some of the modifications necessary to compensate for anatomical variations. Collection of semen by electrical stimulation is not commonly used in birds. Unlike the other two methods which require behavioral cooperation by the bird, electroejaculation is possible in reproductively active birds without prior conditioning when properly restrained.

Fertility from artificial insemination in captive non-domestic-birds has been good. Although some spermatozoal morphology has been reported, most aspects of morphology are not useful in predicting fertility. However, spermatozoal head length in the crane may have a positive correlation with fertility. Nevertheless, insemination with the largest number of live spermatozoa is still the best guarantee of fertile egg production.

INTRODUCTION

As the number of bird species threatened with extinction increases, aviculturists and conservationists have joined forces in a concerted effort to breed certain endangered birds in captivity to increase their numbers (Martin, 1975). Captive breeding of endangered birds has received much attention in recent years owing to the impressive results of a few breeding programs (Cade, 1975; Kear, 1975; Erickson, 1968), but as a conservation technique captive breeding has not been without criticism (Miller, 1953; Zimmerman, 1974). Although the potential of captive propagation for restoring depleted populations is obvious, it is true that many pioneering attempts have been less than successful. One of the primary reasons for poor initial success has been the considerable difficulty that non-domestic bird breeders have encountered in inducing captive reproduction. Hediger (1965) summarized the major reasons for poor breeding performance in captive animals and, as a group, birds seem to show nearly all the types of problems he identifies.

In general, captive birds can be brought into full physiological reproductive condition if proper proximate and ultimate factors such as photoperiod, food, space requirements, protection from excessive disturbance and satisfactory nesting structures are available. Since many factors are species specific (Immelmann, 1971), we will not attempt to review the various manipulations that have been used to bring captive birds into reproductive condition. Unfortunately, even in a carefully controlled environment, where all physical variables are maintained at optimal values for breeding, many birds still fail to reproduce. Many of these difficult species achieve almost full reproductive condition, at least physiologically, but subtle behavioral problems associated with confinement interfere with the final stages of the breeding cycle. Often a pair will engage in all of the preliminary aspects of breeding (including courtship and nest building) but fail to copulate. In most instances where the birds proceed through all the early stages of the breeding cycle, the females will lay infertile eggs (Cade, 1975; Temple & Cade, 1977). In raptors, problems with males seem to be more common than with females, and many breeders have devised elaborate ways to encourage inhibited males to achieve their full reproductive potential (Meng, 1972; Weaver & Cade, 1974).

We believe that natural pairing and mating is the most desirable way to obtain fertile eggs, but when it is impossible or impractical to devise procedures that will permit this to occur alternative techniques must be used. Artificial insemination provides a practical means for circumventing many of the frequently encountered problems with infertility in captive birds. It can be used to supplement natural matings if copulation is absent or incomplete, for instance when:

1. the male of a captive pair engages in all stages of the breeding cycle except natural copulation;
2. behavioral problems such as inappropriate sexual imprinting, extreme aggressiveness, extreme subordination, or confinement-induced stress inhibit natural pairing;
3. crippling or anatomical disability precludes natural copulation;
4. poor spermatozoal concentration in the semen results in poor fertility.

Artificial insemination can be used to satisfy special requirements of a breeding program if:

1. one female must be bred with more than one male to produce pedigree progeny with a diverse genetic constitution;

2. the male and female are geographically separated but both are reproductively active;
3. a shortage of males requires dispersion of semen to produce fertility in all available females;
4. frozen semen is the only source of spermatozoa;
5. hybrid birds are required but natural mating is not possible;
6. quarantine or some other disease hazard precludes transfer of birds.

Here we review and compare the various artificial insemination techniques that have been used with wild birds. We cover both semen collection and insemination and, where possible, compare the results with natural pairings. We draw primarily on results with two groups of non-domestic birds with which we have considerable experience: raptors and cranes.

GENERAL SCOPE OF ARTIFICIAL INSEMINATION WORK WITH NON-DOMESTIC BIRDS

Most attempts with artificial insemination in non-domestic birds have been quite recent, despite the fact that artificial insemination has been used routinely with poultry for decades following the pioneering work by Quinn & Burrows (1936). Artificial insemination techniques have been successfully applied to such non-domestic groups as ducks and geese (Johnson, 1954; Watanabe, 1957; Kinney & Burger, 1960; Lake, 1962; Pingel, 1972; Skinner, 1974), pigeons and doves (Owen, 1941), cranes (Gee, 1969; Archibald, 1974), peafowl (A. T. Leighton, pers. comm.), hawks (Berry, 1972; Temple, 1972; Corten, 1973), falcons (Bird, Lagüe & Buckland, 1976; Boyd, Boyd & Dobler, 1977; Cade *et al.*, in preparation), eagles (Hamerstrom, 1970; Grier, 1973), curassows and turkeys (G. A. Greenwell, C. Emerick & P. Cohen, pers. comm.) and cassowaries (C. Pickett & J. Mollen, pers. comm.).

In almost all non-domestic birds it is important to remember that reproduction is highly seasonal; these birds will only be in reproductive condition for a brief portion of each year. Therefore, successful artificial insemination in non-domestic birds must coincide with the reproductive period in each species. In general, males and females are synchronized so that both achieve their full reproductive potential during the same brief time period. But some management programs, especially photoperiod modification, can upset this synchrony and compensatory programs may be necessary (Nestor, Chamberlin & Renner, 1970).

In general, artificial insemination in non-domestic birds can be grouped into three categories: cooperative, massage and electroejaculation. These categories reflect different degrees of cooperation that both the male and female birds show toward the human handler during semen collection and insemination.

TECHNIQUES USED FOR ARTIFICIAL INSEMINATION IN NON-DOMESTIC BIRDS

Cooperative Semen Collection from Males

Cooperative artificial breeding procedures seem to have been pioneered with non-domestic birds. Although attempts have been made to collect domestic fowl semen through natural mating, males were not trained to deposit semen on their handlers (Smyth, 1968). Basically, cooperative semen collection depends on a male bird voluntarily ejaculating semen that the human handler can collect and use to inseminate a female. In most instances, this is accomplished with birds that have been sexually imprinted on humans. Human imprinting in birds is a complex phenomenon that we will not cover in detail here. In general, it is not uncommon for birds that have been hand-reared and kept in close association with humans and isolated from conspecifics, to socialize and to express sexual behavior towards their handlers. Male birds will attempt to court their handlers, and if encouraged, the birds will usually attempt to mount the handler and go through copulatory movements. If repeated frequently, this type of behavior leads to full copulatory activities with the bird actually ejaculating semen.

The pioneering experiments on the use of sexually imprinted males as semen donors for artificial insemination were performed on birds of prey. The handling techniques of falconry encouraged these raptors to regard human beings as social and, later, sexual partners. Hamerstrom (1970) was one of the first to attempt this technique, but her experiments with golden eagles, *Aquila chrysaetos*, were unsuccessful. Temple (1972) was able to collect semen from a sexually imprinted red-tailed hawk, *Buteo jamaicensis*, that voluntarily ejaculated while mounting his gloved hand. Berry (1972) did the same with goshawks, *Accipiter gentilis*, and Grier (1973) succeeded with golden eagles. Recently, Boyd & Boyd (1976) achieved similar results working with sexually imprinted prairie falcons, *Falco mexicanus*, and J. H. Enderson (pers. comm.) with peregrine falcons, *Falco peregrinus*. In all these cases, the unrestrained males voluntarily ejaculated semen that the handler could collect in or on a suitable receptacle. Sexually imprinted male cranes have also been used for artificial insemination, but their long legs and large size make it impractical to allow them actually to mount a handler and ejaculate voluntarily. Hence, semen is usually collected from imprinted as well as non-imprinted cranes by using the massage method (Gee, 1969; Archibald, 1974). Although the degree of cooperation by the bird for the massage method is less than that necessary for the cooperative collection procedure, the best semen samples are obtained most frequently from birds trained to respond.

Cooperative Insemination of Females

Cooperative insemination techniques can also be used with females. Sexually imprinted females respond to their handlers by assuming copulation postures and by everting their oviducts. To inseminate such birds, no restraint is required. The handler merely deposits semen in the cloaca or oviduct of the receptive bird.

As with collection of semen from imprinted males, most of the work with inseminating imprinted females has been done with birds of prey. Temple (1972), Berry (1972), Grier (1973), Boyd *et al.* (1977), J. H. Enderson (pers. comm.), and Cade *et al.* (in preparation) all succeeded in artificially inseminating unrestrained, receptive female raptors.

Massage Semen Collection from Males

Semen collection by the massage technique must be used on males that are reproductively active but not necessarily sexually imprinted on humans. Collection of semen from uncooperative, non-domestic birds has been achieved by the massage method and is essentially the same as that used for poultry (Quinn & Burrows, 1936; Frank, 1961; Lake, 1962; Smyth, 1968). Using this technique, semen has been collected from wild greater sandhill cranes, *Grus canadensis tabida*, at the time of capture in their native habitat (Gee, 1972).

The basic manipulations of the males are identical to those described by Quinn & Burrows (1936). The bird is restrained while the operator manually stimulates the abdomen and vent; in most cases, this actually results in ejaculation of some semen, but to complete the process semen must be squeezed from the distal ends of the vasa deferentia. However, in many birds no semen sample is obtained without erectile tissue response to the preceding manipulation. With only slight modifications, this procedure has been used with ducks, peacocks, finches, canaries (Bonadonna, 1939; A. T. Leighton, pers. comm.), pigeons and doves (Owen, 1941), waterfowl (Johnson, 1954; Lake, 1962; Pingel, 1972; Skinner, 1974), pheasants (Smyth, 1968), quail (Wentworth & Mellen, 1963), falcons (Bird & Buckland, 1976; Cade *et al.*, in preparation), hawks (Corten, 1973), cranes (Gee, 1969; Archibald, 1974), curassows and turkeys (G. A. Greenwell *et al.*, pers. comm.), and cassowaries (C. Pickett, pers. comm.).

Most of the modifications of the basic technique are the result of special handling or restraining methods dictated by long legs, sharp talons, long beaks and different body sizes. As an example, we describe in detail the techniques that were developed in 1969 at the Patuxent Wildlife Research Center, United States Fish and Wildlife Service, for collecting semen from cranes. The male is caught, guided into the nearest corner and held there. The assistant cradles the bird between his legs with the bird's head toward the corner and the bird's breast against but not between the assistant's thighs. The assistant grasps the bird's legs by the shank and strokes gently several times in a circular inward and downward direction. The operator immediately behind the bird and facing the assistant, strokes the back several times, from the postdorsal region to the interpelvic tail region, and then down to the postlateral region below the tail. The bird responds to this stimulation by pushing forward against the assistant's thighs, emitting a low vocal growl, and raising its tail. The tail is then forced back with the left hand and the abdominal and sternal regions stroked with the right hand. Next, the cloaca is grasped dorsally by the thumb and the index fingers of the left hand (usually the cloaca would have responded to the preceding stimulation by a partial eversion). In the right hand a small glass funnel plugged with wax is held between the fingers for semen collections. Often spontaneous ejaculation occurs and the first drop of semen is collected on the lip of the funnel. Occasionally, the bird lifts its legs off the ground when it ejaculates and must be supported by the thighs. The final steps (abdominal and cloacal massage) are repeated several times and the remaining semen is expressed from the vent with fingers of the left hand. The entire process from stimulation by the operator to collection of the semen is accomplished in five to ten seconds.

Although most birds are manipulated in the same way, in some, preparations other than those already mentioned may be necessary before collecting semen. In waterfowl the same steps are taken to stimulate the male, but prior to ejaculation, early in the stimulating process, the phallus is everted (Skinner, 1974). After stimulation semen can be collected at the base and tip of the phallus with a small funnel, tube or suction device. Careful handling is advised because the phallus, often termed a penis in waterfowl, is very delicate and injury may result from rough treatment (Smyth, 1968). In other birds with copulatory organs, similar procedures would be desirable but can be difficult with such birds as the ratites. The quail contain a cloacal gland that produces a white frothy substance that, though not detrimental to spermatozoa, is removed prior to semen collection (Smyth, 1968). Other birds such as the ratites contain waxy or greasy substances in the vent that may need to be cleaned out before semen is collected.

Massage Insemination of Females

Non-cooperative females, like the males, must also be restrained during the insemination procedure. As with the collection of semen from males, the procedure used to inseminate females of non-domestic species is essentially the same as the procedures used routinely with poultry (Quinn & Burrows, 1936). Females are restrained and held in such a way that the cloacal region is fully exposed and accessible. Semen can then be deposited into the cloaca or directly into the oviduct. It is desirable to place the semen in the oviduct (Smyth, 1968). The distal end of the oviduct can be everted by placing firm pressure on the female's abdomen and the walls of the cloaca. This can be a somewhat difficult task with larger species, but with some practice even birds as large as eagles and cranes can be managed. In some birds such as waterfowl, eversion of the oviduct is difficult or impossible and may result in excessive stress on the bird. In these cases the oviduct can be located by palpation or use of a speculum, and the syringe or inseminating device guided into the oviduct (Johnson, 1954; Watanabe, 1957; Olver, 1971). In cranes, the response to the stimulation of the massage technique results in visual exposure of the vagina if the dorsal wall of the vent separating the urodeum from the proctodeum is pushed up. With practice the inseminating syringe can be inserted into the vagina during the few seconds it is observed. Archibald (1974) has successfully inseminated imprinted female sandhill cranes, *Grus canadensis*, that readily assumed copulation posture and everted their oviducts in response to massage stimulation and handling. Placing semen directly in the everted oviduct has been the preferred insemination technique with most non-domestic birds (Bird *et al.*, 1976; Boyd *et al.*, 1977), although satisfactory results have been obtained by simply depositing semen in the cloaca (Gee, 1969; Temple, 1972; Berry, 1972; Grier, 1973; Archibald, 1974). Gee found that fertility rates in the sandhill crane in excess of 80% were possible from cloacal placement of semen if inseminations were frequent (at least twice each week), accomplished with adequate numbers of sperm (about 50 000), were begun two or three weeks prior to the first egg, and were completed within a few hours after every oviposition.

Collection of Semen by Electroejaculation

Electroejaculation is not a common way to collect avian semen but has been used successfully for years in domestic mammals (Almquist, 1968). Although uncommon today, it may become a preferred method of semen collection from non-domestic birds if the advantages apply to other birds as well as they do with domestic ducks and geese (Serebrovski & Sokolovskaja, 1934; Watanabe, 1957). Both of the preceding methods of semen collection (cooperative and massage) require training and cooperation from the birds to obtain good semen samples. Even though semen can be collected from uncooperative birds with the massage method, to obtain large volumes of contaminant-free semen several weeks of conditioning is required. Watanabe (1957), using electro-ejaculation without prior training, collected duck semen that contained a greater number of spermatozoa and was released in larger quantity than that collected by the conventional massage technique. The semen collected produced fertility comparable to that collected by the massage method, and no adverse effects were observed in the drakes. Since many wild birds are difficult to handle and thus present greater difficulty for conventional semen collection, electroejaculation could be a useful way to collect the semen quickly and reduce the stress associated with handling.

Handling of Semen Specimens

Since semen volumes collected from non-domestic birds are small in comparison to those from chickens and turkeys (Table I), most of the semen is used for insemination. However, a simple semen evaluation helps to obtain good fertility rates from artificial insemination by the elimination of poor males or of samples containing conditions detrimental to spermatozoal survival (Lake, 1962; Smyth, 1968). In our laboratories a small portion of each semen sample is stored in a capillary tube for laboratory observation. It is examined in the capillary tube under the microscope for the presence of motile spermatozoa, and then a 5% eosin–10% nigrosin stained smear is made to determine the proportion of live spermatozoa (Burrows & Quinn, 1939). The entire semen evaluation can be completed within minutes of collection and the stained smears used later for morphological studies (Gee, 1969).

OBSERVATIONS ON COLLECTION OF SEMEN FROM NON-DOMESTIC BIRDS

We will first evaluate semen collection techniques on the basis of factors known to have important bearing on fertility rates, namely, duration of production, quantities and quality of semen obtained, and semen storage.

Duration of Semen Production

Unlike males of many domestic species which are in almost continual breeding condition, males of non-domestic birds are highly seasonal in semen production. Using cooperative semen collection, Grier (1973) was able to collect semen from two male golden eagles over time spans of 70 to 111 days. Using forced-collection techniques with the American kestrel, Bird & Buckland (1976) reported a similar duration of semen production of 73·6 days and a range of 48 to 101 days. Using cooperative collection techniques, Berry (1972) was only able to obtain semen from a male goshawk for a period of less than 10 days, while Corten (1973), using forced-collection techniques, obtained semen from a goshawk over a 60-day period. Temple has found that semen could be collected from a male red-tailed hawk over a period of at least 60 days when forced-collection techniques were used. When collections from the same bird were made using cooperative techniques, the period was shortened to only about 21 days. Sandhill and whooping cranes, *Grus americana*, begin active semen production several weeks prior to egg production and usually continue to produce semen as long as eggs are laid (G. F. Gee, unpublished). Most sandhill cranes produce semen for 80 to 90 days, when collected by the massage method, and eggs for 50 to 60 days when collected as laid.

These results suggest that semen can be collected over a longer period of time using forced-collection techniques. Presumably cooperative males are actually producing spermatozoa over a longer period of time than the relatively brief period when they voluntarily ejaculate semen. Both Temple (1972) and Berry (1972) have noted that voluntary ejaculation of semen corresponded temporally with the time of oviposition in females. Also, with cranes at the Patuxent Wildlife Research Center and at the International Crane Foundation (Gee, 1969; Archibald, 1974), and with geese (Johnson, 1954), the response of both sexes to the massage method is greater during the period of oviposition.

Quantities of Semen Obtained

Representative data on the volume of semen obtained from various birds are given in Table I. The quantity of semen that can be collected from a male varies with body size but may also vary with the type of collection technique used. Temple (1972) found that a male red-tailed

TABLE I

Semen volumes collected from various species of birds

Species	Range of volumes	References
Domesticated birds		
Canary	$\cong 10\ \mu l$	J. G. Griffith (pers. comm.)
Chicken	0·5–0·8 ml	Smyth (1968)
Duck	0·3 ml	Watanabe (1957)
Goose	10–600 μl	Johnson (1954)
Japanese quail	10 μl	Smyth (1968)
Pigeon	10–20 μl	Owen (1941)
Ring-necked pheasant	50–100 μl	Smyth (1968)
Turkey	0·2–0·3 ml	Smyth (1968)
Non-domesticated birds		
American kestrel	14 μl	Bird *et al.* (1976)
Brewer's blackbird	$\cong 10\ \mu l$	Wolfson (1960)
Eclectus parrot	50–100 μl	G. F. Gee & F. Beall (unpublished)
Goshawk	20–30 μl	Berry (1972)
House finch	$\cong 10\ \mu l$	J. G. Griffith (pers. comm.)
Prairie falcon	50–100 μl	Boyd *et al.* (1977)
Red-tailed hawk	0·1 ml	Temple (1972)
Swamp sparrow	$\cong 10\ \mu l$	Wolfson (1952)
Sandhill crane	10–200 μl	G. F. Gee (unpublished)
Whooping crane	10–200 μl	G. F. Gee (unpublished)
Wood thrush	$\cong 10\ \mu l$	Wolfson (1960)
Wattled cassowary	1–5 ml	C. Pickett (pers. comm.)

hawk voluntarily ejaculated semen volumes averaging 0·1 ml. Boyd *et al.*
(1977) obtained semen volumes averaging 0·1 ml from a prairie falcon
that voluntarily ejaculated whereas Cade *et al.* (in preparation), using
forced-collection techniques, obtained semen volumes of only 30–50μl.
Bird *et al.* (1976) obtained from 10 to 59 μl of semen (average 13·8 μl)
from kestrels using a forced-collection method. Gee (1971) found that
semen volume from 10 greater sandhill cranes collected for three
consecutive seasons was small, 30 μl, when collected by the massage
technique.

Other sources of variation in semen volume have been noted. Almost
all investigators have noted that semen volumes decrease markedly
when collections, forced or cooperative, are taken more than once each
day (Temple, 1972; Bird *et al.*, 1976; Cade *et al.*, in preparation). Semen
volume often varies throughout the breeding season. Temple (1972),
Berry (1972), Grier (1973), Grier, Berry & Temple (1973) and Boyd *et
al.* (1977) have all noted that early and late in the breeding season,
volumes are lower than at the height of breeding. However, smaller
volumes occurring during the height of the reproductive season are
often the result of a bird copulating or attempting to copulate with birds
or other objects in the pen.

Even in non-domestic birds that are in full breeding condition,
semen is not obtained every time collection is attempted. The cooperative collection technique generally results in a lower percentage of
attempted collections actually yielding semen. Berry (1972) was only able
to obtain semen from a goshawk in 15 (8%) of 117 attempts using the
cooperative technique; Temple (1972) obtained semen from a red-tailed
hawk in 30% of the collections using the cooperative technique, and Grier
(1973), working with cooperative golden eagles, obtained semen in 216
(46%) of 468 attempts.

Forced-collection techniques appear to result in a higher percentage
of successful collections. Bird *et al.* (1976), using the massage technique,
were able to obtain measurable quantities of semen from their American
kestrels, *Falco sparverius*, in 74% of the collection attempts. G. F. Gee
(unpublished) found that for 25 sandhill cranes, 86% of the forced
semen collection attempts were successful in 1977. After removal of the
few (10%) unresponsive male cranes from these calculations, 99% of the
semen collections were successful.

Quality of Semen Obtained

Many characteristics of a semen sample are known to influence its later
potential for fertilizing eggs. Among these are: concentration and
motility of spermatozoa, morphology of spermatozoa and contamination
with foreign matter (Burrows & Quinn, 1939; Lake, 1962; Smyth, 1968).
When dealing with non-domestic birds, sample sizes are usually so small
that few meaningful statistical comparisons can be made between semen

samples with different characteristics; only a few attempts have been made.

Concentration and motility of spermatozoa in semen

Several factors are known to influence the concentration of spermatozoa in a semen sample. Bird *et al.* (1976) found that in kestrels from which semen was collected by massage techniques, the concentration of spermatozoa was correlated positively with the age of the male, size of the male and volume of semen collected. They also found that concentrations of spermatozoa in semen samples varied through the breeding season, tending to be highest in the first half of the period of semen production.

Most investigators (Temple, 1972; Grier, 1973; Bird *et al.*, 1976; Cade *et al.*, in preparation) have noted that regardless of the method of collection, frequent collections reduce the concentration of spermatozoa. Temple could find no improvement in sperm concentrations of red-tailed hawk semen with collections spaced longer than one day apart. Bird *et al.* (1976) working with large falcons found that intervals between collections of at least two days produced higher concentrations of spermatozoa. Although crane inseminations seldom exceed two per week (Archibald, 1974), at Patuxent we have collected four ejaculates per week from a few birds for most of one season—Monday, Tuesday, Thursday and Friday—without a noticeable decline in semen quantity or quality.

The collection technique may influence the concentration of spermatozoa in a semen sample. Temple obtained semen from the same red-tailed hawk using both cooperative and forced methods, but concentrations of spermatozoa were higher in samples obtained by the cooperative technique. Semen volume may be increased and spermatozoal concentration decreased owing to the release of transparent fluid during the massage of the lymph folds and erectile tissues in birds. When pressure is not applied to the vent during semen collection attempts in chickens, the semen contains fewer contaminants and a higher density of spermatozoa (Kamar, 1958).

Most studies in non-domestic birds have shown a consistent positive correlation between the concentration of spermatozoa and the proportion of spermatozoa showing vigorous progressive motility (Temple, 1972; Grier *et al.*, 1973; Bird *et al.*, 1976; Cade *et al.*, in preparation). In raptors, the motility of spermatozoa in semen samples is low early and late in the breeding season and highest during the middle of the season (Bird *et al.*, 1976).

Although seasonal differences in spermatozoal motility were not observed, semen from captive sandhill cranes that had a high percentage of living, active spermatozoa and a high spermatozoal concentration give excellent fertility rates after insemination (Table II). These cranes produce much less concentrated semen than domestic fowl. Spermatozoal

TABLE II

The relationship between semen quality and fertility in artificially inseminated cranes

Semen quality[a]	Number of eggs		Fertility (%)
	Laid	Fertile	
<2	8	4	50
2·5–3·0	7	4	57
3·5–4·0	23	16	70
4·5–5·0	16	12	75
5·5–6·0	23	20	87

[a] Mean spermatozoal score[b] of last two inseminates preceding oviposition by more than one day.

[b] Spermatozoal score = spermatozoal concentration rating[c] + live cell rating[d].

[c] The spermatozoal concentration rating was determined by examining a sample of semen collected in a capillary tube prior to insemination and scoring as follows: 0, semen containing few spermatozoa; 1, semen containing spermatozoa throughout the fluid but in insufficient numbers to line the inner surface of the tube; 2, semen containing sufficient spermatozoa to line much of the inner surface of the tube; 3, semen containing many packed spermatozoa lining the inner surface of the tube.

[d] The live cell rating was determined by examining semen stained with 5% eosin–10% nigrosin and scoring as follows: 0, less than 60% live cells; 1, 60–79% live cells; 2, 80–89% live cells; 3, 90% or more live cells.

concentration in good quality crane semen is 200–300 million spermatozoa ml^{-1}. Chicken semen averages 3500 million ml^{-1} (Smyth, 1968), and it is produced in larger quantities (chicken 0·5 to 0·8 ml; crane 0·02 to 0·15 ml). One crane ejaculate may not contain the minimal number of spermatozoa necessary for fertility with the artificial insemination technique described. Events in the subsequent development of the artificial insemination program at Patuxent Wildlife Research Center also indicated that spermatozoa contained in one ejaculate may be inadequate for good fertility; for example:

1. more frequent inseminations improved fertility in pairs with a previously low fertility;
2. in cranes known to copulate successfully throughout the season fertility resulted only after artificial insemination;
3. inseminations of the female of a previously infertile sandhill crane pair with whooping crane semen produced three chicks, one hybrid and two sandhill cranes.

Contamination of semen

Semen samples that are contaminated with urates, fecal matter or blood have poorer fertilizing capacities than clean samples (Smyth, 1968). Semen samples obtained by cooperative techniques are always quite pure whereas semen obtained by massage techniques usually contains some contamination. Even with gentle stimulation and squeezing of the cloacal region that is a necessary part of the massage technique, urates from the ureters, fecal matter from the rectum, lymph from the erectile tissues or blood from surface vessels in the cloacal region can be released in the cloaca along with semen. Much depends on the amount of force necessary to obtain the semen, the tameness and responsiveness of the bird and other conditions associated with capture and handling.

Bird *et al.* (1976) found that contaminants averaged 68% of the area in microscopic fields of semen samples obtained from kestrels using the massage technique. Temple (1972) found that red-tailed hawk semen collected by the massage technique had six times the amount of contaminants as semen from the same bird collected by the cooperative technique. With cranes, the amount of contamination in the sample is inversely related to the degree of cooperation with the handlers (G. F. Gee, unpublished). Some cranes release semen without cloacal massage, while others require several vigorous milking strokes to the dorsal lip of the vent; samples obtained by milking usually contain urates in addition to seminal fluids.

Morphology of spermatozoa

Few investigators have attempted to study the morphology of spermatozoa in semen samples collected from non-domestic birds. In studying spermatozoal morphology in semen collected from captive sandhill cranes, Gee identified six distinct cell types (Fig. 1).

All of these cell types were located in live fresh semen samples as well as in the stained smears. On the study slides 87% of all cells appeared to be alive; these were made up as follows: $57 \pm 18 \cdot 4\%$ ($\pm$s.d.) were morphologically complete, of average size and without membrane distortions (N), $17 \cdot 5 \pm 9 \cdot 4\%$ were bent at sharp angles in the head piece, body or mid-piece (B), $6 \pm 4 \cdot 3\%$ had bodies of larger than average diameter (S), $1 \pm 0 \cdot 2\%$ had more than one tail and/or bodies $1 \cdot 5$ or more times larger than the normal cells (G) and $5 \cdot 2 \pm 5\%$ were partially or totally trapped in a spherical body (DL). The remaining 13% of cells were dead; $11 \cdot 6 \pm 12 \cdot 1\%$ appeared swollen and slightly greater than average length (D) and $1 \cdot 5 \pm 2 \cdot 8\%$ appeared abnormal in ways similar to one or more of the abnormal groups in the live cells listed above (Fig. 1). The relative proportions of cell types were constant throughout the reproductive season but early each season a greater number of abnormal dead spermatozoa ($P \leqslant 0 \cdot 01$) were collected (Table III).

No correlations could be found between proportions of cell types in a semen sample and subsequent fertilizing capacity. However, spermatozoal head length in the sandhill crane is positively related to fertility. Subspecific differences in head length are also significant: Florida, *Grus canadensis pratensis*, $13 \cdot 3 \pm 0 \cdot 33$ μm ($\pm$s.e.); greater, *G. c. tabida*, $13 \cdot 8 \pm 0 \cdot 33$ μm; lesser, *G. c. canadensis*, $11 \cdot 3 \pm 0 \cdot 52$ μm and Mississippi, *G. c. pulla* $12 \cdot 1 \pm 0 \cdot 52$ μm (Sharlin, 1976).

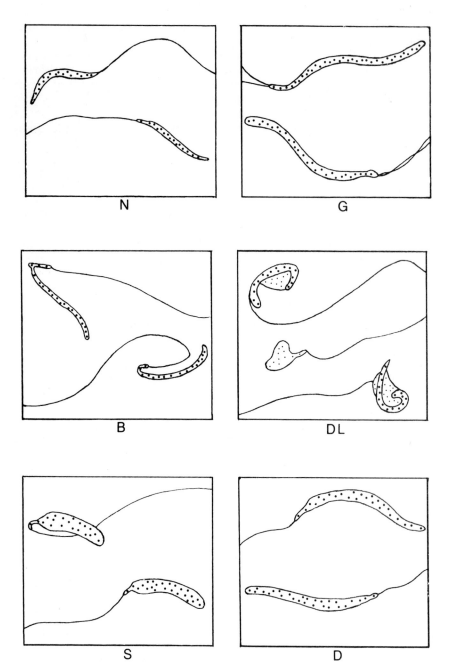

FIG. 1. Spermatozoa of the greater sandhill crane illustrating abnormal forms. N, Normal; B, bent; S. swollen; G. giant; DL, droplet; D, dead.

Table III

Spermatozoal types (%) in the semen of the greater sandhill crane

Trial	Season	Normal	Bent	Live Swollen	Giant	Droplet	Dead Normal	Abnormal
1	Early	69	14	5	1	3	7	2
	Mid	50	18	6	1	11	13	2
	Late	59	12	7	0·5	4	16	2
2	Early	53	18	6	1	6	14	2
	Mid	58	21	6	0·5	5	9	1
	Late	51	18	8	1	6	15	1
3	Early	61	18	7	1	4	7	2
	Mid	64	18	8	0·5	4	6	0·5
	Late	58	16	4	0·5	3	18	0·5
$\bar{x}$	Early	61	17	6	1	4	9	2
	Mid	57	19	7	1	7	9	1
	Late	56	15	6	1	4	16	1
Total	$\bar{x}$	58	17	6	1	5	12	1

Corten (1973, 1974) discussed the morphology of spermatozoa in semen collected from goshawks. He described different cell types but could find no correlation between cell types and other characteristics of the semen. Bird *et al.* (1976) noted increased numbers of abnormal spermatozoa in kestrel semen samples collected early in the breeding season and in samples from young birds.

Comparison of semen quality from wild and captive birds

In an effort to determine whether spermatozoal quality was normal in captive sandhill cranes, Gee and biologists from the Idaho Wildlife Service caught wild greater sandhill cranes in Idaho and collected semen from them in 1971 and 1972. Although six cranes were captured in 1971 and eight in 1972, only six semen samples were collected, and only one of these was definitely from a sexually active male. Semen from the productive male contained 90% live cells, and 360×10^6 spermatozoa ml^{-1} of semen. Owing to the small number of samples, a statistical comparison of semen from the wild cranes with semen from captive birds was not possible, but the few samples obtained did not appear to differ in any way from semen collected from captive cranes held at Patuxent Wildlife Research Center.

Storage of Semen After Collection

Relatively few of the studies of artificial insemination in non-domestic birds have required the storage of semen for extended periods of time between collection and insemination of a female. In most cases, semen has been deposited in the female within one or two hours of collection (Gee, 1969; Temple, 1972; Berry, 1972; Grier, 1973; Archibald, 1974; Bird *et al.* 1976; Boyd *et al.*, 1977). Delays of this duration have little influence on the motility of spermatozoa in golden eagles and goshawks (Corten, 1973; Grier, 1973). Bird *et al.* (1976) found that kestrel spermatozoa retained their motility at 4–10°C for 12 hours, whereas at room temperatures motility was not retained as long.

TABLE IV

Motility of red-tailed hawk spermatozoa under different storage conditions

Strage treatment[a]	Percentage of spermatozoa remaining motile
1 hour at 24°C	80–100
4 hours at 24°C	50–60
6 hours at 24°C	10–20
12 hours at 24°C	0–10
24 hours at 24°C	0
1 hour at 5°C	80–100
4 hours at 5°C	70–90
6 hours at 5°C	40–50
12 hours at 5°C	30–40
24 hours at 5°C	10–20
36 hours at 5°C	<10

[a] Held in open capillary tubes.

Table IV shows data that Temple obtained from experiments with semen of the red-tailed hawk. In agreement with Bird's findings, semen remained in better condition for a longer period of time at lower temperatures. G. F. Gee & G. S. Archibald (unpublished) found that crane semen from the Patuxent Wildlife Research Center shipped by air retains good motility for at least eight hours if kept at 4–10°C and open to the air, while at room temperature most spermatozoa in clean samples cease to be active in three to four hours. A few crane spermatozoa can retain motility at room temperature for 12 hours, especially those trapped in the spherical body. However, crane spermatozoa in contaminated semen may be totally immotile in 15 min. The United States Fish and Wildlife Service and the United States Agricultural Research Service have begun a joint study on cryogenic preservation of crane semen, and a report on that work is in this symposium (Sexton & Gee, this volume p. 89).

OBSERVATIONS ON INSEMINATION OF FEMALE NON-DOMESTIC BIRDS

The ultimate objective of artificial insemination is to obtain fertile eggs, and much of the success in achieving this goal depends on how a semen sample is introduced into the reproductive tract of the female. The two alternative techniques, placement of semen in either the cloaca or the oviduct, can lead to different fertility rates.

Comparison of Insemination in Cloaca and in Oviduct

It is generally agreed that placing semen directly into the oviduct leads to a higher rate of fertility than placing semen into the cloaca (Smyth, 1968). Although breeders of non-domestic birds accept this conclusion, there is often a reluctance physically to handle a delicate, often irreplaceable female in the rather forceful way that is necessary to evert the oviduct. As a result, most inseminations of non-domestic birds have relied on placement in the cloaca rather than the oviduct. Table V

TABLE V

Comparison of fertility rates following inseminations in the oviduct or cloaca

| Species | Percentage fertility following inseminations in | | Reference |
	Oviduct	Cloaca	
Peregrine falcon	73	27	Cade *et al.* (in preparation)
Prairie falcon	50	0	Boyd *et al.* (1977)
Red-tailed hawk	66	33	S. A. Temple (unpublished)

summarizes data on fertility rates that resulted from different insemination routes. Apparently, the lower fertility rates that accompany cloacal inseminations can be overcome if inseminations are performed very frequently, at least in sandhill cranes (Table VI).

Timing of Inseminations and Duration of Fertilizing Ability

A fertilized egg can result only if viable spermatozoa are present in the reproductive tract of the female at or very shortly after the time of ovulation. This means that an effective insemination must precede oviposition by at least the number of hours that it takes an egg to pass through the oviduct. For example, Boyd *et al.* (1977) found that

TABLE VI

Frequency of insemination related to crane egg fertility

Frequency[a]	Number of eggs	Number of fertile eggs	Percentage fertility
1	11	5	45
2	12	8	67
3	50	34	68
4	16	14	88
Total	89	61	68

[a] Number of inseminations in a 10-day period.

insemination of prairie falcons 50 hours before oviposition failed to fertilize the egg, but insemination 60 hours before oviposition did fertilize the egg. Species-specific times must certainly exist, but so far only Boyd *et al.* (1977) have precise data for a non-domestic bird.

Also to be considered when planning the timing of an insemination are the time periods over which spermatozoa can retain their fertilizing ability in the reproductive tract of the female. Spermatozoa of some domestic birds are known to retain fertilizing ability for extended time periods, e.g. up to 45 days in turkeys (Nalbandov & Card, 1943), hence many eggs can be fertilized by a single insemination. All of the available data suggest that most non-domestic birds are incapable of retaining spermatozoa for such extended periods.

Bird *et al.* (1976) found that artificially inseminated female American kestrels could continue to produce fertile eggs for an average of 8·1 days after the last insemination. Boyd *et al.* (1977) found that female prairie falcons did not produce fertile eggs more than 194 hours after the last insemination. Grier (1973) obtained a fertile egg from a golden eagle nine days after an insemination, and S. A. Temple (unpublished) obtained a fertile egg from a female red-tailed hawk six days after an insemination.

In sandhill cranes the timing of inseminations during the egg-production cycle influenced fertility rate (G. F. Gee, unpublished). When cranes were inseminated a few hours after oviposition, the fertility of subsequent eggs was greater (82%) than when birds were inseminated two to three days prior to oviposition (66%). Fertility of the next three eggs of the same post-oviposition inseminated group was also high (82%), but inseminations were continued during the interval between eggs. Eggs that were laid the day after insemination were not included in these comparisons. The eggs would have been in the oviduct and beyond the stage when the semen deposited could affect fertility.

Effects of Artificial Insemination on Egg Production

Many breeders of non-domestic birds have feared that the physical handling and the associated trauma that is necessary for forced artificial insemination would cause a disruption of the female's laying cycle. The available observations suggest that these fears may be exaggerated. In raptors, there have been no reports of unusual laying patterns among birds that had been inseminated using forced artificial insemination techniques (Bird *et al.*, 1976; Boyd *et al.*, 1977; Cade *et al.*, in preparation). In eight cranes inseminated one year but not the next, G. F. Gee (unpublished) found that artificial insemination did not appear to interfere with egg production; $7 \cdot 9 \pm 1 \cdot 6$ (mean $\pm$ s.d.) eggs per bird were obtained in the artificial insemination program and $7 \cdot 1$ eggs per bird when not in the program. Apparently, most non-domestic birds are quite tolerant of the handling that accompanies artificial insemination; at least, it does not seem to disrupt their reproductive cycle.

Comparison of Fertility Rates From Artificial Insemination and Natural Pairing

The fertility of eggs in the wild is greater than fertility obtained in captivity. Although the percentage of fertile eggs can vary considerably in special circumstances, about 95% of all eggs laid by birds in their native habitat are fertile (Skutch, 1976). In 207 crane eggs taken from the wild for incubation at the Patuxent Wildlife Research Center, 91% were fertile. Without artificial insemination, most cranes hatched from these wild eggs failed to produce fertile eggs when they matured, a situation common in other captive crane flocks (Archibald, 1974). From the few fertile crane pairs, 52% of 114 eggs have been fertile—considerably less than from the eggs collected in the wild. Often in captive raptors the first clutch of eggs laid in captivity are infertile (Cade, 1975), and even when they are fertile, captive egg production is not as good as in the wild. The South American snail kites, *Rostrhamus s. sociabilis*, at the Patuxent Wildlife Research Center lay several clutches of eggs every year but only 72% of the eggs have been fertile. Of 604 eggs laid by the Aleutian Canada geese, *Branta canadensis leucopureia*, at the Patuxent Wildlife Research Center, only 60% have been fertile.

CONCLUSIONS

Captive breeding of non-domestic birds has increased dramatically in this century, and production of young often exceeds that produced by the same number of birds in their native habitat. However, a considerable number of difficulties are commonly encountered with some animals. When infertility is a problem, artificial insemination can be a useful method to improve production. Artificial insemination programs with non-domestic birds are relatively recent, but several notable successes have been documented, especially with cranes and raptors.

Fertility from artificial insemination in captive non-domestic birds has been good but less than that achieved by birds in their native habitat. Semen quality has not been studied extensively in non-domestic birds owing to the small volume of semen obtained. In general, insemination with the largest number of live spermatozoa is still the best guarantee of fertile egg production. If the sample is small or the concentration of spermatozoa is low, frequent inseminations may be required to achieve good fertility.

ACKNOWLEDGEMENTS

We thank G. S. Archibald, F. Beall, D. M. Bird, L. L. Boyd, T. J. Cade, P. J. M. Corten, G. F. Greenwell, J. W. Grier, J. G. Griffith and C. W. Pickett for the use of unpublished data. The work of S. A. Temple on artificial insemination of raptors was conducted as part of the falcon breeding program at Cornell University's Laboratory of Ornithology with support from the Natural Science Foundation, National Audubon Society, United States Fish and Wildlife Service, World Wildlife Fund and contributions to the Peregrine Fund.

We are grateful to Mr James Stephenson for his valuable assistance in development of the insemination technique and Dr Rod Drewein for his capture of the wild cranes. Thanks are extended to Dr Ray C. Erickson, Mr Glen Smart, Mr Gene Cowan and many other individuals responsible for the acquisition and maintenance of the cranes at the Patuxent Wildlife Research Center.

REFERENCES

Almquist, J. O. (1968). Dairy cattle. In *The artificial insemination of farm animals*: 105–106. Perry, E. J. (ed.). New Brunswick, N. J.: Rutgers University Press.

Archibald, G. S. (1974). Methods for breeding and rearing cranes in captivity. *Int. Zoo Yb.* **14**: 147–155.

Berry, R. B. (1972). Reproduction by artificial insemination in captive American goshawks. *J. Wildl. Mgmt* **36**: 1283–1288.

Bird, D. M. & Buckland, R. B. (1976). The onset and duration of fertility in the American kestrel. *Can. J. Zool.* **54**: 1595–1597.

Bird, D. M., Laguë, P. C. & Buckland, R. B. (1976). Artificial insemination vs. natural mating in captive American kestrels. *Can. J. Zool.* **54**: 1183–1191.

Bonadonna, T. (1939). Artificial insemination of birds. *Proc. 7 World Poult. Congr. Expos.*: 79–82.

Boyd, L. L. & Boyd, N. S. (1976). Hybrid falcon. *Hawk Chalk* **14**: 53–54.

Boyd, L. L., Boyd, N. S. & Dobler, F. C. (1977). Reproduction of prairie falcons by artificial insemination. *J. Wildl. Mgmt.*

Burrows, W. H. & Quinn, J. P. (1939). Artificial insemination of chickens and turkeys. *Circ. U.S. Dep. Agric.* No. 525: 1–12.

Cade, T. J. (1975). Falcon farming. *Anim. Kingd.* **78**: 3–9.

Cade, T. J., Weaver, J. D., Platt, J. B. & Burnham, W. A. (In preparation). *The propagation of large falcons in captivity.*

Corten, P. J. M. (1973). Artificial insemination methods and equipment. *Captive Breed. diurnal Birds Prey* **1(4)**: 3–5.

Corten, P. J. M. (1974). Further experience, progress and success with AI in birds of prey during 1974. *Captive Breed. diurnal Birds Prey* **1(5)**: 12–13.

Erickson, R. C. (1968). A federal research program for endangered wildlife. *Trans. N. Am. Wildl. Nat. Resour. Conf.* **33**: 418–433.

Frank, A. H. (1961). Artificial insemination in livestock breeding. *Circ. U.S. Dep. Agric.* No. 567: 1–71.

Gee, G. F. (1969). Reproductive physiology of the greater sandhill crane. *A. Progr. Rep. Admin. Rep., Patuxent Wildl. Res. Cent.* **1969**: 245–247.

Gee, G. F. (1971). Reproductive physiology of the greater sandhill crane. *A. Progr. Rep. Admin. Rep., Patuxent Wildl. Res. Cent.* **1971**: 275–277.

Gee, G. F. (1972). Reproductive physiology of the greater sandhill crane. *A. Progr. Rep. Admin. Rep., Patuxent Wildl. Res. Cent.* **1972**: 111–112.

Grier, J. W. (1973). Techniques and results of artificial insemination with Golden eagles. *Raptor Res.* **7**: 1–2.

Grier, J. W., Berry, R. B. & Temple, S. A. (1973). Artificial insemination with imprinted raptors. *Jl. N. Am. Falconers' Assoc.* **11**: 45–55.

Hamerstrom, F. (1970). *An eagle to the sky.* Ames: Iowa University Press.

Hediger, H. (1965). Environmental factors influencing the reproduction of zoo animals. In *Sex and behavior*: 319–354. Beach, F. A. (ed.). New York: Wiley and Sons.

Immelmann, K. (1971). Ecological aspects of periodic reproduction. In *Avian biology* **1**: 342–389. Farner, D. S. & King, J. R. (eds). New York and London: Academic Press.

Johnson, A. S. (1954). Artificial insemination and duration of fertility in geese. *Poultry Sci.* **33**: 638–640.

Kamar, G. A. R. (1958). The collection of cock's semen without milking the copulatory organ. *Poultry Sci.* **37**: 1382–1385.

Kear, J. (1975). The breeding of endangered wild fowl as an aid to their survival. In *Breeding endangered species in captivity*: 49–60. Martin, R. D. (ed.). London and New York: Academic Press.

Kinney, T. & Burger, R. E. (1960). A technique for the insemination of geese. *Poultry Sci.* **39**: 230–232.

Lake, P. E. (1962). Artificial insemination of poultry. In *The semen of animals and artificial insemination*: 331–355. Maule, J. P. (ed.). Farnham Royal, England: Commonwealth Agr. Bur.

Martin, R. D. (1975). (ed.). *Breeding endangered species in captivity.* New York and London: Academic Press.

Meng, H. (1972). Breeding peregrine falcons in captivity. *Hawk Chalk* **11**: 40–44.

Miller, A. H. (1953). The case against trapping California condors. *Audubon Mag.* **55**: 261–262.

Nalbandov, A. V. & Card, L. E. (1943). Effects of stale sperm on fertility and hatchability of chicken eggs. *Poultry Sci.* **22**: 218–226.

Nestor, K. E., Chamberlin, V. D. & Renner, P. A. (1970). Lighting affects semen production in turkeys. *Ohio Rep. Res. Devlmt* **55**: 119–120.

Olver, M. D. (1971). Artificial insemination and duration of fertility in Chinese geese. *Agroanimalia* **3**: 79–86.

Owen, R. D. (1941). Artificial insemination of pigeons and doves. *Poultry Sci.* **20**: 428–431.

Pingel, H. (1972). Artificial insemination in duck breeding. *Wiss. Z. Humboldt-Univ. Berl. (Math.-Nat.)* **21**: 179–181.

Quinn, J. P. & Burrows, W. H. (1936). Artificial insemination of fowls. *J. Hered.* **27**: 31–37.

Serebrovski, A. S. & Sokolovskaja, I. I. (1934). Electorejakuljacia u Ptic. *Prob. Zhivotn., Mosc.* **5**: 57. (*Anim. Breed. Abstr.* **3**: 73–74.)

Sharlin, J. S. (1976). *Sperm head length as a predictor of fecundity in the sandhill crane.* M.S. Thesis: University of Maryland.

Skinner, J. L. (1974). Infertility and artificial insemination. In *Raising wild ducks in captivity*: 147–152. Hyde, D. O. (ed.). New York: E. P. Dutton and Co., Inc.

Skutch, A. F. (1976). The rate of reproduction. In *Parent birds and their young*: 433–446. Austin and London: University of Texas Press.

Smyth, J. R., Jr. (1968). Poultry. In *The artificial insemination of farm animals*: 258–300. Perry, E. J. (ed.). New Brunswick, N.J.: Rutgers University Press.

Temple, S. A. (1972). Artificial insemination with imprinted birds of prey. *Nature, Lond.* **237**: 287–288.

Temple, S. A. & Cade, T. J. (1977). The Cornell University falcon program. In *World conference on birds of prey, Vienna 1975, report of proceedings*: 353–369. Chancellor, R. D. (ed.). London: International Council for Bird Preservation.

Watanabe, M. (1957). An improved technique of the artificial insemination in ducks. *J. Fac. Fish. Anim. Husb., Hiroshima Univ.* **1**: 363–370.

Weaver, J. D. & Cade, T. J. (1974). Special report on the falcon breeding program at Cornell University. *Hawk Chalk* **13**: 31–42.

Wentworth, B. C. & Mellen, W. J. (1963). Egg production and fertility following various methods of insemination in Japanese quail (*Coturnix coturnix japonica*). *J. Reprod. Fert.* **6**: 215–220.

Wolfson, A. (1952). The cloacal protuberance—a means for determining breeding condition in live male passerines. *Bird Banding* **23**: 159–165.

Wolfson, A. (1960). The ejaculate and the nature of coition in some passerine birds. *Ibis* **102**: 124–125.

Zimmerman, D. R. (1974). Captive breeding: boon or boondoggle. *Nat. Hist. Mag.* **83**(10): 6–8, 10–16, 19.

Symp. zool. Soc. Lond. (1978) No. 43, 73–80

Artificial Insemination of Falcons

L. L. BOYD

Washington State University, Pullman, Washington, USA

SYNOPSIS

A technique for conditioning falcons to ejaculate semen voluntarily has been combined with a standard method for artificially inseminating poultry to produce hybrids between species, backcrosses from a hybrid male to its parent species and a hybrid by outcrossing to a third species. This technique for semen collection has provided a comparison of semen production between a prairie falcon and a hybrid prairie x peregrine. The sequence of fertile eggs laid after insemination has provided information on the longevity of spermatozoa in the oviduct and on the optimal time for insemination.

INTRODUCTION

The artificial insemination of some domesticated species of birds has become a standard method of breeding. Various techniques have been used in the application of this procedure to raptors.

The conditioning of males to ejaculate semen voluntarily, and of females to accept insemination voluntarily has been described as cooperative artificial insemination, originally achieved with the golden eagle, *Aquila chrysaetos* (Hamerstrom, 1970; Grier, 1973), the American goshawk, *Accipiter gentilis* (Berry, 1972) and the red-tailed hawk, *Buteo jamaicensis* (Temple, 1972). Prairie falcons, *Falco mexicanus*, have also been produced by artificial insemination of semen from a cooperative male using the poultry technique for everting the oviduct of the female (Boyd, Boyd & Dobler, 1977).

Most of the emphasis on reproducing raptors in captivity has been the pairing of birds for natural breeding. This method facilitates incubation, brooding and feeding of young, but many of these pairs fail to copulate. Standard methods of artificial insemination have been successfully used with kestrels, *F. sparverius* (Bird, Laguë & Buckland, 1976) and also in the large falcons, e.g. *F. peregrinus*, *F. mexicanus* and *F. rusticolus* (Cade & Weaver, 1976; S. L. Bapteste & W. A. Burnham, pers. comm.).

The cooperative male has proven to be an interesting and advantageous asset to breeding experiments. Conditioned males respond to their human handlers as if they were their natural mates. A single male produces enough semen to inseminate several females, and the quality of semen for artificial insemination is probably equal to that in natural insemination.

The hybridization of a saker falcon, *F. cherrug*, and a peregrine falcon by natural mating (Morris & Stevens, 1971) was probably the first in captivity. The creation of these beautiful and interesting specimens also raised the question of fertility in such hybrids.

A male hybrid falcon produced by artificially inseminating a female peregrine (Middle East origin) with prairie falcon semen (Boyd & Boyd, 1975) provided an opportunity to test fertility, and by conditioning it for voluntary semen ejaculation it has been possible to breed it to the prairie falcon, peregrine falcon and merlin, *F. columbarius richardsoni*. Three additional hybrids of the prairie x peregrine, two males and one female, which are included in this report, have been produced since the first of this cross in 1975. The female has not yet laid, therefore the study of the fertility of this hybrid has so far been limited to the male.

SEMEN COLLECTION

The hybrid male was handled in the same manner as the male prairie falcon (Boyd *et al.*, 1977) except that an exogenous hormone treatment was not used. By hand raising it from hatching age and maintaining social contact with it until it was sexually mature (i.e. the first spring after hatching), the hybrid accepted its handler as a mate and voluntarily produced viable semen. As with the prairie falcon, the mutual ledge display and food transfers were the most prominent courtship expressions. Solicitation for copulation was accomplished by kneeling on the floor, back to the falcon, while approximating the soliciting wail of a female prairie falcon. Once the falcon was responding, the handler donned a specially designed hat for semen collection. The hat was of a dense cotton fabric and tightly fitted to prevent it from rocking during the bird's vigorous copulatory activity. A closed cell foam rubber tube, wrapped around the hat just below the crown, formed a soft brim for the bird to copulate on in any direction he faced. The brim was cemented to the hat by a bead of clear silicone rubber sealant which was smoothed while still soft into a concave gutter around the inside edge of the rubber ring. Semen was collected from the gutter with a capillary tube attached to a tuberculin syringe. Semen collections were made early in the morning, at midday, and in the evening.

Semen production started at relatively high level during the hybrid's first season (Fig. 1). This was probably because some time was necessary after spermatozoal production actually commenced for the hybrid to develop enough sexual drive to attempt copulation. During its second season (Fig. 1) there was a gradual increase to maximum semen production and then a decline, which is probably a more accurate representation of the normal reproductive cycle. The prairie falcon (Fig. 1), by comparison, maintained a much lower level of semen production.

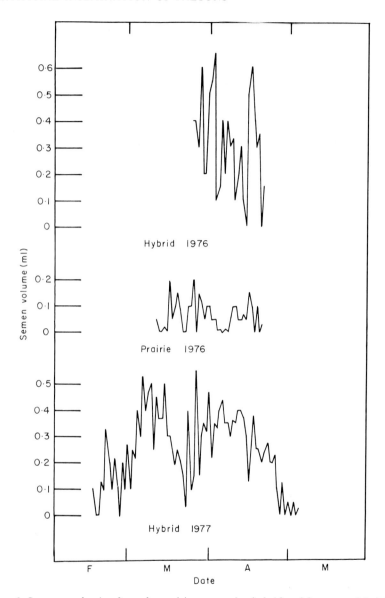

FIG. 1. Semen production from the prairie x peregrine hybrid and from a prairie falcon.

SEMEN CHARACTERISTICS

The semen produced by the hybrid appeared visually at times to have a low spermatozoal concentration, but proved to be comparable to that of the prairie falcon (average ejaculate of 0·1 ml containing 320 000 to 2 million spermatozoa). The spermatozoal count of falcons, in comparison to domestic fowl, is very low. An average ejaculate from chickens and turkeys ranges from 500 to 1500 million spermatozoa (Lake, 1975). This low count is not characteristic of all raptors, since that of the Harris hawk, *Parabuteo unicinctus*, compares favorably with that of domestic fowl, and the shape and size of spermatozoa is uniform (L. L. Boyd, unpublished). An extreme variation of sizes and shapes of spermatozoa was evident in the hybrid, but this also seems to be characteristic of prairie falcons and peregrines.

SEMEN STORAGE

Fresh semen (within one hour of collection) was used for inseminating the peregrine and one of the prairie falcons (sequences A, B, C and D, Fig. 2). Semen from three to 20 hours old, transported uninsulated by carrier pigeon 160 km, was used to inseminate the other three prairie falcons (sequences E, F, G and H, Fig. 2). The merlin was inseminated with semen 14 to 15 hours old which had been transported 600 km by road and packaged in an insulated container with ice-water (sequence I, Fig. 2).

INSEMINATION, FERTILITY AND HATCHING ABILITY

The fertility sequence of eggs laid after insemination is similar to the pattern observed in prairie falcons (Boyd *et al.*, 1977) and in the gyrfalcon x peregrine, *Falco rusticolus* x *Falco peregrinus* (Cade & Weaver, 1976). With fresh semen the minimum insemination time necessary prior to egg-laying to achieve fertility was about 60 hours (first insemination of sequences A and C, Fig. 2). The maximum fertile period from insemination to the last fertile egg was seven days (last insemination of sequence D, Fig. 2). Semen transported by carrier pigeon produced a lower rate of fertility (sequences E, F, G and H, Fig. 2). The maximum fertile period observed in prairie falcons was eight days (Boyd *et al.*, 1977).

All inseminations were made in the manually everted oviduct by restraining the birds and massaging the abdomen. This technique is commonly used for poultry insemination, but it seems to be somewhat more difficult to apply to raptors. Various methods of restraint and manipulation can be used. The method of everting the oviduct consists of applying gentle pressure to the abdomen, while simultaneously spreading the cloacal muscles with the index and middle fingers. This technique has been described in detail by Boyd *et al.* (1977).

Eggs were incubated either naturally by paired falcons, or artificially. The latter consisted of either incubating to full term in an incubator or

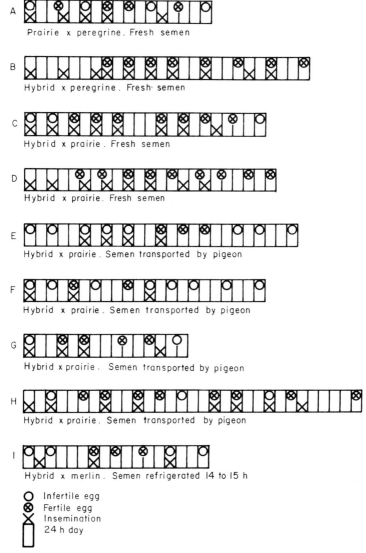

FIG. 2. Egg-laying and egg fertility of falcons inseminated with semen from the prairie x peregrine hybrid.

starting the eggs under a falcon or a chicken and later transferring them to an incubator.

A total of four prairie x peregrine hybrids were produced, one in 1975 and three in 1976 (Table I). There has been, however, a severe reduction in hatching ability of the backcrosses to the peregrine and the

TABLE I

Mortality and fledging success of artificially incubated eggs

	Prairie x peregrine	Hybrid x peregrine	Hybrid x merlin
Eggs set	5	8	3
Eggs broken	0	1	1
Embryo death 1–10 days	0	2	0
Embryo death 10–20 days	0	4	1
Embryo death 20 days to term	1	1	0
Chicks hatched	4	0	1
Chicks fledged	4	0	1

prairie falcons. Eight fertile eggs were produced by backcrossing the hybrid male to the peregrine. None of these eggs hatched (Table I).

Three fertile eggs were produced from the hybrid x merlin cross. One hatched and fledged into a beautiful falcon (Table I). The egg that hatched was incubated for the first 11 days under a domestic pigeon and then transferred to an artificial incubator until hatching at 30 days. Incubation of another egg was commenced under a chicken and the egg was transferred to the artificial incubator at three days, but it became dehydrated and the embryo died at 12 days of incubation. One egg was accidentally broken in the incubator.

A total of only nine healthy offspring from 33 fertile eggs have been produced by backcrossing the hybrid male to prairie falcons. A considerable increase in hatching ability was apparently achieved by natural incubation (Table II).

TABLE II

Mortality and fledging success of artificially and naturally incubated eggs from hybrid x prairie falcon

	Artificial	Natural
Eggs set	24	9
Eggs broken	0	0
Embryo death 1–10 days	13	4
Embryo death 10–20 days	2	1
Embryo death 20 days to term	4	0
Chicks hatched	5	4
Chicks fledged	5	4

HYBRID CHARACTERISTICS

The genetic influence of the peregrine in the prairie x peregrine hybrid
was more obvious in anatomy than in plumage. In juvenile plumage all
of these hybrids have been uniform in their markings. This has also been
true of other prairie x peregrine crosses using a North American peregrine, *F.p. pealei* (S. L. Bapteste, pers. comm.). In adult plumage the
hybrid developed a coloration more typical of the peregrine. The peregrine parentage in this hybrid differs anatomically from northern forms
of the peregrine with a relatively long wing, short tail and heavier
sternum and shoulder girdle (Dementiev, 1957). The result is a heavier
wing loading than that of many other forms of the peregrine, which
causes a quite distinctive flight style. This was the most obvious peregrine trait of the hybrid.

The backcrosses to the prairie falcon have shown an extreme variation in both plumage and anatomical characteristics. They range from
dark to very light in plumage, to the extent that some are lighter than
prairie falcons and others darker than the original hybrids. The anatomical characteristics favoring either the prairie falcon or the peregrine
often, but not always, correspond to plumage of the appropriate type.
There are obvious peregrine characteristics in every backcross.

The hybrid x merlin cross generally resembles a merlin. The most
obvious large falcon characteristics are much heavier tarsi and feet than
the merlin. In weight (300 g) it is barely larger than a female merlin.
Because of the large size difference between the prairie x peregrine
hybrid and a merlin, this cross is tentatively being considered a male. It
has demonstrated a strong flight capability and appears to be as vigorous
as the original hybrid (C. H. Schwartz, pers. comm.).

CONCLUSION

Hybridization between species has occurred in all experiments thus far
within the genus *Falco*. All of these crosses have involved the peregrine
including saker x peregrine, prairie x peregrine and gyrfalcon x peregrine. The nearly cosmopolitan distribution of the peregrine and the
relative geographic isolation of other falcon species suggests that the
peregrine may be an ancestral form common to many or all. Further
experiments in the hybridization and fertility of falcon species could
prove interesting in illuminating the evolution of falcons.

Backcrosses from the male prairie x peregrine hybrid with the female
prairie falcon have been produced, but with a substantial reduction in
hatching ability of fertile eggs. Full-term embryos have resulted from
backcrossing the male prairie x peregrine hybrid with the female peregrine, but the lack of any offspring in this case may be partially due to
incubation conditions since none of these eggs were incubated by falcons.

An outcross between the male prairie x peregrine hybrid with a female merlin was produced from only three fertile eggs. Further experiments are necessary to determine whether this cross might be more viable than the backcross to related species.

ACKNOWLEDGEMENTS

I would like to thank Professor G. A. Laisner and S. Laisner for the financial support which made this report possible. I am grateful for the cooperation and assistance of R. H. Young, C. H. Schwartz, A. L. Gardner and T. B. Stralser. I would also like to thank the WSU Poultry Department for assistance in providing feed for falcons. I appreciate the encouragement and assistance extended by Drs J. R. King and L. B. Kirschner in preparing this manuscript. Full support in these experiments has been provided by N. S. Boyd which has been essential to the progress of this work.

REFERENCES

Berry, R. B. (1972). Reproduction by artificial insemination in captive American goshawks. *J. Wildl. Mgmt* **36**: 1283–1288.
Bird, D. M., Laguë, P. C. & Buckland, R. B. (1976). Artificial insemination vs. natural mating in captive American kestrels. *Can. J. Zool.* **54**: 1183–1191.
Boyd, L. L. & Boyd, N. S. (1975). Hybrid falcon. *Hawk Chalk* **14**: 53–54.
Boyd, L. L., Boyd, N. S. & Dobler, F. C. (1977). Reproduction of prairie falcons by artificial insemination. *J. Wildl. Mgmt* **41**: 266–271.
Cade, T. J. & Weaver, J. D. (1976). Gyrfalcon x peregrine hybrids produced by artificial insemination. *Jl N. Am. Falconers' Ass.* **15**: 42–47.
Dementiev, G. P. (1957). On the shaheen. *Ibis* **99**: 477–482.
Grier, J. W. (1973). Techniques and results of artificial insemination with golden eagles. *Raptor Res.* **7**: 1–12.
Hamerstrom, F. (1970). *An eagle to the sky.* Ames: Iowa State University Press.
Lake, P. E. (1975). Gamete production and the fertile period with particular reference to domesticated birds. *Symp. zool. Soc. Lond.* No. 35: 225–244.
Morris, J. & Stevens, R. (1971). Successful cross breeding of a peregrine tiercel and a saker falcon. *Jl N. Am. Falconers' Ass.* **10**: 10–13.
Temple, S. A. (1972). Artificial insemination with imprinted birds of prey. *Nature, Lond.* **237**: 287–288.

Symp. zool. Soc. Lond. (1978) No. 43, 81–88

Artificial Insemination: A Practical Method for Genetic Improvement in Ring-Necked Pheasants

J. R. CAIN

Texas Agricultural Experiment Station, Texas A and M University, College Station, Texas, USA

SYNOPSIS

Attempts at genetic improvement with ring-necked pheasants have proceeded slowly, primarily owing to colony mating systems conventionally used. It would be desirable to develop specialized lines of pheasants for a meat market or wild-like birds for hunting preserves. A cage management system was developed which allowed pedigree offspring production from pheasant hens. This system permitted the development of two lines of pheasants, one selected for high rate of egg production and another for wide neck rings. Females were housed in single cages, in cock–hen pair cages or in colony floor pens. Caged hens laid significantly better than floor birds, probably because of reduced social stress. Pair-caged hens laid 55·3 eggs, single caged hens laid 53·1 eggs and floor pen hens laid 41·7 eggs per hen. Light intensity affected the onset of egg-laying, but not the total number of eggs laid. Mortality averaged about 5% for this study.

Artificial insemination (AI) techniques used for domestic poultry were modified for use with pheasants. These cocks required more stimulation on the rump and pelvic regions than did chickens. After about three weeks of training, the roosters were sufficiently acquainted with the routine not to be distressed. Pheasant cocks rarely gave more than 0·25 ml semen. While pheasant hens required more manipulation to evert the oviduct than chicken hens, egg production in the hens subjected to AI was not apparently different from pair-caged hens.

Artificial insemination of caged pheasant hens with cock semen yielded acceptable fertility rates for genetic studies. Mean fertility from single pair matings (86·9%) was only slightly better than from artificially inseminated hens (80·4%) during a 20-week reproductive season. For AI, pheasant semen was diluted at a rate of two parts commercial turkey diluent to one part semen. Each hen received about 100 million spermatozoa per week. The duration of fertility following a single insemination averaged 11·4 days. It is anticipated that these studies will provide a model for other species of pheasants.

INTRODUCTION

Although ring-necked pheasants are not native to North America, they currently inhabit much of the continent. These birds are a mixture of Chinese, Korean and Manchurian ring-necks with some influence from the English blackneck pheasant. Regardless of the species mix, the North American ring-necked pheasant is generally referred to as *Phasianus colchicus* (Baxter & Wolfe, 1973). Ring-necked pheasants are raised commercially in North America, primarily for release on hunting preserves. An increased demand for commercially produced pheasants has encouraged research on this species. Current management practices on most pheasant farms involve keeping birds in large outdoor pens in an attempt to simulate a natural habitat to some degree. While this system may be efficient for small numbers of breeder females, it is inefficient for large commercial pheasant farms.

Early attempts to confine breeder hens in small pens were generally unsuccessful (Breittenbach & Meyer, 1959). Some commercial farms reported poor success with breeder hens confined to cages because of low fertility and high mortality. In contrast, others reported no differences between egg production rates from pheasant hens in suspended colony cages and those from hens in floor pens (Adams, Kahrs & Deyoe, 1968; Vandepopuliere et al., 1969). Neither of these cage systems offered the advantage of obtaining offspring of known parentage. To supply the needs of the commercial game bird industry, namely, birds with rapid feather growth, efficient feed conversion and desirable physical characteristics, it is necessary to apply selection pressures to the breeding stock. Such controlled breeding can only be achieved with a system of caging the hens individually.

Single-sire families, in which one cock is mated with several hens, each of which is kept separate for purposes of identifying the parentage of its eggs, provide the greatest potential for genetic studies. One method of obtaining these matings would be the use of artificial insemination (AI), a standard practice in the poultry industry. There has been little research reported on the use of AI as a genetic tool in breeding ring-necked pheasants. Asmundsen & Lorenz (1955) reported fertility of 16·5% when crossing pheasants and turkeys using AI. Shaklee & Knox (1954) obtained a fertility of 3·5% when crossing male pheasants with female cornish chickens using AI. Since AI of chickens has been used extensively since 1936 (Burrows & Quinn, 1939) we felt it might have advantages for use with ring-necked pheasants. Studies were initiated to compare the egg production and fertility rates of artificially inseminated, caged pheasant hens with those from naturally mated females housed in cages and in floor pens. The objective of these studies was to develop a means whereby captive reared stocks of pheasants can be genetically improved. It was anticipated that studies with common ring-necked pheasants would provide a model for some of the rarer species of Phasianus.

METHODS AND MATERIALS

Cage Management Studies

Studies investigating cage management of pheasant breeder hens as a means of maintaining pedigree egg records for genetic experiments have been conducted at the Texas Agricultural Experiment Station during the past three years. A random-bred flock of ring-necked pheasants were wing banded and randomly assigned to one of three types of cages. The hens were either housed individually in standard chicken laying cages or with a cock in standard turkey laying cages.

One-half of the single hen cages had sheet metal tops which were expected to reduce head injuries. The average light intensity in the metal-top cages was 4·5 (range 1·1–8·6) lux, while in the wire-top cages it was four times greater, averaging 17·9 (range 6·1–22·6) lux. A fourth group of breeder hens was housed in four floor pens (16·25 m^2 each) at a density of 0·25 m^2 bird^{-1}; these served as controls. Egg production and mortality were recorded daily for a 20-week period.

Artificial Insemination Studies

In conjunction with the cage management investigation, a study was initiated to compare the fertility of eggs from artificially inseminated hens with those from naturally mated hens. Roosters used for semen collection were housed individually in the wire-floored turkey cages. Twenty-seven randomly chosen females housed in single hen laying cages were inseminated weekly with pooled pheasant semen for 15 weeks. The fertility of these hens was compared to that of 24 pair-mated hens housed in turkey laying cages located adjacent to the single hen cages. Artificial daylength for this experiment was initiated at 14·5 hours day^{-1} and increased 0·5 hour day^{-1} month^{-1} after peak production was reached. All birds were watered and fed *ad libitum*. Eggs were collected daily and stored in an egg cooler at 12·5°C until set for hatching. All undeveloped eggs were broken and checked for fertility.

Semen was collected weekly from pheasant cocks into 3-ml glass vials using standard poultry ejaculation techniques. Pheasant cocks generally required more stimulation on the rump and pelvic regions than did chicken roosters. After about three weeks of training, pheasant cocks were sufficiently acquainted with the routine not to be distressed. Each male rarely gave more than 0·25 ml of semen per collection. The mean concentration of spermatozoa per unit volume of semen was determined at the outset to be 5×10^6 spermatozoa mm^{-3}. Based on this concentration, semen samples were thenceforth diluted (1:2) with commercial turkey semen extender (Minnesota Turkey Growers Association) with the resultant concentration of approximately 100×10^6 spermatozoa 0·1 ml^{-1}. Each female was then inseminated weekly with 0·1 ml of diluted semen intravaginally using a glass tuberculin syringe.

Genetic Studies

Using the system of single hen cage management, which permitted accurate pedigree records to be obtained, and the technique of AI, studies on genetic selection for specialized lines of pheasants were possible. To determine if some characteristics of pheasants are inherited in a similar manner to domestic fowl, two traits were chosen which have been well defined genetically in chickens. Egg production was investigated because a demand exists for pheasant hens which lay more eggs, and while heritability estimates for this trait are low in chickens, genetic progress for increased numbers of eggs has been successful.

Since heritability estimates for feathering characteristics are generally considered to be high in chickens, a study to determine the inheritance of the width of the neck ring of pheasant cocks was also initiated. Sixteen randomly selected males (control group) and 16 cocks with wide neck rings were selected as sires. Females were chosen randomly and housed in laying cages. When the 271 offspring obtained from the 32 matings were 21 weeks of age, the width of their neck rings was measured to the nearest millimeter with micrometer calipers. Measurements were made at the widest part of the ring on the right and left side, and at the midline of the back. Similar measurements were made on the sires and realized heritability values were calculated for these feathering traits.

RESULTS AND DISCUSSION

Cage Management Studies

The average number of eggs laid per hen by pheasants in the floor pens was significantly lower than that laid by their caged counterparts, whereas no differences were observed between the caged bird groups (Table I). Hens housed in metal-top cages receiving low light intensity (mean of 4·5 lux) began laying slightly later than those housed in the wire-top chicken cages with four times greater light intensity (mean of 17·9 lux). The total number of eggs produced during 20 weeks by each

TABLE I

The effects of various breeder management systems on eggs per hen and mortality in ring-necked pheasants

Treatment	N	Mean number of eggs/hen $\pm$ s.e.	Percentage mortality
Wire-top cages (s)	51	$53·1 \pm 3·7^a$	$3·9^a$
Metal-top cages (s)	54	$57·2 \pm 3·7^a$	$9·3^a$
Wire-top cages (p)	21	$55·3 \pm 7·5^a$	$14·3^a$
Floor pens (c)	104	$41·7 \pm 3·2^b$	$3·8^a$

s, Single hen; p, paired birds; c, colony pens.
[a,b] Means followed by the same letter do not differ significantly at the 0·05 confidence level.

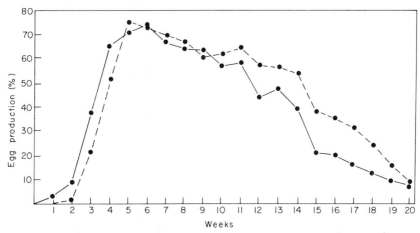

FIG. 1. Comparison of mean weekly egg production rates ([No. of eggs week^{-1}/No. of hens × 7] × 100%) from pheasant hens housed in two types of cages providing different light intensities. ————Wire-top cages, – – – – – metal-top cages.

group did not differ significantly, however (Fig. 1). A similar effect was noted in chickens where pens with a light intensity of 1·1 lux retarded sexual maturity for one week when compared to pullets receiving 3·2 to 32·3 lux (Dorminey, Parker & McCluskey, 1970). Turkeys receiving 18 lux were also sexually retarded when compared to those subjected to a light intensity of 33 or 51 lux (Nestor & Brown, 1972). The present study suggested that pheasants have light intensity requirements slightly higher than chickens but lower than turkeys.

No significant differences in mortality rate existed between breeder hens in cages and those kept in floor pens (Table I). Eggs from hens in single cages were numbered, recorded and used for other studies. We found cage management to be a successful alternative to floor management which allowed us to make genetic gains not possible in colony mating systems.

Artificial Insemination Studies

Fertility in the group of hens artificially inseminated with diluted pheasant semen was significantly lower than in the pair-mated group at the onset of the experiment (Fig. 2). During the remainder of the study, however, the fertility of eggs from artificially inseminated hens was about the same as that obtained from pair matings. The early poor fertility may have been due to inexperience on the part of both the pheasant cocks and the inseminators. The decrease in fertility during Week 11 of the experiment was due to a lack of insemination in the preceding week. At this time an attempt was made to study the duration

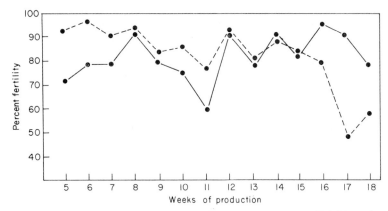

FIG. 2. Fertility of pheasant eggs from two mating systems. ——— Artificial insemination, – – – – caged pair matings.

of fertility following an insemination. Consequently, no artificial insemination was performed on 50% of the hens during the tenth week. Thus, the elapsed time between inseminations was at least two weeks (or until three successive infertile eggs were laid) for some hens. This caused the mean fertility for the flock to be decreased drastically. The mean duration of fertility in those hens not inseminated was found to be 11·4 days (range three to 15 days). The decline in fertility in the pair-mated group at the end of the production period was probably caused by some males becoming sterile owing to hot weather, whereas all artificially inseminated hens continued to receive semen. The semen volume per male did decrease during the month of July, again presumably owing to the high ambient temperature. This observation is supported by Ritter (1962) who claimed that low semen volume had no effect on fertility.

The mean fertility rate for the season in the AI group was significantly lower ($P < 0.05$) than that for the pair mating and floor pen groups (Table II). The fertility obtained in this study might have been higher if undiluted semen had been used. A single insemination of 100

TABLE II

Comparison of fertility in natural matings versus artificial insemination in ring-necked pheasants

	Number of hens	Number of males	Number of eggs set	Number of eggs fertile	Percentage fertility
AI	27	12	843	678	80·43
Pair	24	24	1018	885	86·94
Pen	208	28	419	376	89·74

million spermatozoa has been shown to maintain high fertility for a period of 10 days in chickens (Munro, 1938). From a practical standpoint, it was felt that weekly inseminations of pooled, extended semen provided an acceptable fertility rate to be a useful tool in a breeding program with ring-necked pheasants.

Genetic Studies

A line of ring-necked pheasants selected for increased number of eggs laid per pheasant hen has undergone two generations of selection. The mean number of eggs laid per hen by the unselected population was 42·2. The next generation produced $55·2 \pm 4·1$ eggs (mean ± s.e.) and the third generation laid an average of $62·7 \pm 3·6$ eggs per hen. Initial selection of 20% was based on individual hen performance, but the subsequent generation provided sufficient data to permit 20% selection pressure based on family production rates. A realized heritability for egg number from caged pheasant hens was determined to be 0·21 based on selection gains made between the second and third generations.

Male ring-necked pheasants exhibited extreme variations in the width of the white neck ring. A comparison of the mean neck ring width for the progeny and sires of each group is presented in Table III.

TABLE III

Mean neck ring widths for two lines of ring-necked pheasant cocks

Lines	Right side (mm)	Back (mm)	Left side (mm)
Control; sires	$20·7 \pm 2·0^a$	$8·1 \pm 0·4^c$	$20·0 \pm 1·9^a$
Wide ring; sires	$29·4 \pm 1·9^b$	$12·8 \pm 0·6^d$	$28·4 \pm 1·8^b$
Control; progeny	$22·8 \pm 1·8^a$	$7·8 \pm 0·4^c$	$22·6 \pm 1·1^a$
Wide ring: progeny	$28·5 \pm 1·5^b$	$10·8 \pm 0·5^d$	$28·0 \pm 1·3^b$

[a,b,c,d] Means ±s.e. with different superscripts differ significantly at the $P < 0·05$ level.

Those offspring from the selected sires had significantly ($P < 0·05$) wider neck rings than those from the randomly selected sires. The realized heritability values for these neck-ring width measurements were very high, ranging from 0·88 to 1·0.

The limited studies have shown that genetic information obtained with poultry can be used to predict gains in selection experiments with ring-necked pheasants. Improvements in egg production and feathering were demonstrated with this species and using *Phasianus colchicus* as a model, it can be predicted that other species of the genus *Phasianus* would respond similarly.

REFERENCES

Adams, A. W., Kahrs, A. J. & Deyoe, C. W. (1968). Effects of cage confinement and lighting schedules on performance of ring-necked pheasant breeders. *Poultry Sci.* **47**: 1025–1026.

Asmundsen, V. S. & Lorenz, F. W. (1955). Pheasant-turkey hybrids. *Science, N.Y.* **121**: 307–308.

Baxter, W. L. & Wolfe, C. W. (1973). Life history and ecology of the ring-necked pheasant in Nebraska. *Publ. Nebraska Game Parks Commn* No. 1970.

Breittenbach, R. P. & Meyer, R. K. (1959). Effect of incubation and brooding on fat, visceral weights and body weight of the hen pheasant (*Phasianus colchicus*). *Poultry Sci.* **38**: 1014–1026.

Burrows, W. H. & Quinn, J. P. (1939). Artificial insemination of chickens and turkeys. *Circ. U.S. Dep. Agric.* No. 532: 1–12.

Dorminey, R. W., Parker, J. E. & McCluskey, W. H. (1970). The effects of light intensity on leghorn pullets during the development and laying periods. *Poultry Sci.* **49**: 1657–1661.

Munro, S. S. (1938). Fowl sperm immobilization by a temperature media interaction and its biological significance. *Q. Jl exp. Physiol.* **27**: 281–291.

Nestor, K. E. & Brown, K. I. (1972). Light intensity and reproduction of turkey hens. *Poultry Sci.* **51**: 117–121.

Ritter, B. (1962). Artificial insemination in turkeys. *Turkey Producer.* May: 11.

Shaklee, W. E. & Knox, C. W. (1954). Hybridization of the pheasant and the fowl. *J. Hered.* **45**: 183–190.

Vandepopuliere, J. M., Kealy, R. D., Greene, D. E. & Williamson, J. L. (1969). The influence of environment and housing on pheasant, quail and chukar reproduction. *Poultry Sci.* **48**: 1885.

Symp. zool. Soc. Lond. (1978) No. 43, 89–95

A Comparative Study on the Cryogenic Preservation of Semen from the Sandhill Crane and the Domestic Fowl

T. J. SEXTON and G. F. GEE

Agricultural Research Service, Avian Physiology Laboratory, Beltsville, Maryland, USA
Patuxent Wildlife Center, Laurel, Maryland, USA

SYNOPSIS

Recent findings on the cryogenic preservation of semen from the crane, *Grus canadensis pratensis* and the domestic fowl, *Gallus domesticus*, are compared. Highest levels of post-thaw motility for crane semen (55%) were obtained when semen was diluted 1:1 with the Beltsville Poultry Semen Extender (BPSE) and held for 30 min at 5°C before it was equilibrated with 4% dimethyl sulfoxide (DMSO) for 15 min. In contrast, post-thaw motility for fowl spermatozoa was highest (60%) when semen was diluted 1:3 with BPSE and held for 60 min at 5°C before it was equilibrated with 4% DMSO for 60 min. Post-thaw motility of spermatozoa of both species was highest when the following freezing rates were used: 1°C per min from +5 to −20°C, 50°C per min from −20 to −80°C, then plunging into liquid nitrogen which resulted in a rate of 160°C per min from −80 to −196°C. One of four crane eggs resulting from insemination with frozen-thawed semen was fertile, whereas 27 of 55 fowl eggs were fertile, but this difference may have been due largely to fewer spermatozoa being inseminated into the female crane than into the fowl.

INTRODUCTION

Methods of preserving semen of domestic species have been the subject of intensive reasearch for many years (Neville, 1975). However, despite the vast amount of research effort that has been devoted to frozen semen in farm animals, the only practical application has been in cattle. Although fowl semen can be preserved cryogenically in a manner similar to that for other species, the fertility is too low to be considered for commercial application.

Among the vanishing wildlife all over the world are many non-domesticated avian species, some on the verge of extinction (Allen, 1974). One of several ways to ensure their survival is the captive main-tenance of small breeding nuclei in zoos and animal parks (Martin, 1975). However, under these conditions, maintaining a proper genetic mixture to prevent the problems associated with inbreeding is often difficult. One solution would be to have the capability of storing semen from endangered species in the frozen state for redistribution in years to come.

Although a great deal of information has been obtained on the cryogenic preservation of semen from several domesticated avian species (fowl, turkey, duck, goose) only the work on the fowl has resulted in the production of an adequate number of progeny through artificial breeding with frozen semen (e.g. Watanabe, 1967; Lake, 1970; Mitchell & Buckland, 1976; Sexton, 1976). Therefore, in the absence of a specific method, we have evaluated the freeze-preservation of semen from the crane, *G. c. pratensis*, using a modified procedure developed by Sexton (1976) for freezing semen of the domestic fowl, *G. domesticus*.

MATERIALS AND METHODS

Experiment 1

Five independent trials were conducted sequentially to determine optimal dilution rate, holding time at 5°C in both the presence and absence of dimethyl sulfoxide (DMSO), and freezing rate for maximum spermatozoal motility after freezing and thawing. Semen samples were collected by abdominal massage from at least three males of each species, pooled immediately after collection and diluted with the Beltsville Poultry Semen Extender (BPSE). Details of the composition and physical properties of the BPSE have been reported by Sexton (1977). All semen samples were placed in a water bath at 5°C immediately after dilution.

Trial 1 was designed to determine the optimal dilution rate before freezing. One part semen was diluted at room temperature with either 1, 3 or 5 parts BPSE before being processed and frozen and thawed (Steps 2–6, Table I).

TABLE I

Scheme for investigating the processing of semen to obtain the best recovery of post-thaw motility of crane and fowl spermatozoa[a]

Step 1. Dilution rate at 25°C	1 : 1
Step 2. Holding time at +5°C prior to addition of DMSO	30 min
Step 3. Addition and level of DMSO	4% v/v
Step 4. Equilibration time at +5°C in the presence of DMSO	15 min
Step 5. Freezing rate	
Step 5a. +5 → −20°C	1°C per min
Step 5b. −20 → −80°C	50°C per min
Step 5c. −80 → −196°C	160°C per min
Step 6. Thawing temperature	+2°C

[a] Figures in the right-hand column were derived from an initial study with the fowl and used as a basis for working out the optimal method for each species in the present study. Substitutions for these parameters were made in succession in the various trials—see Materials and Methods.

Trial 2 was designed to determine the optimal period for holding diluted semen at 5°C before the addition of DMSO. On the basis of the results of Trial 1, crane semen was diluted 1:1, and fowl semen was diluted 1:3. Both samples were then held at 5°C for 15, 30 or 60 min before being processed further (Steps 3–6, Table I).

Trial 3 was designed to determine the optimal level of DMSO. Semen was diluted as described for Trial 2 and held at 5°C for 30 and 60 min for crane and fowl semen, respectively, before the addition of DMSO. Selection of these holding periods was based on the results of Trial 2. Levels of DMSO were added to cooled diluted semen to a final concentration of 2, 4 or 6% (v/v) and frozen and thawed (Steps 4–6, Table I).

Trial 4 was designed to determine the effect of time of equilibration at 5°C with DMSO before freezing. On the basis of the results of Trials 1–3 the following optimal procedures were used in this trial; crane semen was diluted 1:1 and held at 5°C for 30 min before the addition of DMSO to a final concentration of 4%, whereas fowl semen was diluted 1:3 and held at 5°C for 60 min before addition of DMSO to a final concentration of 4%. After DMSO was added, both semen samples were stored at 5°C for a further 15, 30, 45 or 60 min before being frozen according to the procedure outlined in Table I.

Trial 5 was designed to determine the effect of freezing rate from +5 to −20°C. Steps 1–3 (Table I) used in this trial were as for Trial 4, with crane semen being equilibrated (Step 4, Table I) for 15 min and fowl semen for 60 min. Semen of both species was frozen from +5 to −20°C (Step 5a, Table I) at either 1, 5, or 10°C per min. The freezing rates from −20°C and the thawing temperature are outlined in Table I.

In order to assess the most efficient treatments, a drop of diluted semen was removed in all trials before freezing (after Step 4, Table I) and after thawing (Step 6, Table I) and placed on a clean microscope slide for estimating the progressive motility of spermatozoa. Progressive motility was evaluated microscopically at a magnification of ×430 by examining several areas of the slide and estimating on a scale from 0 to 100 the percentage of spermatozoa moving in a forward motion.

Experiment 2

On the basis of the results of Experiment 1, the optimum procedure for each species (outlined in Table II) was used to assess the fertilizing capacity of crane and fowl spermatozoa after freezing and thawing.

Semen that had been frozen in 0·5 ml plastic straws and stored in liquid nitrogen for one to four days was thawed by placing the straws in ice water at 2°C. Immediately upon thawing, hens (one crane and five white leghorns) were artificially inseminated intravaginally at intervals from April to early June. The crane hen was inseminated every fourth day with 10 million spermatozoa, whereas the fowl hens were inseminated every seventh day with 50 million spermatozoa.

TABLE II

Effect of various treatments before freezing on the motility of crane and fowl spermatozoa (Experiment 1)[a]

| | | | Progressive motility (%)[b] | | | |
| | | | Pre-freeze | | Post-thaw | |
Trial	Treatment		Crane	Fowl	Crane	Fowl
		1:1	83	90	40[c]	43[c]
1	Dilution	1:3	83	93	15	50
	(semen:diluent)	1:5	87	90	20	35
		15	85	80	40[c]	15
2	Time held at 5°C	30	85	83	45[c]	45
	(min)	60	83	87	5	60[c]
		2	92[c]	92[c]	5	30
3	Level of DMSO	4	80[c]	85[c]	45[c]	60[c]
	(% v/v)	6	40	70	5	40
		15	80[c]	87	55[c]	50
4	Equilibration time	30	60	80	35	50
	at 5°C	45	55	93	30	50
	(min)	60	45	95	20	60[c]

[a] Composite summary of four independent trials conducted sequentially.
[b] Mean of three replicates.
[c] Within each trial, values with superscript differ from those without ($P < 0.05$).

Eggs were collected daily and incubation was commenced within three days. Fertility was estimated by candling on Day 7 of incubation and expressed as the percentage of viable embryos. Any questionable eggs were broken and examined macroscopically for embryonic development.

Each trial in Experiment 1 was replicated three times. Percentage data for both experiments were transformed to $\arcsin\sqrt{\%}$ and significance was determined at the 5% level of probability by analysis of variance.

RESULTS AND DISCUSSION

Table II gives a composite summary of the effect of dilution, time held at 5°C, level of DMSO and equilibration time at 5°C on the motility of crane and fowl spermatozoa. Pre-freeze motility values for both crane and fowl spermatozoa were unaffected by dilution and/or time held at 5°C. However, the pre-freeze motility values for crane and fowl spermatozoa were significantly depressed by the addition of DMSO at levels greater than 4% ($P < 0.05$). Equilibration times greater than 15 min resulted in a decrease in the pre-freeze motility of crane spermatozoa whereas the motility of fowl spermatozoa before freezing was unaffected by equilibration time.

Because of the limited scope of this study, it is difficult to explain why there was a species difference in the motility response of spermatozoa to dilution and equilibration at 5°C with DMSO. However, a number of speculations warrant discussion. First, consideration must be given to the inherent differences between crane and fowl semen with respect to spermatozoal density and its relationship to the "dilution effect" phenomenon. The average ejaculate from cranes used in this study contained seven million spermatozoa whereas that from the fowl contained about 1000 million. Therefore, crane spermatozoa were subjected to a more dilute environment than fowl spermatozoa. Although it is well documented that fowl spermatozoa are susceptible to the dilution effect phenomenon (Wales & White, 1961), no such information is available regarding crane spermatozoa. However, crane spermatozoa may react similarly to dilution and may undergo certain functional changes (in motility, metabolism, etc.) resulting in a permanent loss in viability. Secondly, there may be a difference in cell membrane permeability to DMSO because shorter equilibration periods were necessary with crane semen. Sexton (1975) reported that Krebs cycle intermediates of low molecular weight readily penetrated turkey spermatozoa but were unable to penetrate easily into fowl spermatozoa. It is also interesting to note that turkey spermatozoa, like those of the crane, require short equilibration periods (15 min) in the presence of DMSO if maximum levels of fertility are to be maintained (T. J. Sexton, unpublished).

Although the dilution rate and the time semen was held at 5°C in the absence of DMSO had no significant effect on the pre-freeze motility of spermatozoa of either species, the recovery of spermatozoal motility after thawing was affected by these treatments. Dilution rates greater than 1:1 and 1:3 significantly depressed post-thaw motility values for crane and fowl spermatozoa respectively ($P < 0.05$, Table II). Crane semen held at 5°C for 15 or 30 min before the addition of DMSO resulted in higher post-thaw motility than semen held for 60 min. Conversely, higher post-thaw motility of fowl spermatozoa was achieved when the semen was held for 60 min. Similarly, a short equilibration period (15 min) for crane semen in the presence of DMSO resulted in maximum recovery of post-thaw motility, whereas a slightly longer period (60 min) was best for fowl semen.

It is generally agreed that the sub-zero temperature range which most affects the viability of cells is that associated with the evolution of the latent heat of fusion (transition from liquid to solid state). With fowl and crane semen this occurred between +5 and −20°C. The post-thaw motility data for various freezing rates between +5 and −20°C are shown in Table III (Trial 5, Experiment 1), and indicate that, of the three

TABLE III

Effect of various freezing rates from +5°C to −20°C on the post-thaw motility of crane and fowl spermatozoa (Experiment 1)[a]

Trial	Freeze rate (°C per min)	Progressive motility (%) Crane	Fowl
	1	55[b]	60[b]
5	5	5	43
	10	5	25

[a] Mean of three replicates.
[b] Values with superscript differ from those without ($P < 0.05$).

freezing rates studied, 1°C per min gave the best revivals of spermatozoa for both species ($P < 0.05$).

Frozen-thawed crane spermatozoa retained their fertilizing capacity but with only four eggs the fertility rate could not be accurately assessed (Table IV). The low value (25%) may partly be explained by the difference in the number of spermatozoa inseminated after thawing in each species. Lake (1970) reported that one reason for the low fertility of frozen fowl semen was the insemination of insufficient spermatozoa surviving the freezing procedure. In this study, the number of spermatozoa inseminated weekly in the crane was 15–20 million compared with 50 million in the fowl. This difference in the number of spermatozoa inseminated was unavoidable because of the inherent differences in spermatozoal density in the ejaculate, as discussed previously. For physical reasons it was impossible to inseminate a sufficient volume of diluted semen into the female crane to equalize the number of spermatozoa. In this study, in order to inseminate 50 million fowl spermatozoa, 0·2–0·3 ml of diluted semen was required. Insemination volumes greater than 0·3 ml result in mechanical problems; specifically, leakage from the vagina during insemination of the fowl. Although the degree of semen leakage from the vagina with respect to the volume of

TABLE IV

Fertilizing capacity of frozen and thawed crane and fowl spermatozoa

Species	No. of eggs examined	Fertility (%)[a]
Crane	4	25
Fowl	50	55

[a] Viable embryos on Day 7 of incubation.

semen inseminated was not observed in the crane it probably would be a significant factor because the insemination volume needed in the female crane to equalize the number of spermatozoa inseminated in the fowl would be 0·5–0·6 ml.

Although the absolute fertility values were drastically different between the species, the difference in the overall recovery of the fertilizing capacity after thawing was not as great. For example, the average fertility of cranes artificially inseminated with fresh semen diluted with the BPSE is approximately 60% whereas with the fowl, fertility levels of 95% are to be expected. Therefore, the actual recovery of the fertilizing capacity of crane spermatozoa was 42% as compared to 57% for the fowl.

In summary, the data in this study indicate that semen from the crane can be frozen using a modified procedure developed for the fowl, and progeny can be expected from artificial breeding.

REFERENCES

Allen, T. B. (1974). *Vanishing wildlife of North America*. Washington, D.C.: National Geographic Society.

Lake, P. E. (1970). The storage of fowl spermatozoa in liquid nitrogen. *Proc. 14th Wld Poultry Congr., Madrid* 1: 349–351.

Martin, R. D. (1975). *Breeding endangered species in captivity*. New York and London: Academic Press.

Mitchell, R. L. & Buckland, R. B. (1976). Fertility of frozen chicken semen after intravaginal and intrauterine inseminations using various concentrations and equilibration times of dimethylsulfoxide and a range of freezing and thawing rates. *Poultry Sci.* 55: 2195–2200.

Neville, W. J. (1975). Recent advances in semen preservation. *Jl S. Afr. Vet. Med. Ass.* 46: 311–323.

Sexton, T. J. (1975). Utilization of Krebs cycle intermediates by chicken and turkey spermatozoa. *Poultry Sci.* 54: 1815 (Abstract).

Sexton, T. J. [1976]. Studies on the fertility of frozen fowl semen. *VIII Int. Congr. Anim. Reprod. Artif. Insem.* [Krakow] 4: 1079–1082.

Sexton, T. J. (1977). A new poultry semen extender, 1. Effect of extension on the fertility of chicken semen. *Poultry Sci.* 56: 1443–1446.

Wales, R. G. & White, I. G. (1961). The viability of fowl spermatozoa in dilute suspension. *Aust. J. biol. Sci.* 14: 637–645.

Watanabe, M. (1967). Studies on deep-freezing preservation of chicken semen. *J. Fac. Fish. Anim. Husb., Hiroshima* 7: 9–23.

Symp. zool. Soc. Lond. (1978) No. 43, 97–126

A Review of Techniques of Semen Collection in Mammals

P. F. WATSON

Royal Veterinary College, London, England

SYNOPSIS

The variety of methods by which viable spermatozoa have been obtained from mammals reflects not only the varying ease with which animals may be handled and trained, but also the objectives for which the spermatozoa are obtained. For preservation and artificial insemination the sample must have an adequate motility and sperm count, and must be free of gel or coagulum if this is normally present. A knowledge of the characteristics of the normal ejaculate is therefore of assistance in devising a suitable collection technique.

The ideal method combines reliability in obtaining a consistently good semen sample at regular intervals with safety for both animals and personnel involved. The extent to which the available methods of semen collection fulfil these requirements are considered.

Erection and ejaculation may be induced by stimulating sensory neural pathways involved in the ejaculatory reflex or by artificially stimulating motor nerves supplying the internal accessory organs and urethra. A wide range of sensory receptors is associated with these reflexes, but sensations from the penis are of paramount importance. An ejaculate most closely resembling the normal ejaculate will be produced as this range of stimuli is approximated, but some difficulties are experienced in collecting semen by methods which employ natural stimuli, especially with non-domestic animals. When these methods are inappropriate artificial stimulation may offer a realistic alternative. It is also possible to obtain viable spermatozoa for the purposes of preservation and artificial insemination from the tail of the epididymis and vas deferens of animals shortly after death.

The applications of the artificial vagina, electroejaculation, manual stimulation and other methods of semen collection to domestic and non-domestic animals are reviewed. Emphasis is laid on the advantages and disadvantages of each method and its suitability for use in non-domestic animals.

INTRODUCTION

Methods of semen collection appropriate to man and domestic and laboratory animals have been refined over many years as a result of the demand for semen for both fundamental and applied reproductive studies. In contrast, the development of suitable methods for non-domestic animals has been haphazard and slow because, until recently, there has been no such demand for semen. With the growing interest in the possibilities of artificial breeding in these animals, and with a developing awareness of the importance of comparative studies in reproduction to an understanding of reproductive processes in general, semen collection from a number of non-domestic species has now been reported. In this paper, criteria are considered for the collection of spermatozoa for preservation and artificial insemination. (The term "semen" will be restricted to samples in which spermatozoa are suspended in secretions derived from the accessory organs). An attempt has been made to draw together the available information on spermatozoal collection techniques in nondomestic mammals; the work on the domestic species is not documented in detail, although it has been cited in appropriate areas. For further details of semen collection methods for domestic animals, the reader is referred to First (1971) and Maule (1962), while for laboratory animals Hafez (1970) should be consulted.

For the purposes of semen storage and artificial insemination the main requirement is a sufficient number of mature, viable spermatozoa. The composition of the seminal fluid component of the sample is of relatively little consequence because, generally, artificial media will be used to dilute the semen. For fundamental studies, however, the composition of the semen must resemble that of the normal ejaculate and deviations from this seriously limit the validity of the studies. Thus the choice of method of semen collection will be influenced by the objectives for which the semen is obtained. In all cases, however, cleanliness in collection is necessary to avoid contamination of the sample with extraneous material. Whatever method is chosen, glassware for collection should always be warmed to prevent temperature shock to the spermatozoa (Mann, 1964).

Standardization of semen assessment procedures would aid comparisons between results from different laboratories. In the human subject some progress has been made (Eliasson, 1971) but in animals, assessment procedures are very diverse. Moreover, the simplest procedures depend on subjective assessments which cannot be compared between laboratories with any degree of precision. Published accounts of semen collection should include details of the collection method, the volume of the ejaculate, the spermatozoal concentration, the percentage of motile spermatozoa, the percentage of spermatozoa resistant to vital staining (Campbell, Dott & Glover, 1956) and the percentage of abnormal spermatozoa. In addition, some statement should be made about the presence of coagulum.

Seminal Coagulation and Gel Formation

The semen of rodents and primates coagulates immediately after ejaculation forming a cervical or vaginal plug, while the semen of the boar, the stallion and the rabbit contains a variable proportion of gel (Mann, 1964). Rodger & White (1975) observed that the seminal plasma of macropod marsupials also coagulates. For semen preservation it is necessary to obtain a suspension of spermatozoa free of gel or coagulum.

This problem has been approached in several ways. In rabbit semen, Mukherjee, Johari & Bhattacharya (1951) found large numbers of spermatozoa in the washed gel which could be released by incubation at 37°C in normal saline. Clearly, the common practice of discarding the gel in rabbit semen involves spermatozoal losses. The gel of stallion semen can be separated during collection by filtration (Komarek *et al.*, 1965) but again, spermatozoal losses were found to be considerable (Pickett, Gebauer *et al.*, 1974). Boar semen is usually collected in fractions through a cotton gauze filter to obtain the sperm-rich fraction undiluted by accessory secretions and free of gel (First, 1971).

Electroejaculation in the boar, however, apparently stimulates the release of only the sperm-rich fraction containing no gel (Dziuk, Graham, Donker *et al.*, 1954; Vera Cruz, 1959; Campbell & Lingham, 1965). Moreover, occasional reports of electroejaculation in rats and mice claim that non-coagulated semen was obtained (Birnbaum & Hall, 1961; Snyder, 1966; Kalasiewicz & Wolanski, 1970) but no common stimulus pattern was responsible. In primates the coagulum resulting from electro-ejaculation has been found to constitute a variable portion of the ejaculate (Weisbroth & Young, 1965; Kraemer & Vera Cruz, 1969; Van Pelt & Keyser, 1970; Hardin, Liebherr & Fairchild, 1975), and the proportion declined with repeated ejaculation (Bennett, 1967).

Surgical removal of the coagulating glands was highly effective in rats and chinchillas (Scott & Dziuk, 1959; Lawson, Krise & Sorensen, 1967; Weir, 1970) but less so in monkeys (Greer, Roussel & Austin, 1968). An alternative approach, the use of enzymes to digest the seminal coagulum, has been studied in guinea-pigs (Freund, 1958) and primates (Hoskins & Patterson, 1967; Roussel & Austin, 1967; Settlage & Hendrickx, 1974b). Both chymotrypsin and trypsin have been shown to be highly effective in dissolving the coagulum at concentrations which were not severely toxic to spermatozoa.

The Ideal Semen Collection Method

Certain broad criteria can be established by which to assess any semen collection procedure. The ideal method should fulfil the following requirements.

1. The technique must be applicable to the live animal without risk to its health and welfare. Any interference with a non-domestic animal carries with it some risk to the animal. The possibility of excitation and subsequent death from circulatory shock (Harthoorn, 1976) must be balanced against the risks of sedation and anaesthesia. The collection procedure itself should not involve risk of traumatic injury to the animal, and the possibility of disease transmission must be eliminated by attention to hygiene.

2. The technique must be capable of repetition at frequent intervals. The frequency of collection of semen will vary according to the demand, but in principle any technique that cannot be repeated at least once per week is not acceptable.

3. The semen obtained must be of normal physical and chemical composition. One of the main difficulties surrounding the

application of semen collection techniques to the non-domestic species is that a knowledge of the "normal" ejaculate is frequently lacking. This severely limits the interpretation of results from such techniques as electroejaculation. Comparisons with related domestic species are fraught with difficulties because the domestic animal has often been selected for reproductive performance and produces an ejaculate untypical of the non-domestic group of species it is taken to represent. Moreover, criteria of normality for the ejaculate are notoriously difficult to define. The poor correlation between semen assessment procedures and fertility has been noted by several authors for the bull (Bishop *et al.*, 1954; Foote, 1975), the ram (Hulet & Ercanbrack, 1962; Hulet, Foote & Blackwell, 1964) and the human (Eliasson, 1971; Rehan, Sobrero & Fertig, 1975). For most non-domestic species the problem has yet to be investigated but the solution promises to be just as elusive. While considerable variation in normal semen constituents exists between species (Mann, 1964) the occurrence of abnormal elements (e.g. blood cells, pus cells or urine) in the semen render the collecting method unacceptable if their presence results from the technique employed.

4. The technique must involve no risk to personnel. Lack of attention to the safety of personnel involved in semen collection from a non-domestic animal can only bring such endeavours into disrepute. It should be obvious that this is of paramount importance in selecting an appropriate method.

CLASSIFICATION OF METHODS OF SEMEN COLLECTION

Spermatozoa have been obtained by a variety of different methods, and before discussing the various methods it may be helpful to consider a classification.

The erection and ejaculation reflexes comprise a complex sequence of neural events involving both the divisions of the autonomic nervous system, and somatic afferent and efferent fibres (see Martin, this volume, p. 127). The role of the autonomic nervous system in erection and ejaculation has been reviewed by Bell (1972). Erection results from the action of the parasympathetic nervous system in those species having a large amount of penile erectile tissue, but in species possessing a fibro-elastic penis, efferent fibres to somatic muscles are involved. Emission of spermatozoa and seminal fluid from the ampullae and accessory glands into the pelvic urethra is brought about by the action of sympathetic nerve fibres, while ejaculation, the propulsion of semen along the urethra to the exterior, is achieved by striated muscles innervated by somatic efferent fibres.

It is well known that these reflexes may be modified in domestic animals as a result of conditioning to semen collection techniques. Similar conditioned reflexes have also been observed in non-domestic animals (Scott & Dziuk, 1959; Kraemer & Vera Cruz, 1969; Jainudeen *et al.*, 1971; Lambiase, Amann & Lindzay, 1972; Fussell, Franklin & Frantz, 1973; Settlage & Hendrickx, 1974a; Richardson & Wenkoff, 1976).

The principle of the classification presented in Table I is the extent of involvement of the nervous system. Methods near the head of Table I involve a full range of sensory input, the resulting ejaculation being reflex. For masturbation and manual stimulation the sensory input may be restricted to sensation from the penis but ejaculation is still reflex. Artificial ejaculation methods, on the other hand, bypass the sensory input and ejaculation results from direct stimulation of motor nerves, although some reflex activity may still be present (see p. 109). Erection may not accompany ejaculation. The non-ejaculatory methods do not depend on neural activity except perhaps at a local level.

TABLE I

A classification of spermatozoal collection methods

Ejaculatory	Natural	Recovery of semen *post coitum* (including condom or vaginal liner)
		Artificial vagina
		Masturbation
		Manual stimulation
		Electrovibration and penile electrical stimulation
	Artificial	Electroejaculation
		Drug-induced ejaculation
Non-ejaculatory		Massage of internal organs *per rectum*
		Epididymal recovery

METHODS OF SEMEN COLLECTION

Methods Involving Ejaculation

Recovery of semen *post coitum*

Recovery of semen after natural mating is probably the oldest of the methods of collection, its use being restricted to those species with intravaginal semen deposition. Polotzoff (1928) used a sponge inserted into the vagina prior to coitus in order to recover stallion spermatozoa. A similar technique, mentioned by Fernandez-Baca & Calderon (1967), to obtain alpaca spermatozoa resulted in contamination with vaginal secretions, together with the loss of considerable numbers of spermatozoa. To avoid this loss, semen has been collected directly by aspiration from the anterior vagina in horses (Swire, 1962) and mink (Ahmad, Kitts & Krishnamurti, 1975). Spermatozoa have also been obtained by collecting the fluids draining from the urinogenital sinus after mating of captive Asiatic elephants (Landowski & Gill, 1964). These methods yield large numbers of spermatozoa but still result in semen contaminated by vaginal secretions. That such secretions have an adverse effect on spermatozoal survival was noted by Ahmad *et al.* (1975). Wallace-Haagens, Duffy & Holtrop (1975) found that the mean survival time of spermatozoa in the human vagina was about two and a half hours and commented that this period was comparable with earlier findings in laboratory rodents.

An alternative approach has been to use either a vaginal liner or a condom, thus avoiding contact of the spermatozoa with the vagina. However, the vaginal liner described by Fernandez-Baca & Calderon (1967) for the alpaca caused some inhibition of coital reflexes and also caused damage to the penis. Semen collected by condom from stallions was considerably more contaminated with preputial debris than that collected by artificial vagina and the method offered no advantages (Nishikawa, 1959).

Eliasson (1971) recommends that human semen samples be collected by masturbation rather than by condom, since condoms are frequently impregnated with spermicides.

The considerable problems encountered in obtaining an uncontaminated semen sample with any of these approaches far outweigh the possible advantages of obtaining a natural ejaculate. In view of these considerations, this method should only be used if other methods are inappropriate to the species in question.

The artificial vagina

Collection of semen by means of an artificial vagina (AV) is the method most commonly practised for cattle, sheep, horses and rabbits. The animal is allowed to mount either a "teaser" female or a dummy and the penis is directed into the AV. Semen is ejaculated into a collecting vessel attached to the AV.

The designs of AVs suitable for different species are all based on a common pattern (Fig. 1). A rigid tube houses a flexible rubber liner which is reflected back over the ends of the tube to form a water-tight seal. One end is attached directly or indirectly to a collecting vessel of suitable dimensions. Warm water is put into the space between the liner and the tube, and the inner cavity is lubricated. The internal temperature of the AV should be several degrees higher than body temperature; the optimal range varies with both species and individual (Faulkner & Pineda, 1975).

Modifications of this basic design have been made to suit individual species and a range of AVs is shown in Fig. 2. For the boar it is necessary to simulate the locking of the spiral tip of the glans penis into the cervix of

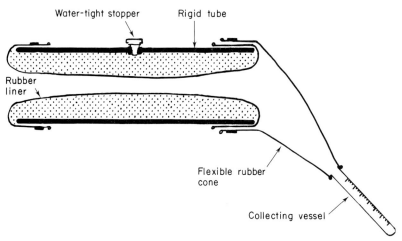

Fig. 1. The construction of an artificial vagina.

the sow. This is achieved either by means of a spiral wire over the tapered end of the AV or by grasping the tip of the penis through the rubber sheath between the AV and collecting vessel, the fingers thus simulating the cervix. In addition, the water jacket is connected to a rubber bulb by means of which pressure fluctuations can be delivered to the liner. Details of the construction of the AVs for individual domestic species may be found in Maule (1962) or Faulkner & Pineda (1975). A pulsation type of AV for the dog was described by Harrop (1954) who found, using this AV, there was no need for an oestrous bitch as a teaser, and better semen samples could be obtained.

Collection by means of an AV is dependent upon the animal having a libido sufficiently high to overcome inhibition of sexual behaviour associated with the presence of man. A low libido in some species, e.g. *Bos indicus*, is widely recognized, and has led some workers to consider electroejaculation as an alternative method of semen collection (Rollinson & Nunn, 1962). The use of an oestrous female as a teaser animal, frequent changes of teaser, or even urine from an oestrous female (Krzywiński, 1976) has often been found to be beneficial in promoting sexual behaviour, particularly in the period of training.

The training (conditioning) of males to serve an AV has been described for the ram (Miller, 1961; Salamon & Lindsay, 1961), the boar (Rowson, 1962; First, 1971) and the stallion (Richardson & Wenkoff, 1976). Agricultural animals become accustomed to having the penis directed into an AV held by an attendant; domesticated camels (Rakhimzhanov, 1971; Khan & Kohli, 1973), reindeer (Dott & Utsi, 1971) and red deer (Krzywiński & Jaczewski, this volume, p. 271 have also been trained in this manner. Fussell *et al.* (1973) conditioned chimpanzees to

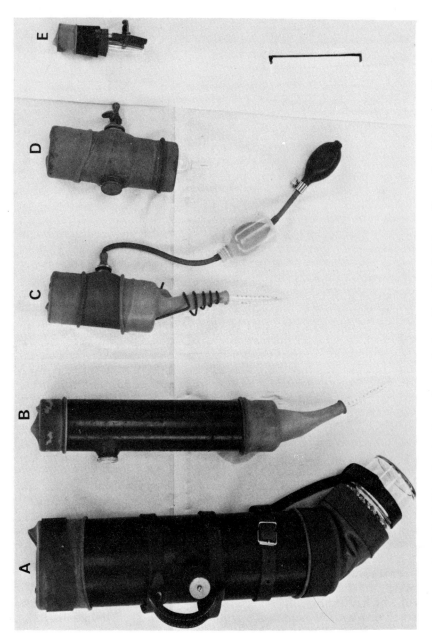

FIG. 2. Artificial vaginas for domestic animals. A. stallion, B. bull, C. boar, D. ram, E. rabbit (scale = 15 cm).

respond to apple juice so that when it was offered at the cage bars erection occurred. A special designed AV was then manually pumped on the penis. Hardin *et al.* (1975) described an AV of standard pattern suitable for chimpanzees and collected semen which was 50% gel indicating that a normal ejaculate was probably obtained. A successful pregnancy followed insemination of this semen. For red deer (Krzywiński, 1976; Krzywiński & Jaczewski, this volume, p. 271) and alpaca (Fernandez-Baca, 1975) the AV was strapped between the hind legs of a dummy so that a collection could be made without the need for an attendant. Krzywiński (1976) solved the problem of semen running back out of the collecting vessel by constricting the entrance of the vessel to retain the semen.

In general, no great changes have been made to the basic design of the AV for collecting from non-domestic or semi-domestic animals. Khan & Kholi (1973) used a modified pig AV complete with pressure fluctuation to stimulate ejaculation in the camel, whereas others have used a short bull AV (Abdel-Raouf & El-Naggar, 1964; Rakhimzhanov, 1971). Dott & Utsi (1971) used a ram AV for reindeer, and Krzywiński (1976) made use of a bull AV for red deer. For domestic cats Scott (1970) described an AV, made from a pasteur pipette bulb, which was used regularly with success (Sojka, Jennings & Hamner, 1970).

Several authors have examined the losses of semen in the collecting apparatus (Foote & Heath, 1963). Pickett, Gebauer *et al.* (1974) and Seidel & Foote (1969) suggested that such losses could be reduced by using an artificial vagina suited to the penile length of individual bulls. If, as they suggest, the length of the AV is critical within a species, their finding is likely to be of even more significance when a standard AV is employed for a range of species. Perhaps the design principle of a variable length rabbit AV (Bredderman, Foote & Yassen, 1964) could be adapted for larger models.

In domestic bulls, so-called teasing of the bull (preventing mounting, or allowing mounting but not permitting ejaculation) is claimed to improve the semen quality (Hale & Almquist, 1960). In Aberdeen Angus and Hereford bulls such sexual preparation improved the initial motility of spermatozoa (Foster, Almquist & Martig, 1970) but had no effect on their survival after freezing (Martig, Almquist & Foster, 1970). These authors comment that sexual preparation was less effective for beef bulls than for dairy bulls. In stallions, teasing decreased the period between exposure to the mare and ejaculation and increased the semen volume but had no effect on the number of spermatozoa released (Pickett, Voss & Gebauer, 1973). Sexual restraint has been found to increase the number of spermatozoa in the ejaculate of buck rabbits (Macmillan & Hafs, 1967) but no reports exist of the effectiveness of sexual preparation in non-domestic animals.

The use of drugs injected prior to semen collection to increase the yield of spermatozoa has been studied by a number of workers. A single dose of 6 iu oxytocin increased the spermatozoal output in bulls

(Milovanov, Bereznev & Gorohov, 1962). A similar response was found in rams (Knight, 1974) and rabbits (Kihlström & Melin, 1963). Hafs, Louis & Stellflug (1973) reported that prostaglandin $F_{2\alpha}$ injected two and four hours prior to semen collection increased the number of spermatozoa ejaculated in rabbits.

Providing the animals can be managed without risk to themselves or to attendants, the use of an AV allows the collection of a clean "normal" ejaculate. However, semen collection by means of an artificial vagina is suited only to domestic and semi-domestic animals. The need for close proximity of man to a fully conscious animal, and the need for a period of training preclude the use of this method for dangerous animals.

Masturbation and manual stimulation

In this paper, the term "masturbation" is used to connote self-stimulation of the penis. Masturbation has been observed in a number of species, especially when the male animal is isolated from sexual contact. Walton (1960) suggests that it is more commonly observed in stallions and bulls than in boars and rams; in primates it constitutes part of the normal pattern of behaviour. In most of the non-human primates the ejaculate is commonly consumed and cannot, therefore, be readily collected. However, by offering a reward to chimpanzees, Martin, Graham & Gould (this volume, p. 249) regularly obtained an ejaculate. In man, masturbation is the method generally recommended since other methods (interrupted intercourse, condom) lead to contaminated, or otherwise unsatisfactory, semen samples (Eliasson, 1971).

Stimulation of the penis by an attendant to produce reflex erection and ejaculation is a technique restricted mainly to the dog and the boar. The method normally involves no stimulatory apparatus and the ejaculate may be collected through a funnel into a collecting vessel.

In the dog, partial erection may be stimulated either by manual pressure behind the glans penis in a sexually experienced dog or by the presence of an oestrous bitch (Kirk, 1970). The prepuce is then drawn back over the bulbus glandis and a firm grasp is taken of the penis behind the bulb, simulating the vulval "lock" of natural coitus. Ejaculation will then commence. In natural mating the dog adopts a position facing in the opposite direction to the bitch early during ejaculation, thus deflecting the penis posteriorly (Grandage, 1972). Deflection of the penis can be achieved manually during collection, producing a more complete ejaculate (W. E. Allen, pers. comm.). Boucher, Foote & Kirk (1958) found that collection was generally easier, and better samples were produced, if an oestrous bitch was present. It is generally held that this method of semen collection is preferable to the use of an AV, although Harrop (1954) claimed that the AV gave better quality semen samples than manipulation. Later workers found the reverse to be the case and showed that the latex rubber of the artificial vagina was detrimental to dog spermatozoa (Boucher et al., 1958). Creed (1964) successfully induced ejaculation in hand-reared red foxes using Harrop's AV combined with manual stimulation, but failed with wild-caught foxes.

Semen has been collected by manual stimulation from both blue and silver-black foxes (Pomytko, Bautina & Vladimirova, 1972; Aamdal, Nyberg & Fougner, 1976) and timber wolves (Seager, Platz & Hodge, 1975). Not all dog foxes respond to manual collection procedures; 35 out of 40 blue foxes yielded semen but silver-black foxes were found to be less easy to collect (Aamdal, Fougner & Nyberg, this volume, p. 241). An alternative method, electroejaculation, was more frequently successful in producing semen (Pomytko, Bautina et al., 1972) but the semen was of poorer quality (Bautina, 1974). Pregnancies have been reported following insemination of spermatozoa collected by manual stimulation of both foxes (Aamdal et al., 1976; Vladimirov & Pomytko, 1976) and wolves (Seager et al., 1975).

Boars are readily trained to mount a dummy and erection occurs spontaneously so that the penis protrudes from the prepuce. Semen may be collected with an AV (see above) or by manual stimulation. In the latter method the spiral tip of the glans penis is grasped in the fingers to simulate the cervix of the sow, and ejaculation will occur so long as a firm grasp is maintained; usually a rubber glove is worn. Ejaculation in both the boar and the dog takes several minutes, and the ejaculate, which is of relatively large volume, is produced in fractions. Thus it is possible with both the AV and manual stimulation to obtain the sperm-rich (second) fraction free from dilution by the accessory secretions ejaculated before and after, simply by changing the collecting vessel at the appropriate moment.

King & Macpherson (1973) reported that the hand collection method in boars yielded semen not significantly different in major characteristics from that yielded by AV. In another study, better semen was obtained with the gloved hand method than with either of two patterns of AV (Cerovsky & Slechta, 1973). Neither method has been applied to non-domestic pigs.

Few reports exist of manual stimulation applied to other mammals, but Megale (1968) described a method of inducing erection in quiet, unsedated bulls by massaging the glans penis within the preputial sheath. Erection was frequently followed by ejaculation, but normal volumes could only be obtained by introducing the erect penis into an AV. The method was also claimed to be applicable to rams and goats. In one report an Asiatic elephant was induced to ejaculate in the presence of an oestrous female on several occasions by stroking of the pectoral region together with verbal encouragement from its keeper (Jainudeen et al., 1971). The authors stress the importance of the regular handler but it is questionable whether more than a few male elephants could be trained in this way.

Electrovibration and electrical stimulation of the penis

The use of an electrical vibrator cup applied to the glans penis to induce ejaculation in men was described by Sobrero, Stearns & Blair (1965). Reflex ejaculation without full erection or orgasm occurred in a number of patients with ejaculation problems, while all normal subjects produced a normal ejaculate. Schellen (1968), employing this method, obtained complete success with nine out of 11 patients incapable of ejaculation during coitus or masturbation, and the period of stimulation leading to ejaculation decreased with experience. Applying electrovibration to dogs, Schefels (1969) found that 22 out of 34 animals ejaculated and the semen quality was excellent.

Mastroianni & Manson (1963) described an electrical method of inducing ejaculation in conscious macaque monkeys. A monophasic AC stimulus of 20–40 volts was intermittently applied between two electrodes, one placed at the base of the penile shaft and the other held near the frenulum of the glans penis. Settlage & Hendrickx (1974a) found that monkeys became conditioned to this procedure and showed no signs of distress. Several workers have used this method and some have claimed that the ejaculate quality is better than with more conventional electroejaculation techniques (see Hendrickx et al., this volume, p. 219). However, Hendrickx et al. (this volume) do not consider these reported differences to be of any real significance. The method is included in this section because it produces reflex ejaculation by stimulation of sensory nerve fibres in the penis, and is thus more akin to electrovibration than to conventional electroejaculation (see below).

Electroejaculation

Electrode arrangements and stimulation patterns. Electroejaculation is of particular relevance to non-domestic animals since it can be employed on anaesthetized subjects, and there is now a growing literature on this application. In principle, the method involves the stimulation of the nerves supplying the reproductive organs by means of a weak electric current.

Modern developments of the technique date from the work of Gunn (1936) who found that when a series of stimuli was applied between electrodes placed in the lumbar musculature and in the rectum of a ram, erection and ejaculation resulted. However, he observed that the considerable degree of stimulation of somatic motor nerves which also occurred, subsequently resulted in temporary ataxia. Later developments have led to the incorporation of both electrode poles on a rectal probe, resulting in a considerable reduction of stimulation of somatic musculature. The earliest rectal probes had a single pair of circular electrodes (Laplaud & Cassou, 1948; Nichols & Edgar, 1964) or multiple circular electrodes of alternating polarity (Thibault, Laplaud & Ortavant, 1948; Healey & Sadleir, 1966). This latter modification enabled the stimulus to be applied over a greater area of rectal wall. Better results, however, have been reported using probes with electrode strips parallel to the long axis (Marden, 1954; Ball & Furman, 1972; Warner, Martin & Keeling, 1974).

Extraneous stimulation has been further reduced in bulls and rams by eliminating the dorsally placed electrode and using only lateral and ventral strip electrodes; better ejaculatory responses were also obtained (Ball, 1976). Figure 3 shows a range of rectal probes of varying sizes and designs.

The probe should fill the rectum to maintain electrical contact between the electrodes and the rectal wall. Directing the probe against the ventral wall of the rectum in the region of the ampullae, seminal vesicles and prostate gland is often attended by greater success, suggesting the importance of direct stimulation of the underlying glands. It is generally believed that electroejaculation does not involve stimulation of reflexes, but Hovell *et al.* (1969) observed serial contractions of the ampullae in an anaesthetized ram following the periods of stimulation. Moreover pulsations in the urethra can frequently be palpated in conscious rams between the stimulation periods (P. F. Watson, unpublished). These observations suggest that in the ram, at least, the response is partly reflex in both the conscious and anaesthetized state (see also Martin, this volume, p. 127).

An alternative to the rectal probe is the use of electrodes introduced into the rectum on the fingers of the operator's gloved hand (Rowson & Murdoch, 1954; Dowling, 1961; Watson, 1976). By this means, the stimulus can be applied to the region of the seminal vesicles and ampullae with much greater precision.

A variety of stimulator circuits have been used delivering either an alternating current sine wave or square wave, or direct current pulses. It is not the purpose of this paper to review the detailed variations in probe designs or stimulator circuitry used for electroejaculation. Further details may be found in the review by Ball (1976) and the paper by I. C. A. Martin (this volume, p. 127).

Advantages and disadvantages of electroejaculation. Apart from the previously mentioned advantage of its application to anaesthetized animals, electroejaculation requires no prior training on the part of the subject. Nevertheless, better results were reported with successive attempts on the same animals indicating some degree of conditioning (Scott & Dziuk, 1959). Some individuals are completely refractory to stimulation (Ball, 1976) or may become so for a few hours following collection (Emmens & Robinson, 1962). Stimulation of the sciatic nerves may cause haemorrhage and bruising in somatic muscles with subsequent stiffness (Ball & Furman, 1972). In the present author's experience, however, this is generally minimal, and no permanent damage results (Dziuk, Graham & Petersen, 1954; Marden, 1954).

A number of workers report the problem of contamination of the ejaculate with urine, since the neural pathways controlling bladder function run in close proximity to those causing emission and ejaculation (Semans & Langworthy, 1938). In mice (Scott & Dziuk, 1959) and marmosets (Watson, unpublished) urine contamination occurred when

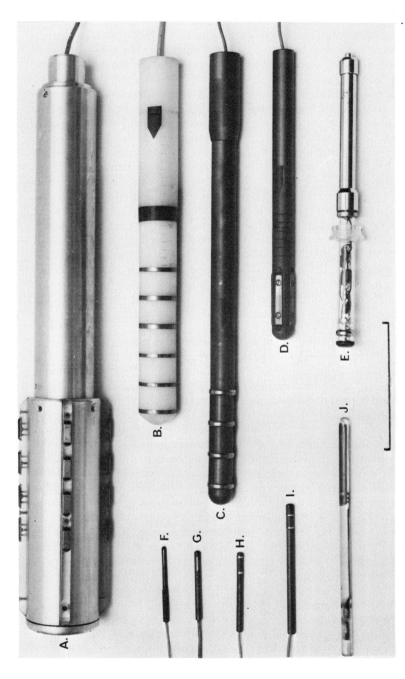

Fig. 3. A range of rectal probes for electroejaculation. Probes D, F, G and J have longitudinal strip electrodes, probes E, H and I have a single pair of ring electrodes while probes B and C have multiple ring electrodes of alternate polarity. Probe A (designed for elephants) has four groups of short strip electrodes, permitting stimulation at varying levels within the rectum (see Jones, Bailey & Skinner, 1975) (scale = 15 cm).

the voltage exceeded the minimum necessary for ejaculation or when the electrodes were displaced anteriorly. Urine contamination appears to be particularly problematical in the cat family (Seager, 1976). These observations suggest that there may be less anatomical separation between the nerve pathways controlling urination and ejaculation in the cat family and/or that the nerve fibre thresholds may be similar.

Characteristics of the electroejaculate. A wide variability in semen quality has been observed with electroejaculation (Lawson, Krise *et al.*, 1967; Roth & Smidt, 1970). The ejaculate produced by electric stimulation in the ram and the bull is characteristically of larger volume, lower spermatozoal count and higher pH than that collected with an AV but the total number of spermatozoa obtained is similar (Mattner & Voglmayr, 1962; Salamon & Morrant, 1963; Hulet *et al.*, 1964; Austin *et al.*, 1968; Foster *et al.*, 1970). In the boar, however, an animal with a normally large semen volume ejaculated in fractions, only the sperm-rich fraction is released by electroejaculation (Dziuk, Graham, Donker *et al.*, 1954; Vera Cruz, 1959; Campbell & Lingham, 1965). With respect to freeze-preservation or fertility it seems that no differences have been detected between bull semen collected either by electroejaculation or AV (Colleary & Ehlers, 1964; Andrews, 1967; Martig *et al.*, 1970; Clarke, Hewetson & Thompson, 1973). Similarly for ram semen, Lapwood, Martin & Entwistle (1972) found no significant difference in fertility between the two collection methods, but Quinn, Salamon & White (1968) reported that ram spermatozoa obtained by electroejaculation were less resistant to cold shock and freezing.

Applications of electroejaculation. Semen has been obtained from a large number of mammalian species from several different orders. The species range from marsupials to man and from mice to elephants. The literature on the application of electroejaculation in the bull and the ram is extensive but there are no references to electroejaculation in the stallion, the buffalo and the rabbit probably because in these species collection of semen by AV presents little difficulty. Table II summarizes the literature on successful applications in non-agricultural mammals.

TABLE II

Successful applications of electroejaculation in non-domestic mammals

Animal	Anaesthetic or sedative	Reference
Marsupialia		
Euro, *Macropus robustus*	(post-mortem)	Sadleir (1965)
Red kangaroo, *Megaleia rufa*	(post-mortem)	Sadleir (1965)
Brush-tailed possum,	Ether	Howarth (1950)
Trichosurus vulpecula	Halothane	Rodger & White (1975)

TABLE II—*continued*

Animal	Anaesthetic or sedative	Reference
Primates		
Tree-shrew, *Tupaia glis*	unanaesthetized	Roussel & Austin (1968)
Marmoset, *Callithrix jacchus*	alphaxalone/ alphadolone	P. F. Watson (unpublished)
Black-capped capuchin monkey, *Cebus apella*	phencyclidine	Roussel & Austin (1968)
Agile mangabey, *Cercocebus galeritus*	phencyclidine	Roussel & Austin (1968)
African green (Grivet) monkey, *Cercopithecus aethiops*	phencyclidine	Roussel & Austin (1968)
Celebes monkey, *Cynopithecus niger*	unanaesthetized	Weisbroth & Young (1965)
	phencyclidine	Weisbroth & Young (1965)
Patas monkey, *Erythrocebus patas*	phencyclidine	Roussel & Austin (1968)
Stumptailed monkey, *Macaca arctoides*	unanaesthetized	Mastroianni & Manson (1963)[a]
	phencyclidine	Roussel & Austin (1968)
Crab-eating monkey, *Macaca fascicularis*	unanaesthetized	Valerio *et al.* (1969)[a]
	phencyclidine	Roussel & Austin (1968)
Japanese monkey, *Macaca fuscata*	unanaesthetized	Weisbroth & Young (1965)
	phencyclidine	Weisbroth & Young (1965)
Pigtail monkey, *Macaca nemestrina*	unanaesthetized	Mastroianni & Manson (1963)[a]
	unanaesthetized	Weisbroth & Young (1965)
	phencyclidine	Gilman (1969)
	phencyclidine	Roussel & Austin (1968)
	phencyclidine	Weisbroth & Young (1965)
Rhesus monkey, *Macaca mulatta*	unanaesthetized	Hoskins & Patterson (1967)[a]
	unanaesthetized unanaesthetized unanaesthetized unanaesthetized unanaesthetized	Mastroianni & Manson (1963)[a]; Settlage & Hendrickx (1974a)[a]; Valerio *et al.* (1969)[a]; Weisbroth & Young (1965)
	phencyclidine	Roussel & Austin (1968)
	phencyclidine	Weisbroth & Young (1965)
Baboon, *Papio* sp.	phencyclidine	Kraemer & Vera Cruz (1969)
	"Imalgène"	Francois *et al.* (1975)

TABLE II—*continued*

Animal	Anaesthetic or sedative	Reference
Squirrel monkey, *Soimiri sciureus*	unanaesthetized methoxyfluorane phencyclidine	Roussel & Austin (1968) Lang (1967) Bennett (1967)
Gelada baboon, *Theropithecus gelada*	phencyclidine	Roussel & Austin (1968)
Gibbon, *Hylobates lar*	phencyclidine	Roussel & Austin (1968)
Chimpanzee, *Pan troglodytes*	phencyclidine phencyclidine or ketamine	Roussel & Austin (1968) Warner *et al.* (1974)
Orang-utan, *Pongo pygmaeus*	phencyclidine or ketamine	Warner *et al.* (1974)
Lowland gorilla, *Gorilla gorilla*	phencyclidine or ketamine	Warner *et al.* (1974)
Edentata		
Nine-banded armadillo, *Dasypus novemcinctus*	?	J. D. Roussel (pers. comm. to Austin, this Symposium)
Rodentia		
Squirrel, *Sciurus* sp.	?	Miljkovic, Zivanovic & Buric (1966)
Golden hamster, *Mesocricetus auratus*	?	Miljkovic *et al.* (1966)
Mouse, *Mus musculus*	unanaesthetized Ether	Scott & Dziuk (1959) Snyder (1966)
Rat, *Rattus rattus*	? unanaesthetized	Miljkovic *et al.* (1966) Birnbaum & Hall (1961)
	unanaesthetized	Lawson, Krise *et al.* (1967)
	unanaesthetized hexobarbitone	Scott & Dziuk (1959) Kalasiewicz & Wolanski (1970)
Guinea-pig, *Cavia porcellus*	? unanaesthetized unanaesthetized	Miljkovic *et al.* (1966) Freund (1958, 1969) Scott & Dziuk (1959)
Chinchilla, *Chinchilla laniger*	unanaesthetized unanaesthetized	Healey & Weir (1967) Lawson & Sorensen (1969)
Carnivora		
Dog, *Canis familiaris*	pentobarbitone	Christensen & Dougherty (1955)
Silver fox, *Vulpes* sp.	?	Pomytko, Bautina *et al.* (1972)

TABLE II—*continued*

Animal	Anaesthetic or sedative	Reference
Polar bear, *Thalarctos maritimus*	ketamine/ acepromazine	Seager (1974)
Ferret, *Mustela putorius*	tiletamine/ zolasepam	Shump, Aulerich & Ringer (1976)
Mink, *Mustela vison*	?	Miljkovic *et al.* (1966)
	unanaesthetized	Aulerich *et al.* (1972)
	phencyclidine	Aulerich *et al.* (1972)
Cougar, *Felis concolor*	phencyclidine/ acepromazine	Seager (1976)
Cat, *Felis catus*	unanaesthetized	Scott (1970)
	ketamine/ acrepromazine	Platz *et al.* (1976)
Ocelot, *Felis pardalis*	phencyclidine/ acepromazine	Seager (1976)
Temminck's golden cat, *Felis temminckii*	phencyclidine/ acepromazine	Seager (1976)
Lynx, *Lynx lynx*	phencyclidine/ acepromazine	Seager (1976)
Bobcat, *Lynx rufus*	phencyclidine/ acepromazine	Seager (1976)
Clouded leopard, *Neofelis nebulosa*	phencyclidine/ acepromazine	Seager (1976)
Lion, *Panthera leo*	phencyclidine/ acepromazine	Seager (1976)
Spotted leopard, *Panthera pardus*	phencyclidine/ acepromazine	Seager (1976)
Bengal tiger, *Panthera tigris tigris*	phencyclidine/ acepromazine	Seager (1976)
Proboscidea		
African elephant, *Loxodonta africana*	etorphine/ acepromazine	Jones (1973); Jones *et al.* (1975)
Artiodactyla		
Camel, *Camelus bactrianus*	?	Anon. (1961)
Alpaca, *Lama pacos*	unanaesthetized	Fernandez-Baca & Calderon (1967)
American bison, *Bison bison*	etorphine/ chlorpromazine	P. F. Watson (unpublished)
Yak, *Bos grunniens*	xylazine	D. M. Jones & P. F. Watson (unpublished)
Red deer, *Cervus elaphus*	unanaesthetized	Jaczewski & Jasiorowski (1973)
	chlorpromazine	Jaczewski & Jasiorowski (1973)

TABLE II—*continued*

Animal	Anaesthetic or sedative	Reference
Black sika deer, *Cervus nippon*	unanaesthetized	P. F. Watson (unpublished)
White-tailed gnu, *Connochaetes gnou*	xylazine	P. F. Watson (unpublished)
Brindled gnu, *Connochaetes taurinus*	xylazine	P. F. Watson (1976)
White tailed deer, *Odocoileus virginianus*	unanaesthetized	Bierschwal *et al.* (1970)
	unanaesthetized	Lambiase *et al.* (1972)
Reindeer, *Rangifer tarandus*	unanaesthetized	Dieterich & Luick (1971)

[a] Penile electrodes.
The nomenclature of the primates follows that given by Napier & Napier (1967).

In man, electrical stimulation accompanied by massage of the seminal vesicles and prostate gland in paraplegic patients produced semen containing motile spermatozoa (Horne, Paull & Munro, 1948; Thomas, McLeish & McDonald, 1975). In non-paraplegic patients with sexual disorders, electrical stimulation evoked neither erection nor ejaculation (Potts, 1957; Rowan, Howley & Nova, 1962). It is difficult to evaluate the technique in humans from these few observations, but it seems likely that electroejaculation could be developed for use in certain clinical conditions.

Restraint. The smaller animals may be electroejaculated unanaesthetized if properly restrained. Suitable restraint devices have been described for small monkeys (Kuehl & Dukelow, 1974) and laboratory rodents (Lawson, Barranco & Sorensen, 1966). Table II also shows the range of volatile and non-volatile anaesthetics and sedatives which have been employed without apparently interfering with electroejaculation (see also Martin, this volume, p. 127). Indeed, occasional reports suggest that sedation may have a beneficial effect (Wells *et al.*, 1966; Aulerich, Ringer & Sloan, 1972). Phenothiazine compounds, however, in dose levels appropriate to non-domestic animals (>1 mg kg^{-1}) may interfere with emission or ejaculation (Warner *et al.*, 1974; Jaczewski *et al.*, 1976). Seager (1976) also observed that acepromazine, at dose levels greater than about 1 mg kg^{-1} in domestic cats, caused an increased incidence of contamination of the ejaculate with urine. In bulls, however, chlorpromazine at low dose levels ($0\cdot044$–$0\cdot88$ mg kg^{-1}) actually facilitated the electroejaculation response and improved the semen quality (Wells *et al.*, 1966). In view of the complex interference with autonomic function by phenothiazine drugs (Byck, 1975), these conflicting observations are not surprising and may be due to the choice of dose level or to species variation in response. Atropine should not be used to control salivation under anaesthesia since several reports indicate that it adversely affects emission and ejaculation (see Martin, this volume, p. 127). In primates undergoing electroejaculation, atropine totally inhibited the ejaculation response in some cases but erection still occurred (Roussel & Austin, 1968; Warner *et al.*, 1974).

Drug-induced ejaculation

Suzuki (1952) collected semen from rats during inhalation anaesthesia with ether or by injection of ethyl alcohol. The method using ether was successful in rats over five months of age, frequently producing samples with motile spermatozoa. A similar technique using halothane, penthrane or chloroform was unsuccessful, but intra-peritoneal injection of high doses of these anaesthetics caused 21 out of 29 rats to ejaculate, although 14 died (Kessler, 1974). This response was presumably due to the non-specific irritant action of these drugs. In mice, the injection of a mixture of a central nervous stimulant (yohimbine) and a hypnotic, 5-butyl-5-bromallyl barbituric acid (pernoston), was found to cause ejaculation (Loewe, 1937, 1938).

A series of erections and ejaculations may be evoked in paraplegic men during the period following intrathecal injection of neostigmine (Guttman, 1973). Pregnancy has been reported following insemination of spermatozoa released by this means (Spira, 1956; Chapelle *et al.*, 1976). The response is part of a generalized autonomic hyper-reflexia, which produced, among other side effects, a severe elevation of blood pressure (Guttman & Walsh, 1971; Rossier *et al.*, 1971). The use of drugs to elicit ejaculation is thus not likely to find ready acceptance for non-domestic animals because of the severity of the side effects produced.

Non-Ejaculatory Methods

Manual expression of the seminal vesicles and ampullae

This technique involves massaging the ampullae and seminal vesicles *per rectum* and can therefore only be conveniently carried out on animals large enough to permit the introduction of a hand into the rectum. A firm but gentle pressure is exerted repeatedly from anterior to posterior over the ampullae, seminal vesicles and root of the penis (Salisbury & VanDemark, 1961). Protrusion of the penis beyond the prepuce is usually seen but semen is not forcibly ejected from the penis. In a recent comparison with electroejaculated semen, that produced by massage was of smaller volume (<4 ml), with a higher spermatozoal concentration and fewer abnormal spermatozoa while motility was similar (Goodwin, 1970). The technique was more satisfactory on younger bulls, and was better for Aberdeen Angus cattle than for Hereford cattle; Santa Gertrudis bulls responded poorly. Semen has also been obtained from water buffalo by this method but the bulls were more difficult to train than domestic cattle (Bhattacharya, 1962). Semen expressed by massage has been frozen and inseminated (Goodwin, 1959). In later experiments (Goodwin, 1970) the method of collection was confounded with storage time at 5°C which invalidated conclusions about the relative preservation qualities.

There are no published reports of the use of this method on other animals, but P. F. Watson (unpublished) failed to express semen by massage from two yak bulls.

Recovery of spermatozoa from the epididymis

Mature fertile spermatozoa are stored in the cauda epididymidis (Bedford, 1975). Surgical cannulation techniques have been developed to obtain epididymal spermatozoa from the vas deferens in the living animal (White, Larsen & Wales, 1959; Amann, Hokanson & Almquist, 1963; Bennett & Rowson, 1963; Wierzbowski & Wierzchos, 1969; Holtz & Foote, 1974; Johnson & Pursel, 1975), but Grove, Bollwahn & Mahler (1968) failed to collect boar spermatozoa adequate for preservation by this method. As an alternative, they resorted to electrical stimulation of the excised cauda epididymidis, and obtained 4–6 ml of fluid containing spermatozoa suitable for preservation. For preservation of spermatozoa from the living animal, semen would normally be obtained by one of the methods discussed earlier. Only after unexpected death of a rare animal might the possibility of recovery of spermatozoa from the epididymis arise.

Recovery of epididymal spermatozoa *post mortem* has been the usual method for small laboratory animals (Hafez, 1970). The method involves either flushing or macerating the cauda epididymidis and vas deferens in a physiological salt solution. With the latter alternative, tissue debris may be separated off leaving a suspension of spermatozoa. Data from non-laboratory animals are not readily available. Pomytko, Bernatskii *et al.* (1972) obtained mink spermatozoa by this means but preservation and fertility were not tested. Duran del Campo (1969) obtained spermatozoa from the epididymis of a bull 15 min after death. The cells were frozen by a standard pelleting method, and two out of 15 cows subsequently conceived to artificial insemination. In another study of four bulls, an average of $8\cdot29\times10^7$ spermatozoa was collected from the epididymides, vasa and ampullae after death, and spermatozoa from all but one bull survived freezing (Waters, 1976). Spermatozoa were obtained from a wild boar *post mortem* and used for insemination of domestic sows with positive non-return to oestrus at one month after insemination (Li *et al.*, 1965). Jones (1973) collected spermatozoa from the excurrent ducts of the testes of wild African elephants within 15 min after death and found good motility after dilution, which was maintained after freezing.

The paucity of such reports may reflect the difficulty in obtaining spermatozoa sufficiently soon after death. Because changes set in rapidly *post mortem* the method does not offer a simple means of obtaining spermatozoa for preservation from a rare non-domestic animal which has died in captivity, although Graham *et al.* (this volume, p. 153) claim that, providing the testes and epididymides are refrigerated at 4°C soon after death, viable spermatozoa may be retrieved up to 48 hours later. However, in view of the poor fertility of ram spermatozoa stored for 24–48 hours in a supporting diluent (Watson & Martin, 1976), it is questionable whether these epididymal spermatozoa would be fertile; this point has never been tested. Moreover, the cause of death of the animal needs to be established. Febrile conditions may affect the quality of the semen, and the possibility of disease transmission via the epididymal fluid should always be considered. The use of this method is thus restricted to animals suffering traumatic injury and detected within minutes of death, a sequence of events uncommon in captive stock.

CONCLUSION

No single method of semen collection fulfils all the criteria of an ideal method. Moreover, the selection of any particular method is, to a large extent, determined by the species being studied. For this reason, several of the methods described are of limited use for non-domestic species except perhaps in exceptional circumstances. Where possible, however, either the use of an artificial vagina or manual stimulation is probably the method of choice, but in those species where neither method is practical, electroejaculation offers a realistic alternative. Further study is still required to adapt these methods to new species in order to obtain maximal release and collection of spermatozoa.

The need to obtain a semen sample may simply be to assess the potential fertility of a male animal as part of an overall assessment of its breeding capacity. On the other hand, the prospects of long-term banking and international exchange of semen extend the range of possibilities. Isolated females could be bred in a co-operative scheme to maximize the breeding potential of captive stock; outcrossing of inbred lines to maintain genetic variation in captive animals would become possible and it would be feasible to introduce new genetic material via semen from wild animals into the gene pool of captive stock. Satisfactory semen collection techniques are a prerequisite for the development of any of these ideas.

REFERENCES

Anon. (1961). Artificial insemination of a Bactrian camel. *Int. Zoo Yb.* **3**: 94.
Aamdal, J., Nyberg, K. & Fougner, J. [1976]. Artificial insemination in foxes. *VIII Int. Congr. Anim. Reprod. Artif. Insem.* [Krakow] **4**: 956–959.

Abdel-Raouf, M. & El-Naggar, M. A. (1964). Studies on reproduction in camels, *Camelus dromedarius*. I. Mating technique and collection of semen. *J. vet. Sci. U.A.R.* **1**: 113–119.

Ahmad, M. S., Kitts, W. D. & Krishnamurti, C. R. (1975). Mink semen studies. I. Liquid preservation and prospect of freezing spermatozoa. *Theriogenology* **4**: 15–22.

Amann, R. P., Hokanson, J. F. & Almquist, J. O. (1963). Cannulation of the bovine ductus deferens for quantitative recovery of epididymal spermatozoa. *J. Reprod. Fert.* **6**: 65–69.

Andrews, L. G. (1967). Collection of bull semen by electroejaculation for evaluation and storage in the field. *Aust. vet. J.* **43**: 490–494.

Aulerich, R. J., Ringer, R. K. & Sloan, C. S. (1972). Electroejaculation of mink, *Mustela vison*. *J. Anim. Sci.* **34**: 230–233.

Austin, J. W., Leidy, R. B., Krise, G. M. & Hupp, E. W. (1968). Normal values for semen collected from Spanish goats by two methods. *J. appl. Physiol.* **24**: 369–372.

Ball, L. (1976). Electroejaculation. In *Applied electronics for veterinary medicine and animal physiology*: 394–441. Klemm, W. R. (ed.). Springfield, Illinois: Charles C. Thomas.

Ball, L. & Furman, J. W. (1972). Electroejaculation of the bull. *Bov. Practitioner*, No. 7: 46–48.

Bautina, E. P. (1974). (Collection of semen and its use in artificial insemination of silver black foxes.) *Razvedenie pushn. zvereĭ krolikov. Vyp.* 2: 96–108. (In Russian.) (*Anim. Breed. Abstr.* **43**: No. 181.)

Bedford, J. M. (1975). Maturation, transport and fate of spermatozoa in the epididymis. In *Handbook of physiology*. Section 7 Endocrinology, **5**. Male reproductive system. Hamilton, D. W. & Greep, R. O. (eds). Washington D.C.: Am. Physiol. Soc.

Bell, C. (1972). Autonomic nervous control of reproduction: Circulatory and other factors. *Pharm. Revs* **24**: 657–736.

Bennett, J. P. (1967). Semen collection in the squirrel monkey. *J. Reprod. Fert.* **13**: 353–355.

Bennett, J. P. & Rowson, L. E. A. (1963). A fistula for the collection of epididymal semen from the bull. *J. Reprod. Fert.* **6**: 61–64.

Bhattacharya, P. (1962). Artificial insemination in the water buffalo. In *The semen of animals and artificial insemination*: 184–204. Maule, J. P. (ed.). Farnham Royal, England: Comm. Agric. Bur.

Bierschwal, C. J., Mather, E. C., Martin, C. E., Murphy, D. A. & Korschgen, L. J. (1970). Some characteristics of deer semen collected by electroejaculation. *J. Am. vet. med. Assn* **157**: 627–632.

Birnbaum, D. & Hall, T. (1961). An electroejaculation technique for rats. *Anat. Rec.* **140**: 49–50.

Bishop, M. W. H., Campbell, R. C., Hancock, J. L. & Walton, A. (1954). Semen characteristics and fertility in the bull. *J. agric. Sci., Camb.* **44**: 227–248.

Boucher, J. H., Foote, R. H. & Kirk, R. W. (1958). The evaluation of semen quality in the dog and the effect of frequency of ejaculation upon semen quality, libido and depletion of sperm reserves. *Cornell Vet.* **48**: 67–86.

Bredderman, P. J., Foote, R. H. & Yassen, A. M. (1964). An improved artificial vagina for collecting rabbit semen. *J. Reprod. Fert.* **7**: 401–403.

Byck, R. (1975). Drugs and the treatment of psychiatric disorders. In *The pharmacological basis of therapeutics*: 152–200 5th edn. Goodman, L. S. & Gilman, A. (eds). New York: Macmillan Publishing Co. Inc.

Campbell, E. A. & Lingham, S. A. (1965). Artificial insemination of pigs in Australia. I. Training of boars and collection of samples. *Aust. vet. J.* **41**: 147–150.

Campbell, R. C., Dott, H. M. & Glover, T. D. (1956). Nigrosin eosin as a stain for differentiating live and dead spermatozoa. *J. agric. Sci., Camb.* **48**: 1–8.

Cerovsky, J. & Slechta, J. (1973). (The effect of ejaculate collection technique on principle semen characters.) *Zivocisna Vyroba* **18**: 887–892. (In Czech.) (*Anim. Breed. Abstr.* **42**, No. 5401.)

Chapelle, P. A., Jondet, M., Durand, J. & Grossiord, A. (1976). Pregnancy of the wife of a complete paraplegic by homologous insemination after an intrathecal injection of neostigmine. *Paraplegia* **14**: 173–177.

Christensen, G. C. & Dougherty, R. W. (1955). A simplified apparatus for obtaining semen from dogs by electrical stimulation. *J. Am. vet. med. Assn* **127**: 50–52.

Clarke, R. H., Hewetson, R. W. & Thompson, B. J. (1973). Comparison of fertility of bovine semen collected by artificial vagina and electroejaculation from bulls with a low libido. *Aust. vet. J.* **49**: 240–241.

Colleary, C. W. & Ehlers, M. H. (1964). Freezability of spermatozoa obtained with the ejaculator and with the artificial vagina. *J. Dairy Sci.* **47**: 115.

Creed, R. F. S. [1964]. Collection of semen from the red fox. *V Int. Congr. Anim. Reprod. Artif. Insem.* [Trento] **4**: 557–561.

Dieterich, R. A. & Luick, J. R. (1971). Reindeer in biomedical research. *Lab. Anim. Sci.* **21**: 817–824.

Dott, H. M. & Utsi, M. N. P. (1971). The collection and examination of semen of the reindeer. *J. Zool., Lond.* **164**: 419–424.

Dowling, D. F. (1961). Electrical stimulation of ejaculation in the bull. *Aust. vet. J.* **37**: 176–181.

Duran del Campo, A. (1969). (The use of semen from dead bulls.) *Ganadería, Madr.* **27**: 540–541. (In Spanish.)

Dziuk, P. J., Graham, E. F., Donker, J. D., Marion, G. B. & Petersen, W. E. (1954). Some observations in collection of semen from bulls, goats, boars and rams by electrical stimulation. *Vet. Med.* **49**: 455–458.

Dziuk, P. J., Graham, E. F. & Petersen, W. E. (1954). The technique of electro-ejaculation and its use in dairy bulls. *J. Dairy Sci.* **37**: 1035–1041.

Eliasson, R. (1971). Standards for investigation of human semen. *Andrologie* **3**: 49–64.

Emmens, C. W. & Robinson, T. (1962). Artificial insemination in sheep. In *The semen of animals and artificial insemination*: 205–251. Maule, J. P. (ed.). Farnham Royal, England: Comm. Agric. Bur.

Faulkner, L. C. & Pineda, M. H. (1975). Artificial insemination. In *Veterinary endocrinology and reproduction*: 304–341 2nd edn, McDonald, L. E. (ed.). Philadelphia: Lea and Febiger.

Fernandez-Baca, S. (1975). Alpaca raising in the high Andes. *Wld Anim. Rev.* No. 14: 1–8.

Fernandez-Baca, S. & Calderon, W. (1967). (Methods of semen collection from alpaca.) *Revta Fac. Med. vet. Univ. nac., Lima* **18/20**: 13–26. (In Spanish.)

First, N. L. (1971). Collection and preservation of spermatozoa. In *Methods in mammalian embryology*: 15–36 Daniel, J. C. (ed.). San Francisco: W. H. Freeman & Co.

Foote, R. H. (1975). Semen quality from the bull to the freezer: An assessment. *Theriogenology* **3**: 219–235.

Foote, R. H. & Heath, A. (1963). Effect of sperm losses in semen collection equipment on estimated sperm output by bulls. *J. Dairy Sci.* **46**: 242–244.

Foster, J., Almquist, J. O. & Martig, R. C. (1970). Reproductive capacity of beef bulls. IV. Changes in sexual behaviour and semen characteristics among successive ejaculates. *J. Anim. Sci.* **30**: 245–252.

Francois, N., Maury, M., Vacant, J., Cukier, J. & David, G. (1975). Etude expérimentale de l'électroéjaculation chez le babouin. *J. Urol. Nephrol.* **81**: 533–541.

Freund, M. (1958). Collection and liquefaction of guinea pig semen. *Proc. Soc. exp. Biol. Med.* **98**: 538–542.

Freund, M. (1969). Interrelationships among the characteristics of guinea pig semen collected by electroejaculation. *J. Reprod. Fert.* **19**, 393–403.

Fussell, E. N., Franklin, L. E. & Frantz, R. C. (1973). Collection of chimpanzee semen with an artificial vagina. *Lab. Anim. Sci.* **23**: 252–255.

Gilman, S. E. (1969). Relationships between internal resistance, stimulation voltage, and electroejaculation in the pigtail monkey, *Macaca nemestrina*: A preliminary report. *Lab. Anim. Care* **19**: 800–803.

Goodwin, D. E. (1959). Examination of Aberdeen Angus bulls for breeding soundness with comments on pathology. *J. Am. vet. med. Assn* **135**: 556–558.

Goodwin, D. E. (1970). The collection of semen from Aberdeen Angus bulls by massage of the intrapelvic organs. *J. Am. vet. med. Assn* **157**: 831–833.

Grandage, J. (1972). The erect dog penis: A paradox of flexible rigidity. *Vet. Rec.* **91**: 141–147.

Greer, W. E., Roussel, J. D. & Austin, C. R. (1968). Prevention of coagulation in monkey semen by surgery. *J. Reprod. Fert.* **15**: 153–155.

Grove, D., Bollwahn, W. & Mahler, R. (1968). Versuche zur Gewinnung und Tiefgefrierkonservierung von Nebenhodenschwanzsperma beim Eber. *Dt. tierärztl. Wschr.* **75**: 35–38.

Gunn, R. M. C. (1936). Fertility in sheep. Artificial production of seminal ejaculation and the characters of the spermatozoa contained therein. *Bull. C.S.I.R., Aust.* No. 94: 1–116.

Guttman, L. (1973). *Spinal cord injuries.* Comprehensive management and research. Oxford: Blackwell Scientific.

Guttman, L. & Walsh, J. J. (1971). Prostigmine assessment test of fertility in spinal man. *Paraplegia* **9**: 39–51.

Hafez, E. S. E. (ed.) (1970). *Reproduction and breeding techniques for laboratory animals.* Philadelphia: Lea and Febiger.

Hafs, H. D., Louis, T. M. & Stellflug, J. N. (1973). Ductus deferens and ejaculated sperm after $PGF_{2\alpha}$. *J. Anim. Sci.* **37**: 313.

Hale, E. B. & Almquist, J. O. (1960). Relation of sexual behaviour to germ cell output in farm animals. *J. Dairy Sci.* **43** (Suppl.): 145–169.

Hardin, C. J., Liebherr, G. & Fairchild, O. (1975). Artificial insemination in chimpanzees. *Int. Zoo Yb.* **15**: 132–134.

Harrop, A. E. (1954). A new type of canine artificial vagina. *Br. vet. J.* **110**: 194–195.

Harthoorn, A. M. (1976). *The chemical capture of animals.* London: Balliere Tindall.

Healey, P. & Sadleir, R. M. F. S. (1966). The construction of rectal electrodes for electroejaculation. *J. Reprod. Fert.* **11**: 299–301.

Healey, P. & Weir, B. J. (1967). A technique for electroejaculation in chinchillas. *J. Reprod. Fert.* **13**: 585–588.

Holtz, W. & Foote, R. H. (1974). Cannulation and recovery of spermatozoa from the rabbit ductus deferens. *J. Reprod. Fert.* **39**: 89–92.

Horne, H. W., Paull, D. P. & Munro, D. (1948). Fertility studies in human male with traumatic injuries of spinal cord and cauda equina. *New Eng. J. Med.* **239**: 959–961.

Hoskins, D. D. & Patterson, D. L. (1967). Prevention of coagulum formation with recovery of motile spermatozoa from rhesus monkey semen. *J. Reprod. Fert.* **13**: 337–340.

Hovell, G. J. R., Ardran, G. M., Essenhigh, D. M. & Smith, J. C. (1969). Radiological observations on electrically induced ejaculation in the ram. *J. Reprod. Fert.* **20**: 383–388.

Howarth, V. S. (1950). A method for the collection of the secretions of the individual accessory glands in a marsupial (*Trichosurus vulpecula*). *Med. J. Aust.* **1950(i)**: 566–567.

Hulet, C. V. & Ercanbrack, S. K. (1962). A fertility index for rams. *J. Anim. Sci.* **21**: 489–493.

Hulet, C. V., Foote, W. C. & Blackwell, R. L. (1964). Effect of natural and electrical ejaculation on predicting fertility in the ram. *J. Anim. Sci.* **23**: 418–424.

Jaczewski, Z. & Jasiorowski, T. (1973). Observations on the electro-ejaculation of red deer. *Acta theriol.* **19**: 151–157.

Jaczewski, Z., Morstin, J., Kossakowski, J. & Krzywiński, A. [1976]. Freezing of semen of red deer stags. *VIII Int. Congr. Anim. Reprod. Artif. Insem.* [Krakow] **1**: 994–997.

Jainudeen, M. R., Eisenberg, J. F. & Jayasinghe, J. B. (1971). Semen of the Ceylon elephant, *Elephas maximus. J. Reprod. Fert.* **24**: 213–217.

Johnson, L. A. & Pursel, V. G. (1975). Cannulation of the ductus deferens of the boar: A surgical technique. *Am. J. vet. Res.* **36**: 315–317.

Jones, R. C. (1973). Collection, motility and storage of spermatozoa from the African elephant, *Loxodonta africana. Nature, Lond.* **243**: 38–39.

Jones, R. C., Bailey, D. W. & Skinner, J. D. (1975). Studies on the collection and storage of semen from the African elephant, *Loxodonta africana. Koedoe* **18**: 147–164.

Kalasiewicz, J. & Wolanski, Z. (1970) (Electroejaculation method for obtaining rat semen.) *Acta physiol. pol.* **21**: 567–573. (In Polish.)

Kessler, G. (1974). Eine Methode zur Provokation von Ejakulationen bei der Ratte. Kurze Mitteilung. *Z. Versuchstierknde* **16**: 73–75.

Khan, A. A. & Kohli, I. S. (1973). A note on collection of semen from camel with help of an artificial vagina. *Ind. J. Anim. Sci.* **43**: 454–455.

Kihlström, J. E. & Melin, P. (1963). Influence of oxytocin upon some seminal factors in the rabbit. *Acta physiol. scand.* **59**: 363–369.

King, G. J. & Macpherson, J. W. (1973). A comparison of two methods for boar semen collection. *J. Anim. Sci.* **36**: 563–565.

Kirk, R. W. (1970). Dogs. In *Reproduction and breeding techniques for laboratory animals*: 224–236. Hafez, E. S. E. (ed.). Philadelphia: Lea and Febiger.

Knight, T. W. (1974). The effect of oxytocin and adrenaline on the semen output of rams. *J. Reprod. Fert.* **39**: 329–336.

Komarek, R. J., Pickett, B. W., Gibson, E. W. & Lanz, R. N. (1965). Composition of lipids in stallion semen. *J. Reprod. Fert.* **10**: 337–342.

Kraemer, D. C. & Vera Cruz, N. C. (1969). Collection, gross characteristics and freezing of baboon semen. *J. Reprod. Fert.* **20**: 345–348.

Krzywiński, A. [1976]. Collection of red deer semen with the artificial vagina. *VIII Int. Congr. Anim. Reprod. Artif. Insem.* [Krakow] **4**: 1002–1005.

Kuehl, T. J. & Dukelow, W. R. (1974). A restraint device for electro-ejaculation of squirrel monkeys (*Saimiri sciureus*). *Lab. Anim. Sci.* **24**: 364–366.

Lambiase, J. T., Amann, R. P. & Lindzay, J. S. (1972). Aspects of reproductive physiology of male white-tailed deer. *J. Wildl. Mgmt* **36**: 868–875.

Landowski, J. & Gill, J. (1964). Einige Beobachtungen über das Sperma des Indischen Elefanten (*Elephas maximus* L.). *Zool. Gart., Lpz.* **29**: 205–212.

Lang, C. M. (1967). A technique for the collection of semen from squirrel monkeys (*Saimiri sciureus*) by electroejaculation. *Lab. Anim. Care* **17**: 218–221.

Laplaud, M. & Cassou, R. (1948). Recherches sur l'électro-éjaculation chez le taureau et le verrat. *C.r. Séanc. Soc. Biol., Paris* **142**: 726–727.

Lapwood, K., Martin, I. C. A. & Entwistle, K. W. (1972). The fertility of merino ewes artificially inseminated with semen diluted in solutions based on skim milk, glucose or ribose. *Aust. J. agric. Res.* **23**: 457–466.

Lawson, R. L., Barranco, S. & Sorensen, A. M. (1966). A device to restrain the mouse, rat, hamster and chinchilla to facilitate semen collection and other reproductive studies. *Lab. Anim. Care* **16**: 72–79.

Lawson, R. L., Krise, G. M. & Sorensen, A. M. (1967). Electroejaculation and evaluation of semen from the albino rat. *J. appl. Physiol.* **22**: 174–176.

Lawson, R. L. & Sorensen, A. M. (1969). Electroejaculation and evaluation of semen from the chinchilla. *Texas J. Sci.* **21**: 189–193.

Li, V. A., Esjkov, P. A., Dolgih, M. N. & Kaldybaev, S. U. (1965). (The use of semen of a wild boar for artificial insemination of sows of farm breeds.) *Trudy Inst. eksp. Biol. Alma-Ata* **2**: 53–56 (In Russian.) (*Anim. Breed. Abstr.* **36**, No. 1741.)

Loewe, S. (1937). A pharmacological ejaculation test for bioassay of male sex hormone. *Proc. Soc. exp. Biol. Med.* **37**: 483–484.

Loewe, S. (1938). Influence of autonomic drugs on ejaculation. *J. Pharmacol.* **63**: 70–71.

Macmillan, K. L. & Hafs, H. D. (1967). Semen output of rabbits ejaculated after varying sexual preparation. *Proc. Soc. exp. Biol. Med.* **125**: 1278–1281.

Mann, T. (1964). *The biochemistry of semen and of the male reproductive tract.* 2nd edn. London: Methuen and Co. Ltd, New York: John Wiley and Sons Inc.

Marden, W. G. R. (1954). New advances in the electroejaculation of the bull. *J. Dairy Sci.* **37**: 556–561.

Martig, R. C., Almquist, J. O. & Foster, J. (1970). Reproductive capacity of beef bulls. V. Fertility and freezability of successive ejaculates collected by different methods. *J. Anim. Sci.* **30**: 60–62.

Mastroianni, L. & Manson, W. A. (1963). Collection of monkey semen by electroejaculation. *Proc. Soc. exp. Biol. Med.* **112**: 1025–1027.

Mattner, P. & Voglmayr, J. K. (1962). A comparison of ram semen collected by the artificial vagina and by electro-ejaculation. *Aust. J. exp. Agric. Anim. Husb.* **2**: 78–81.

Maule, J. P. (ed.) (1962). *The semen of animals and artificial insemination.* Farnham Royal, England: Comm. Agric. Bur.

Megale, F. (1968). Induction of erection and ejaculation in the bull by local massage. *Cornell Vet.* **58**: 88–89.

Miljkovic, V., Zivanovic, D. & Buric, I. (1966). (Electroejaculation in the mink and some laboratory animals.) *Acta vet., Beogr.* **16**: 363–371. (In Serbo-croat.) (*Anim. Breed. Abstr.* **36**, No. 745.)

Miller, S. J. (1961). Ram management for artificial insemination. In *Proc. Conf. Arti. Breed. Sheep in Aust.*: 163–167. Roberts, E. M. (ed.). Sydney: Univ. N.S.W. Press.

Milovanov, V. K., Bereznev, A. P. & Gorohov, L. N. (1962). (The effect of oxytocin on the reproductive system of male livestock.) *Vestn. Sel-hoz. Nauk. (Mosk.)* **7**(2): 99-103. (In Russian.) (*Anim. Breed. Abstr.* **32**, No. 683.)

Mukherjee, D. P., Johari, M. P. & Bhattacharya, P. (1951). The gelatinous mass in rabbit semen. *Nature, Lond.* **168**: 422–423.

Napier, J. R. & Napier, P. H. (1967). *A handbook of living primates: morphology, ecology and behaviour of non-human primates.* London and New York: Academic Press.

Nichols, G. de la M. & Edgar, D. G. (1964). A transistorized rectal probe for electroejaculating rams. *N.Z. vet. J.* **12**: 145–146.

Nishikawa, Y. (1959). *Studies on reproduction in horses.* Tokyo: Japan Racing Assn.

Pickett, B. W., Gebauer, M. R., Seidel, G. E. & Voss, J. L. (1974). Reproductive physiology of the stallion: Spermatozoal losses in the collection equipment and gel. *J. Am. vet. med. Assn* **165**: 708–710.

Pickett, B. W., Voss, J. L. & Gebauer, M. R. (1973). The effect of teasing on stallion seminal characteristics. *J. Anim. Sci.* **37**: 324.

Platz, C., Follis, T., Demorest, N. & Seager, S. [1976]. Semen collection, freezing and insemination in the domestic cat. *VIII Int. Congr. Anim. Reprod. Artif. Insem.* [Krakow] **4**: 1053–1056.

Polotzoff, V. (1928). Die Spermatozoenproduktion beim Pferde. I. Mitteilung. *Pflug. Arch. ges. Physiol.* **218**: 374–385.

Pomytko, V. N., Bautina, E. P. & Vladimirova, N. V. (1972). (The problem of obtaining good quality semen from silver-black foxes.) *Nauch. Trudy nauchnoissled. Inst. Pushn. Zverov. Krolikov.* **11**: 175–182. (In Russian.) (*Anim. Breed. Abstr.* **42**, No. 2277.)

Pomytko, V. N., Bernatskii, V. G., Kruglova, N. M. & Dubovaya, R. G. (1972). (Semen collection, dilution, storage and artificial insemination in mink.) *Nauch. Trudy nauchnoissled. Inst. Pushn. Zverov. Krolikov.* **11**: 165–170. (In Russian.) (*Anim. Breed. Abstr.* **42**, No. 2278.)

Potts, I. F. (1957). Mechanism of ejaculation. *Med. J. Aust.* **1957(i)**: 495–497.

Quinn, P. J., Salamon, S. & White, I. G. (1968). The effect of cold shock and deep-freezing on ram spermatozoa collected by electrical ejaculation and by an artificial vagina. *Aust. J. Agric. Res.* **19**: 119–128.

Rakhimzhanov, G. R. (1971). (Semen collection from the two-humped camel.) *Vest. sel'.-khoz. nauk, Alma-Ata.* **14** (5): 100–102, 119. (In Russian.) (*Anim. Breed. Abstr.* **40**: No. 3658.)

Rehan, N-E., Sobrero, A. J. & Fertig, J. W. (1975). The semen of fertile men: Statistical analysis of 1300 men. *Fert. Steril.* **26**: 492–502.

Richardson, G. F. & Wenkoff, M. S. (1976). Semen collection from a stallion using a dummy mount. *Can. vet. J.* **17**: 177–180.

Rodger, J. C. & White, I. G. (1975). Electroejaculation of Australian marsupials and analyses of the sugars in the seminal plasma from three macropod species. *J. Reprod. Fert.* **43**: 233–239.

Rollinson, D. H. L. & Nunn, W. R. (1962). Development of artificial insemination in the tropics. In *The semen of animals and artificial insemination*: 358–395. Maule, J. P. (ed.). Farnham Royal, England: Comm. Agric. Bur.

Rossier, A. B., Ziegler, W. H., Duchosal, P. W. & Meylan, J. (1971). Sexual function and dysreflexia. *Paraplegia* **9**: 51–59.

Roth, E. & Smidt, D. (1970). Untersuchungen zur fruzeitigen Spermagewinnung mittels Elektroejakulation bei Deutschen veredetten Landschweinen und Gottinger Miniaturschweinen. *Berl. Munch. tierärztl Wschr.* **83**: 7–11.

Roussel, J. D. & Austin, C. R. (1967). Enzymic liquefaction of primate semen. *Int. J. Fertil.* **12**: 288–290.

Roussel, J. D. & Austin. C. R. (1968). Improved electroejaculation of primates. *J. Inst. Anim. Tech.* **19**: 22–32.

Rowan, A. L., Howley, T. F. & Nova, H. R. (1962). Electroejaculation. *J. Urol.* **87**: 726–729.

Rowson, L. E. A. (1962). Artificial insemination in the pig. In *The semen of animals and artificial insemination*: 263–280. Maule, J. P. (ed.). Farnham Royal, England: Comm. Agric. Bur.

Rowson, L. E. A. & Murdoch, M. I. (1954). Electrical ejaculation of the bull. *Vet. Rec.* **66**: 326–327.

Sadleir, R. M. F. S. (1965). Reproduction in two species of kangaroo, *Macropus robustus* and *Megaleia rufa*, in the arid Pilbara region of Western Australia. *Proc. zool. Soc., Lond.* **145**: 239–261.

Salamon, S. & Lindsay, D. R. (1961). The training of rams for the artificial vagina, with some observations on ram behaviour. *Proc. Conf. Artif. Breed. Sheep in Aust.*: 173–182. Roberts, E. M. (ed.). Sydney: Univ. N.S.W. Press.

Salamon, S. & Morrant, A. J. (1963). A comparison of two methods of artificial breeding in sheep. *Aust. J. exp. Agric. Anim. Husb.* **3**: 72–77.

Salisbury, G. W. & VanDemark, N. L. (1961). *Physiology of reproduction and artificial insemination of cattle*. San Francisco and London: W. H. Freeman and Co.

Schefels, W. (1969). Die Spermagewinnung beim Ruden mit Hilfe eines Vibrators. *Dt. tierärztl. Wschr.* **76**: 289–290.

Schellen, T. M. C. M. (1968). Induction of ejaculation by electrovibration. *Fert. Steril.* **19**: 566–569.

Scott, P. P. (1970). Cats. In *Reproduction and breeding techniques for laboratory animals*: 192–208 Hafez, E. S. E. (ed.). Philadelphia: Lea and Febiger.

Scott, J. V. & Dziuk, P. J. (1959). Evaluation of the electroejaculation technique and the spermatozoa thus obtained from rats, mice and guinea pigs. *Anat. Rec.* **133**: 655–664.

Seager, S. W. J. (1974). Semen collection and artificial insemination in captive wild cats, wolves and bears. *Ann. Proc. Am. Assn Zoo vets* **1974**: 29–33.

Seager, S. W. J. (1976). Electroejaculation of cats (domestic and captive wild Felidae). In *Applied electronics for veterinary medicine*: 410–418. Klemm, W. R. (ed.). Springfield, Illinois: Charles C. Thomas.

Seager, S. W. J., Platz, C. C. & Hodge, W. (1975). Successful pregnancy using frozen semen in the wolf, *Canis lupus irremotus*. *Int. Zoo Yb.* **15**: 140–143.

Seidel, G. E. & Foote, R. H. (1969). Motion picture analysis of ejaculation in the bull. *J. Reprod. Fert.* **20**: 313–317.

Semans, J. H. & Langworthy, O. R. (1938). Observations on the neurophysiology of sexual function in the male cat. *J. Urol.* **40**: 836–846.

Settlage, D. S. F. & Hendrickx, A. G. (1974a). Electroejaculation technique in *Macaca mulatta* (Rhesus monkeys). *Fert. Steril.* **25**, 157–159.

Settlage, D. S. F. & Hendrickx, A. G. (1974b). Observations on coagulum characteristics of the rhesus monkey ejaculate. *Biol. Reprod.* **11**: 619–623.

Shump, A. U., Aulerich, R. J. & Ringer, R. K. (1976). Semen volume and sperm concentration in the ferret (*Mustela putorius*). *Lab. Anim. Sci.* **26**: 913–916.

Snyder, R. L. (1966). Collection of mouse semen by electroejaculation. *Anat. Rec.* **155**: 11–14.

Sobrero, A. J., Stearns, H. E. & Blair, J. H. (1965). Technique for induction of ejaculation in humans. *Fert. Steril.* **16**: 765–767.

Sojka, N. J., Jennings, L. L. & Hamner, C. E. (1970). Artificial insemination in the cat (*Felius catus* L.) *Lab. Anim. Care* **20**: 198–204.

Spira, R. (1956). Artificial insemination after intrathecal injection of neostigmine in a paraplegic. *Lancet* **1956(i)**: 660–671.

Suzuki, Y. [1952]. New ejaculatory responses in the rat and their application as indicators of sexual activities. *II Int. Congr. Physiol. Path. Anim. Reprod. Artif. Insem.* [Copenhagen] **1**: 53–62.

Swire, P. W. (1962). Artificial insemination in the horse. In *The semen of animals and artificial insemination*: 281–297. Maule, J. P. (ed.). Farnham Royal, England: Comm. Agric. Bur.

Thibault, C., Laplaud, M. & Ortavant, R. (1948). L'électroéjaculation chez le taureau. Technique et résultats. *C.r. hebd. Séanc. Acad. Sci., Paris* **226**: 2006.

Thomas, R. J. S., McLeish, G. & McDonald, I. A. (1975). Electroejaculation of the paraplegic male followed by pregnancy. *Med. J. Aust.* **1975(ii)**: 798–799.

Valerio, D. A., Ellis, E. B., Clark, M. L. & Thompson, G. E. (1969). Collection of semen from macaques by electroejaculation. *Lab. Anim. Care* **19**: 250–252.

Van Pelt, L. F. & Keyser, P. E. (1970). Observations on semen collection and quality in macaques. *Lab. Anim. Care* **20**: 726–733.

Vera Cruz, N. C. (1959). Artificial insemination in pigs. I. Semen collection by electrical stimulation in boars. *Philipp. Agricult.* **43**: 225–235.

Vladimirov, A. V. & Pomytko, V. N. [1976]. Artificial insemination of blue foxes. *VIII Int. Congr. Anim. Reprod. Artif. Insem.* [Krakow] **4**: 1090–1092.

Wallace-Haagens, M. J., Duffy, B. J. & Holtrop, H. R. (1975). Recovery of spermatozoa from human vaginal washings. *Fert. Steril.* **26**: 175–179.

Walton, A. (1960). Copulation and natural insemination. In *Marshall's Physiology of reproduction.* **I**, Pt. 2. 3rd edn. Parkes, A. S. (ed.). London: Longmans Green Co.

Warner, H., Martin, D. E. & Keeling, M. E. (1974). Electroejaculation of the great apes. *Annls Biomed. Engin.* **2**: 419–432.

Waters, R. J. [1976]. Sperm recovery *post-mortem* from the bull. *VIII Int. Congr. Anim. Reprod. Artif. Insem.* [Krakow] **1**: 275.

Watson, P. F. (1976). Electroejaculation, semen characteristics and semen preservation of the brindled gnu. *J. Reprod. Fert.* **47**: 123–126.

Watson, P. F. & Martin, I. C. A. (1976). Artificial insemination of sheep: the fertility of semen extended in diluents containing egg yolk and inseminated soon after dilution or stored at 5°C for 24 or 48 hours. *Theriogenology* **6**: 553–558.

Weir, B. J. (1970). Chinchillas. In *Reproduction and breeding techniques for laboratory animals*: 209–223 Hafez, E. S. E. (ed.). Philadelphia: Lea and Febiger.

Weisbroth, S. & Young, F. A. (1965). The collection of primate semen by electroejaculation. *Fert. Steril.* **16**: 229–235.

Wells, M. E., Philpot, W. N., Musgrave, S. D., Jones, E. W. & Brock, W. E. (1966). Effect of method of semen collection and tranquilization on semen quality and bull behaviour. *J. Dairy Sci.* **49**: 500–503.

White, I. G., Larsen, L. H. & Wales, R. G. (1959). Method for the *in vivo* collection of epididymal spermatozoa and for their comparison with ejaculated cells. *Fert. Steril.* **10**: 571–577.

Wierzbowski, S. & Wierzchos, E. (1969). Cannulation of the vas deferens in the boar. *J. Reprod. Fert.* **19**: 173–175.

Symp. zool. Soc. Lond. (1978) No. 43, 127–152

The Principles and Practice of Electroejaculation of Mammals

I. C. A. MARTIN

University of Sydney, New South Wales, Australia

SYNOPSIS

Normal ejaculation is a highly organized sequence of responses to a number of sensory stimuli. It involves several organ systems including both the somatic and autonomic nervous systems. Techniques of electrical stimulation applied for the collection of semen from a wide range of mammalian species are reviewed in this paper. The effectiveness of these stimuli is considered in terms of their efficacy in producing penile erection, emission of spermatozoa and glandular secretions into the urethra, closure of the neck of the urinary bladder and projectile ejaculation of semen, as these are the events which are co-ordinated in natural ejaculation.

The electrical factors important in electroejaculation have been found to be the form and frequency of the pulses and the power delivered by the stimulator. The nature and position of the electrode system in, or on, the animal have also been shown to be critical for satisfactory responses. Circuits used for generating the electrical stimuli have changed substantially with the development of transistors and micro-integrated components. In contrast to the instruments used 30 to 40 years ago when research on electroejaculation began, stimulators constructed recently have been made smaller in size and have a lower power requirement for the generation of electrical pulses. The most commonly used stimuli have been sine wave alternating currents or wave forms derived from this basic signal, but direct current pulses have also been used successfully.

Experiments employing direct current pulses for the electroejaculation of rams are also reported in this paper. Responses have been related to the nature of the electrical fields generated in various regions of the pelvic contents by particular electrode arrangements on bipolar rectal probes.

INTRODUCTION

The earliest reports of systematic experimentation on the electro-ejaculation of mammals appear to be those of Batelli (1922) and Moore & Gallagher (1930) who stimulated guinea-pigs with 60 cycle s^{-1} alternating current (AC) at 5 to 35 V using electrodes placed in the mouth combined with another needle electrode inserted at the base of the skull. Semen was collected, but responses were usually diffuse, often involving urination, defecation and convulsions. Some animals were electrocuted.

Gunn (1936), using rams as test animals, advanced the technique significantly by inserting a needle electrode into lumbar muscles and placing the other electrode in the rectum. The needle was replaced by a disc electrode applied to the shaved skin of the lumbar or sacral regions of the guinea-pig by Dalziel & Phillips (1948) and Freund (1958, 1969).

Laplaud & Cassou (1945, 1948) designed a bipolar rectal electrode system which was used successfully on the bull, ram and boar. Their work established that stimulation of structures in the pelvis was sufficient for emission of semen and involvement of higher spinal cord or of brain stem was not necessary.

Initially, circuits for the generation of electrical stimuli were simple and were usually powered by mains AC electricity. They were, accordingly, limited in the range of stimulating pulses they could deliver and were often hazardous for both the test animal and the operator. Later, a number of more elaborate circuits were used (Benham & Enders, 1941; Dowling, 1961; Freund, 1969). Such instruments usually permitted control of the frequency, amplitude and wave form of the electrical pulses. However they were fragile and operated at a high power requirement relative to the stimuli they generated. More recently, the use of transistors and integrated circuits in the stimulators has allowed the generation of a much greater range of electrical pulses for testing in experiments on electroejaculation. These units are much more compact and robust than the earlier instruments and are frequently powered by low voltage batteries. Thus, they are basically safer than the earlier circuits.

Research on electroejaculation of a range of species of mammals was reviewed by Ball (1976). With the addition of the work by Rodger & White (1975) on Australian marsupials and by Jones (1973) on the African elephant, references exist describing the successful electroejaculation of animals from six mammalian orders (Marsupialia, Primates, Rodentia, Carnivora, Artiodactyla and Proboscidea—see Watson, this volume, p. 97). However most of the research on the technique has used bulls or rams as experimental animals. After the first report by Mastroianni & Manson (1963) of the electroejaculation of the rhesus monkey, there have been numerous reports of semen collection from more than 20 species of simian primates. In addition, Warner, Martin & Keeling (1974) reported the development of a stimulator and rectal electrodes for semen collection from chimpanzees, orang-utans and gorillas. Although the techniques employed differ in detail from species to species, certain factors are basic to all of them. The electrical stimulus must be intermittent and usually of relatively low frequency, e.g. 20 to 80 Hz. Application of continuous direct current (DC) is not effective. The electrical field generated in the animal tissue must be sufficiently powerful to radiate beyond the rectal wall and stimulate nerves and/or smooth muscle. The variables determining this field are the geometry of the probe and the electrodes it carries, the potential differences (voltage) applied across the electrodes and the resistance of the electrode–tissue interfaces.

Success with the technique is also closely related to characteristics of the normal coital behaviour of the species, the tractability of the animal, the duration of natural intromission and ejaculation and whether the semen coagulates after ejaculation. Where the animal can be trained for semen collection using the artificial vagina, this is clearly the method of choice. However, many research and controlled breeding projects involve the collection of semen on a time schedule independent of the mating behaviour of the male. Furthermore it is impractical to attempt to collect semen from most non-domestic species using the artificial vagina.

In general, species which mate and ejaculate quickly and whose semen does not coagulate, e.g. ram and bull, are the easiest to electroejaculate. The native temperament of the species is also important and, again, the general systemic response of the fully conscious, unsedated ram or bull to electroejaculation is less than that of most other species tested. Therefore, tranquillizers and anaesthetics are commonly used for the restraint and relaxation of animals during electroejaculation. Indeed, the great progress recently made in the electroejaculation of primates and carnivores has been due principally to medication of the animal rather than to innovations in the electrical equipment.

This paper is concerned with a comparison of the normal process of ejaculation and responses elicited by electrical stimuli, with the types of electrical stimulus which have greatest efficacy in causing erection of the penis and emission of semen, and with the selection of drugs most suitable for the control of the animal during electroejaculation.

PHYSIOLOGY OF EJACULATION

Copulatory Behaviour of Mammals

Dewsbury (1972) has classified male copulatory behaviour in a wide range of mammalian species according to the characteristics of intromission, thrusting, genital lock and frequency of ejaculation. Species differ widely in the frequency and duration of vaginal penetration by the penis. In some, repeated intromission occurs before ejaculation; in others, ejaculation occurs at a single intromission. During intromission, the male may make repetitive thrusting movements of the penis or ejaculation may occur after a single thrust. Genital lock of the penis and vagina occurs during intromission in some species, most notably among the Canidae. The principal variations in the characteristics of ejaculation are the duration of ejaculation of semen and the frequency of ejaculation. Duration at any copulation can range from seconds, e.g. ram, bull or cat, with a single expulsion of semen, to minutes in the boar or stallion in which time there will be numerous pulsatile ejaculations. After ejaculation, there is usually a refractory period in which the male cannot ejaculate and is usually not interested in copulation. This time is usually characteristic of the species and can be minutes, hours or days. Taking the species mentioned above, rams and bulls will resume mating within minutes and can ejaculate repeatedly in a period of an hour. Stallions, boars and dogs are likely to be refractory for an hour or more.

In general electroejaculation has been applied most successfully to those species in which ejaculation normally occurs quickly after intromission and a single thrust and which have a short refractory period. In species in which ejaculation occurs in a series of propulsive waves, it is usual for variation in composition of semen to occur during ejaculation. The sperm-rich fraction of the ejaculate may be produced early or late in the expulsion of semen. These animals frequently produce an ejaculate of high volume and generally either they are difficult to electroejaculate or no reports of successful electroejaculation exist, e.g. boar, stallion or dog. There is clearly a sequencing, rather than synchronizing, of reflexes in the whole ejaculatory process and this must be considered when developing a technique of electroejaculation for a particular species.

Neurology of Erection and Ejaculation

Marberger (1974), Kedia & Markland (1976) and Lundberg (1976) have described the process of ejaculation in man. Some of the information on erection and ejaculation has come from experimental observations made on animals, but most originates from studies of men with various deficiencies in erection and ejaculation caused by disease, surgery or drugs. The sequence and duration of reflexes associated with erection and ejaculation differ in detail from man to the other species but all mammals studied appear to have the same basic reflexes. In man the process is frequently described as occurring in four phases: erection, emission of semen into the urethra, formation of a pressure chamber in the urethra near the bladder and expulsion of the semen through the urethra. All of these responses are spinal reflexes; the complex sensations described as "orgasm" or "climax" involve the cerebral cortex and are not necessarily synchronous with emission or expulsion of semen.

Table I summarizes the nervous pathways involved in these reflexes. Erection can be stimulated by reflexes arising from input to the cerebral cortex from non-genital tactile sensors and the special senses of smell, sight and hearing. However, spinal reflexes involving stimulation of the genitalia are also important. Parasympathetic fibres in the erigent nerves are the motor elements in this reflex causing vasodilatation in penile arteries and constriction in penile veins. Erection is normally maintained throughout the remainder of the ejaculatory process. Emission, which is the movement of spermatozoa and accessory secretions into the urethra, is caused in particular by continued stimulation of sensory receptors of the genitalia, especially those in the glans penis. The efferent nerve fibres concerned are sympathetic and are grouped in the pelvis as the hypogastric nerve. They cause peristaltic contractions of smooth muscle in the epididymis and vas deferens propelling spermatozoa and fluid to the ampulla. The smooth muscles of the prostate, ampulla and seminal vesicles also contract and the bladder sphincter is stimulated to close tightly. Filling of the upper urethra is then sensed by visceral sensory

TABLE I

Reflexes involved in ejaculation

Phase	Sensory nerves	Central processing	Motor nerves
Erection	Special senses	Brain, spinal cord	Erigent (S2–S4, P)
	Somatic from external genitalia including pudendal N (S2–S4)	Spinal cord	Erigent
Emission	Continued stimulation of external genitalia especially glans penis	Spinal cord	Hypogastric (T10–L3, Sy)
Formation of pressure chamber	Continued stimulation as above	Spinal cord	Hypogastric pudendal pelvic (S2–S4, P)?
Expulsion	Visceral afferents from pelvic urethra	Spinal cord	Somatic motor mainly pudendal

T, Thoracic spinal segment; P, parasympathetic; S, sacral spinal segment; Sy, sympathetic.

fibres in the pudendal nerve causing reflex clonic contractions of the striated bulbo- and ischio-cavernosus muscles together with other muscles of the perineal region.

Several similar nervous pathways are involved in micturition, so that the reflexes associated with the two acts must be channelled quite specifically. Certainly, contamination of semen with urine is the commonest technical problem occurring in electroejaculation. Additionally, it seems unlikely that a technique of electroejaculation could be developed as organized as the natural sequence of reflexes. However, if the electrical stimulus is sufficient to move semen into the urethra, then the animal is likely to expel the contents of the urethra reflexly when stimulation is stopped.

Effect of Drugs on Ejaculation

Compounds affecting the autonomic nervous system have been studied for their effects on ejaculation. Dziuk & Norton (1962), Baker & Dziuk (1964) and Baker *et al.* (1964) examined the effects of the administration of pilocarpine or atropine to bulls, boars and rabbits 20 to 30 min before ejaculation. Atropine depressed the volume of semen produced by boars and the number of spermatozoa ejaculated by bulls and rabbits was reduced. Pilocarpine had no effect on the semen characteristics of boars, but increased the volume of semen and number of spermatozoa per ejaculate from bulls. Guttman (1973) reported that intrathecal injection of neostigmine gave erection and ejaculation in paraplegic men.

These drugs are active at cholinergic nerve endings, suggesting that the parasympathetic nervous system is involved in emission. Kedia & Markland (1976), in reviewing the pharmacological responses of the smooth muscle of male internal genitalia, did not present any data to support this suggestion. They did, however, assemble clear evidence that the motor innervation to the smooth muscle of seminal vesicles and vas deferens was adrenergic, operating through α-adrenergic receptors. Indeed, these neurones appear to be highly sensitive to the anti-hypertensive drug, guanethidine, an α-adrenergic blocking agent. From human clinical reports cited by Marberger (1974) and Kedia & Markland (1976), other anti-hypertensive drugs, such as the *Rauwolfia* alkaloids, and the psycho-active compounds, thioridazine, phenoxybenzamine and chlorpromazine, will either inhibit ejaculation or cause retrograde ejaculation into the bladder.

Currently, except for pilocarpine, there do not appear to be any drugs available which will enhance the ejaculatory response and might facilitate electroejaculation. However, a number of drugs used for the capture and restraint of animals are likely to inhibit ejaculation. Various compounds used for the sedation and anaesthesia of animals in preparation for electroejaculation are reviewed in the next section.

REVIEW OF METHODS USED IN ELECTROEJACULATION

Preparation and Restraint of the Animal

Electroejaculation will be facilitated if the animal is prevented from eating or drinking for several hours before stimulation. Martin & Rees (1962) found that bulls grazing lush pasture gave better responses to electroejaculation if they were brought into yards overnight. Using this management procedure, the bull's rectum was easier to empty before placing the electrode and fewer bulls had full bladders at the time of stimulation. Obviously, if anaesthesia is the method of restraint chosen, food and water should be restricted 8 to 12 hours in advance of semen collection. Although many of the early reports indicated than an enema was administered to clear faeces from the rectum, more recently the rectal probe has been placed initially to induce defecation and then cleansed and repositioned in the evacuated rectum (Ball, 1976).

The form and degree of restraint applied to the animal is determined by its size and whether sedation or anaesthesia is to be used. Bulls and rams are usually electroejaculated without any medication and the animal need only be held in a crush or crate to restrict movement. Martin & Rees (1962) described a simple adaptation of a commercial cattle crush which limited the bull's movement and prevented it from lying down. Gunn (1936) used a counterbalanced, hinged table-top onto which rams could be strapped and variants of his table are still in use. However, rams can be electroejaculated while standing in a race (Edgar, Inkster & MacDairmid,

1956) or simply held down in lateral recumbency by one or two assistants. As Ball (1976) has observed, restraint of the boar for electroejaculation is more difficult than for the ram or bull. Dziuk, Graham, Donker *et al.* (1954) electroejaculated boars while the animals were restrained between a gate and the adjacent wall of the pen. However, anaesthesia of the boar appears to be more satisfactory than mechanical restraint.

Scott & Dziuk (1959) restrained mice, rats and guinea-pigs by tying them to a board and Freund (1969) described a similar method of restraint for guinea-pigs. Seager (1976) outlined a method of restraint of the domestic cat in which the cat was drawn head first into a clear plastic pipe. However, he suggested that light surgical anaesthesia was a more practical method of restraint.

Mastroianni & Manson (1963) were able to restrain smaller monkeys manually for electroejaculation, but other workers have outlined the use of boards or chairs to which monkeys can be strapped (Valerio *et al.*, 1969). However, after Roussel & Austin (1968) demonstrated the efficacy of the tranquillizer, phencyclidine hydrochloride (Sernylan, Parke-Davis), for the control of non-human primates during electroejaculation, this compound has been used widely (Warner *et al.*, 1974). These authors used this compound or ketamine (Vetalar, Parke-Davis) on chimpanzees, orang-utans and gorillas and successfully electroejaculated these species while they were lightly anaesthetized. Additional medication with atropine or with tranquillizers derived from phenothiazine (promazine, chlorpromazine and acetylpromazine) reduced the volume of the semen obtained or, in some cases, the response to stimulation was only erection.

Hovell *et al.* (1969) electroejaculated rams which had been maintained in general anaesthesia for two to three hours using sodium pentobarbitone (Nembutal, Abbott). Anaesthesia was used in that research to immobilize the animal so that cinefluorograms could be obtained. Xylazine (Rompun, Bayer) has been used to sedate zebu bulls before electroejaculation (I. C. A. Martin, unpublished) and Watson (1976) reported the use of the same drug on the brindled gnu. Ball (1976) cited a report of the use of sodium thiamylal (Surital, Parke-Davis) to induce light anaesthesia in boars being electroejaculated. Semen was collected containing large numbers of spermatozoa but volume was reduced and the gel fraction was frequently missing. Semen was collected from a Père David's deer after immobilization with etorphine and acepromazine mixture (Immobilon, Reckitt & Colman) (I. C. A. Martin, unpublished). A rectal electrode and stimulus pattern suitable for rams was applied and semen was ejaculated at the third stimulation. This preparation has been used widely for the capture of wild Artiodactyla so that further use seems warranted. Jones (1973) also reported the use of Immobilon for the restraint of African elephants during electroejaculation.

Seager (1976) has used a combination of acetylpromazine (Acepromazine, Ayerst) and ketamine for the induction of anaesthesia in domestic cats before electroejaculation. He cautioned that the dose of acetylpromazine should be restricted otherwise contamination of the ejaculates with urine occurred frequently.

Aulerich, Ringer & Sloan (1972) used phencyclidine (Sernylan, Parke-Davis) to anaesthetize mink for electroejaculation. Some animals were also stimulated without anaesthesia and comparison of characteristics of the semen obtained showed that this anaesthetic did not affect the quality of the ejaculates collected. A combination of tiletamine hydrochloride and zolasepam (Tilazol, Parke-Davis) was employed by Shump, Aulerich & Ringer (1976) to anaesthetize ferrets before electroejaculation. They then used this technique to ejaculate ferrets as frequently as once per day for four successive days without any evidence of ill-effect to the animals or significant changes in the characteristics of the semen collected (Shump, Shump *et al.*, 1977).

Rodger & White (1975) found that electroejaculation proved to be a satisfactory method of obtaining seminal plasma, but not spermatozoa, from a variety of Australian marsupials. The volume of seminal plasma obtained was greater from anaesthetized animals than from conscious or sedated animals. The marsupials examined (brush-tailed possum, grey kangaroo, red kangaroo, tammar wallaby and long-nosed bandicoot) were also much less affected by the anaesthetics and tranquillizers (acetylpromazine, halothane, phencyclidine and xylazine) tested than eutherian mammals would have been. For several of the compounds, the marsupials required three to five times the effective dose commonly used on the species described above. Denny (1974) prepared a list of dose rates of anaesthetics and tranquillizers used on monotremes and marsupials and, except for the barbiturates, they are generally higher than would be used in eutherian mammals.

Equipment

Stimulators

Most of the stimulators used for electroejaculation have generated sine wave alternating current (Fig. 1, A) in the frequency range of 20 to 80 Hz (Dziuk, Graham & Petersen, 1954; Marden, 1954; Dowling, 1961). Other workers have used circuits which generated square wave pulses (Fig. 1, E) at frequencies ranging from 1000 pulses s^{-1} (Freund, 1969) to 50 pulses s^{-1} (Warner *et al.*, 1974) and 20 pulses s^{-1} (Furman, Ball & Seidel, 1975). There are also a few reports of the use of square wave DC pulses (Martin & Rees, 1962; Nichols & Edgar, 1964) (Fig. 1, F). Figure 1 also shows other variants of the basic wave form. Both half wave rectification (B) and clipping (C) of the sine wave reduced the response of bulls to stimulation (Marden, 1954). Spike waves of AC can be generated by proportional control circuits (D) but do not appear to have been tested for their value as stimuli for electroejaculation. The comparable differentiated wave form from DC pulses (H) has been used (see pp. 138–149) but was less

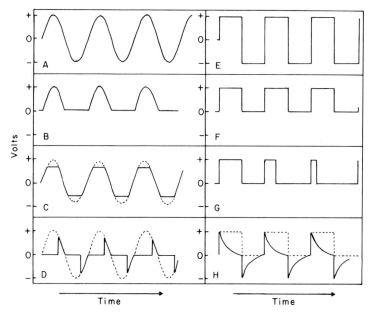

FIG. 1. Characteristics of electrical pulses. A is Sine wave AC. B, C and D are modifications of the sine wave. B is half wave rectified. C is a clipped amplitude sine wave. D is reduction of the sine wave by proportional control. E, F, G and H are square waves. E and F are AC and DC pulses respectively. In G the pulses vary from a mark space ratio of 0·5 (left) to 0·125 (right). H is an example of a differentiated wave form derived from direct current pulses (indicated by dotted line).

satisfactory than the original square wave (F). Square waves can also be varied in duration, i.e. variation in "mark space ratio", and G shows pulses of 0·5, 0·25 and 0·125 relative duration.

Comparisons of the responses to square waves or sine waves have shown the sine form to be more effective (Warner *et al.*, 1974; Furman *et al.*, 1975) in causing emission of semen. However, in these cases, mark space ratios in the range 0·02 to 0·25 were employed. Martin & Rees (1962) demonstrated that reduction of mark space ratio from 0·3 to 0·1 caused a fall in the output of semen from bulls.

Apart from the basic characteristics of the shape and frequency of the electrical pulses, the other electrical factors of voltage applied, resistance and current flow are important. Warner *et al.* (1974) have measured current densities on rectal electrodes during the electroejaculation of the great apes and demonstrated that the electrode–tissue interface resistance of rectal electrodes fell and current density rose during stimulation. The resistances measured lay between 58 and 132 Ω which were comparable to those quoted by Marden (1954) and Martin & Rees (1962) for bulls. Thus a satisfactory stimulator should be capable of delivering power and maintaining wave or pulse characteristics despite low output resistances. The range of real voltages applied to animals may well have been narrower than those quoted in various papers (5 to 45 V) if stimulators could not maintain voltage at higher discharge rates. In terms of efficacy of stimulus and safety of the animal, current flow is probably the most important parameter. Rock (1976) has stated that currents above 100 mA are potentially hazardous and very much lower currents applied directly to the heart could induce fibrillation.

Electrodes

Bipolar rectal electrodes, in a variety of forms adapted to the anatomy of the species to be electroejaculated, have been used most frequently (Ball, 1976). However, electrodes carried into the rectum on the hand of the operator have been useful for preliminary studies of the electro-ejaculation of large animals (Dowling, 1961; Watson, 1976) as they permit exploration for the most responsive regions of the pelvic contents. Valuable information on the segments of the spinal cord principally concerned with erection and emission was obtained by Freund (1969) who placed one electrode in the rectum of the guinea-pig and applied the other with electrode jelly to the skin over the vertebral column in the lumbar and sacral regions. Penile electrodes have been used on monkeys by a number of workers (Mastroianni & Manson, 1963; Hoskins & Patterson, 1967; Settlage & Hendrickx, 1974) but rectal electrodes were employed successfully by Weisbroth & Young (1965). Warner *et al.* (1974) gave details of the construction and use of rectal probes for the electroejaculation of apes.

Two basic designs of rectal probe have been used. In one, the electrodes were placed as rings around the probe (Blackshaw, 1954; Martin & Rees, 1962) and in the other the electrodes were mounted as strips on the probe (Marden, 1954). In the latter design, four longitudinal electrodes 90° apart, and alternately connected in pairs, were mounted on a plastic probe. Warner *et al.* (1974) constructed multi-electrode rectal probes on which various configurations of current flow could be arranged. They found that circular current flow was better than longitudinal, and using this information constructed a probe carrying longitudinal strip electrodes which was essentially a scaled-down version of that produced by Marden (1954) for use in bulls. Ball & Furman (1972) also favoured the use of strip electrodes and published details of a probe for bulls. Of the three electrode strips on it, the two "lateral" electrodes were linked and the "ventral" one was of opposite polarity. This modification in design was claimed to reduce contraction of muscles in the hind limbs during stimulation.

Patterns of stimulation

The stimulus has usually been given by steadily increasing and decreasing the voltage applied in a period of up to 20 s and then, after a brief period of rest of several seconds, repeating the stimulus (Warner *et al.*, 1974; Furman *et al.*, 1975; Ball, 1976). However other workers have set the voltage and applied the stimulus in a fixed proportion of the stimulation cycle. Thus, Freund (1969) controlled the application of the stimulus to guinea-pigs with a mechanical timer which gave a 3 s "on period" and a 12 s "off period". Five cycles of stimulation were given and the guinea-pig usually ejaculated during the first or second "on period". Shump, Aulerich *et al.* (1976) described a similar pattern of stimulation for the ferret of 4 s "on" in a 10 s cycle which was repeated up to 20 times. Chinchillas were also successfully electroejaculated using the same cycle of 4 s application of current in every 10 s up to 15 shocks (Healey & Weir, 1967). Although a cycle of stimulus and rest appears to have been fundamental in all the techniques described, the number of stimuli applied has usually been governed by the response of the animal, and the absolute and relative periods of "on" and "off" have been chosen empirically.

Methods of collection of semen

A container for collection of semen must have certain basic features. It must protect spermatozoa from temperature shock and be easy to replace with a clean container so that various fractions of the ejaculate may be collected or a sample contaminated with urine can be discarded. Also, it should not be potentially injurious to the animal. Furman *et al.* (1975) have described equipment which satisfied these requirements. Most of their container was made from flexible plastics so that, should bulls kick the apparatus, or fall on it, there was no risk of injury.

The semen from a number of species coagulates during, or after, ejaculation. This poses problems in the examination of the ejaculate and there is a risk of retention of coagulated semen plugging the urethra if electroejaculation does not cause complete expulsion of the semen. Scott & Dziuk (1959) reported deaths of rats and mice from uraemia after blockage of the urethra but found that, after removal of the seminal vesicles and coagulating glands, rats could be electroejaculated repeatedly without injury.

Primate semen often coagulates after ejaculation, but the coagulum can be prevented from forming by treatment of the ejaculate with trypsin (Hoskins & Patterson, 1967; Roussel & Austin, 1967). Residual coagulum in the urethra does not appear to cause serious urethral obstruction in primates after electroejaculation and Kraemer & Vera Cruz (1969) mentioned that, in some cases, the penile urethra of the baboon was manually stripped of strings of coagulated seminal material.

Seminal plasma from kangaroos coagulates shortly after collection but semen from the brush-tailed possum clots only occasionally (Rodger &

White, this volume, p. 289). No problems with urethral obstruction after electroejaculation of marsupials were reported by these authors.

Comparison of Semen Collected by Electroejaculation with that Obtained by Other Methods

Apart from studies of rams and bulls, there are few reports of comparisons of the semen collected by artificial vagina with that obtained from electroejaculation. Mattner & Voglmayr (1962), Salamon & Morrant (1963), Hulet, Foote & Blackwell (1964) and Lapwood, Martin & Entwistle (1972) all found that the volume of the ejaculate was higher when rams were collected by electroejaculation but the total number of spermatozoa output per ejaculate did not differ significantly between methods of collection. Austin, Hupp & Murphree (1961) came to the same conclusion in research involving 12 bulls in a balanced experimental design and crossover from one method of collection to the other. Obviously the response of the animal and the quality of semen obtained will be, in part, determined by the training of the animal for collection by the artificial vagina and the efficiency of the particular technique of electroejaculation. However, in all of these reports the number of spermatozoa per electrically stimulated collection was one-quarter, or more, of the estimated daily sperm production rate for that species. Fertility of ram semen was not significantly affected by method of semen collection (Salamon & Morrant, 1963; Lapwood *et al.*, 1972).

One baboon of the group of four studied by Kraemer & Vera Cruz (1969) ejaculated spontaneously on three occasions during preparation for electroejaculation and the characteristics of these ejaculates were compared with 13 ejaculates obtained after electrical stimulation of the same animal. Again the method of semen collection did not significantly affect semen characteristics. Shump, Aulerich *et al.* (1976) compared the mean number of spermatozoa collected per electrically induced ejaculation of the ferret with estimates of number of spermatozoa found *post mortem* in the cauda epididymides and vasa deferentia (Chang, 1965) and concluded that emission of a high proportion of spermatozoa in these regions occurred at electroejaculation.

STUDIES OF THE ELECTROEJACULATION OF RAMS

Introduction

The major factors influencing the output of semen during electro-ejaculation appear to be amplitude of the stimulus, position of the electrodes, duration of periods of stimulation and rest (stimulation cycle) and characteristics of the electrical pulses used (wave form and frequency). In this section, results are presented from experiments which tested various combinations of some variables within each of these factors.

Materials and Methods

Stimulators

Both circuits used in these experiments were designed and constructed in the Department of Veterinary Physiology, University of Sydney.

The circuit for generating pulses of square wave DC was that used by Martin & Rees (1962). It was a discrete component circuit employing 10 transistors grouped in three functional stages of pulse generation (Schmitt trigger), square wave shaping and power output. The second stimulator was designed so that square wave and sine wave AC pulses could be compared. Integrated circuit modules were used extensively in this stimulator which contained a Wien bridge sine wave oscillator driving a voltage pre-amplifier followed by a pair of power amplifiers. Square wave pulses were generated by feeding the output from the Wien bridge into a Schmitt trigger square wave oscillator and then to the voltage pre-amplifier. Both stimulators were powered by lead-acid accumulators (12 V for the DC circuit and 24 V with a central common lead at 12 V for the AC circuit). Output voltage was restricted to 10 V (20 V, peak-to-peak, in the AC circuit) in both cases. All square wave pulses were set at 0·5 mark space ratio for these experiments.

Three bipolar rectal electrodes were used in these experiments and all were the same diameter (20 mm). On the four-ring probe, the electrodes were 50 mm apart and connected so that adjacent rings were opposite in polarity. The pair of electrodes on the two-ring probe were 90 mm apart. The four electrodes on the strip probe were spaced at 90° around the probe and opposite pairs of electrodes had the same polarity. All probes had the same total surface area of electrodes (1500 mm^2).

Experimental animals

Merino rams, ranging in age from two to six years, were used in these experiments. A different group of rams was used for each of the experiments based on Latin Squares. In the course of the 11 experiments, 138 rams were electroejaculated. Where rams were used repeatedly in Latin Square experimental designs, there was an interval varying from three days to one week between collections. All rams were given the full period of stimulation required by the design even if they ejaculated at the first or second stimulus.

Appraisal of semen

The volume of each ejaculate was measured. Concentration of spermatozoa in each ejaculate was derived from the optical density of a sample of the semen diluted 200-fold in buffered formol saline with a calibrated colorimeter (Unicam SP.1300, Pye Unicam).

Experimental design and analyses of data

Latin Squares were frequently used in the experimental designs

(Experiments 1 to 6 and 10). Thus, individual rams were represented in the "rows" of the square and the successive collection days by the "columns".

The index of total output of spermatozoa was 1000 (volume of semen x optical density of diluted semen). Preliminary analyses showed significant heterogeneity of sample variances so that the final analyses of variance were performed on the transform, $\log_{10}(\text{index} + 1)$. Accordingly, geometric means have been calculated for the summaries of results.

Results

Experiments using the square wave DC stimulator and ring electrodes

Mean responses in Experiments 1 and 2 are summarized in Table II. Both experiments were factorial designs conducted as Latin Squares. In the first experiment it was found that responses did not differ significantly when two or four-ring electrodes were used. There was a highly significant difference between the two methods of application of the

TABLE II

Mean output of spermatozoa from rams stimulated with direct current, square wave pulses

Experiment No.	Factors and levels		Mean output of spermatozoa ($\times 10^9$)
1	2^3 factorial design using 8 rams in an 8 × 8 Latin Square		
	Rectal probe	Two-ring electrodes	5·92
		Four-ring electrode	4·27
	Stimulus	10 V, 10 s on, 10 s off	6·94[c]
		Slow rise to 10 V followed by slow fall to 0 in 20 s	3·25[c]
	Frequency	40 pulses s^{-1}	6·59[b]
		80 pulses s^{-1}	3·60[b]
2	2 × 3 factorial design using 6 rams in a 6 × 6 Latin Square. Cycle pattern of stimuli, 20 s on, 40 s off for 4 min		
	Frequency	40 pulses s^{-1}	7·36[a]
		80 pulses s^{-1}	4·09[a]
	Volts	5	3·41[b]
		7·5	5·81[b]
		10	7·96[b]

[a] $P < 0.05$.
[b] $P < 0.01$.
[c] $P < 0.001$.

stimulus. More than twice the number of spermatozoa were collected when the stimulus was switched on at 10 V for 10 s and off for 10 s four times than when the voltage was raised to 10 V in 10 s and then reduced to zero in the next 10 s repeating this cycle four times. Stimuli at 40 pulses s^{-1} gave a significantly greater output of spermatozoa than those delivered at 80 pulses s^{-1}. The same order of response to these frequencies was observed in Experiment 2 in which the two-ring electrode was used. Additionally, the total output of spermatozoa increased significantly as the voltage was raised from 5 to 10 V. In this experiment stimuli were applied rapidly in a cycle pattern of 20 s on and 40 s off for four minutes.

The two-ring electrode was used in Experiments 3 to 7.

Mean responses in Experiments 3 and 4 are shown in Table III. In both experiments, rams were stimulated in a 2-min test period with pulses of 8 V applied rapidly, but the stimulation to rest ratio was varied. It is clear that 20 s on and 40 s off, repeated once, was not satisfactory. There were no significant differences in response to the pulse frequencies tested in these experiments.

Table IV contains the mean responses in Experiment 5 in which pairs of rams were electroejaculated using 8 or 10 V stimuli at 40 pulses s^{-1} applied rapidly. One week elapsed between days of collection in this Latin

TABLE III

Mean output of spermatozoa from rams stimulated with direct current, square wave pulses[a]

Experiment No.	Factors and levels	Mean output of spermatozoa ($\times 10^9$)
3	Frequency	
	20 pulses s^{-1}	3·40
	40 pulses s^{-1}	4·60
	Cycle pattern of stimulation in 2-min test period	
	20 s on, 40 s off	1·26[b]
	10 s on, 20 s off	4·50[b]
	5 s on, 10 s off	5·60[b]
4	Frequency	
	32 pulses s^{-1}	3·55
	40 pulses s^{-1}	3·55
	50 pulses s^{-1}	4·13
	Cycle pattern of stimulation in 2-min test period	
	5 s on, 10 s off	4·32
	5 s on, 5 s off	3·17

[a] All stimuli were 8 V applied through a two-ring electrode rectal probe. Both experiments were 2×3 factorial designs using 12 rams, 2 to each "row" of a 6×6 Latin Square.
[b] $P < 0.01$.

TABLE IV

Experiment 5: mean output of spermatozoa per collection from rams electroejaculated using square wave DC stimuli at 40 pulses s^{-1} applied through a two-ring electrode rectal probe[a]

	Mean output of spermatozoa $(\times 10^9)$	
Factors and levels	Initial stimulation	4 h later
Duration of stimulation		
1 min	3.03^b	2.18
2 min	6.00^b	2.68
Amplitude of stimuli		
8 V	3.97	2.02
10 V	5.06	2.84

[a] Cycle pattern of stimulation in test period was 5 s on and 5 s off. 2^2 factorial design using 8 rams, 2 per "row" of a 4×4 Latin Square.
[b] $P < 0.01$.

Square but rams were stimulated twice, four hours apart, on each collection day. The output of spermatozoa was significantly increased by extension of the stimulation period from 1 min to 2 min. The longer period of stimulation had no significant effect when electroejaculation was repeated four hours later.

Experiment 6 was designed to test the possibility that an initial conditioning stimulus of low voltage would facilitate electroejaculation. Three rams were used in each row of a Latin Square design but responses to treatments did not differ significantly (Table V).

TABLE V

Experiment 6: mean output of spermatozoa per collection of rams electroejaculated with square wave DC pulses of 9 V amplitude and frequency 40 pulses s^{-1a}

Factors and levels	Mean output of spermatozoa $(\times 10^9)$
Preliminary "conditioning" stimulus	
None	4.80
3 V for 1 min	5.39
Cycle pattern of 9 V stimulus in 1-min test period	
2.5 s on, 2.5 s off	5.45
5 s on, 5 s off	4.74

[a] 2^2 factorial design using 12 rams, 3 per "row" of a 4×4 Latin Square. Two-ring electrode rectal probe used.

Table VI shows the sequence in which semen was collected from four pairs of rams by either electroejaculation or artificial vagina (Experiment 7). On a collection day, two ejaculates were collected four hours apart from each ram. Significantly fewer spermatozoa (overall mean, 3.07×10^9) were produced per collection by electroejaculation than by artificial vagina (4.95×10^9) ($P < 0.01$). The sequence of methods of collection did not have any significant effect, but there were highly significant differences in response to electroejaculation between rams ($P < 0.01$).

TABLE VI

Experiment 7: mean output of spermatozoa ($\times 10^9$) per semen collection, made with electroejaculation (EE) or artificial vagina (AV)

	Week			
	1	2	3	4
Ram pair				
a	EE 3·93	EE 3·07	AV 4·74	AV 4·48
b	EE 3·81	AV 5·25	EE 2·77	AV 3·77
c	AV 5·43	EE 1·30	AV 4·64	EE 0·85
d	AV 5·48	AV 5·77	EE 3·93	EE 4·92

Comparison of strip and ring rectal electrodes

The two-ring electrode and the strip electrode rectal probes were compared in Experiment 8 which was, in design, a four-point assay employing two levels of voltage. Nine rams were randomly allocated to each treatment. Stimuli were applied in a cycle of 5 s on in every 10 s for 2 min and frequency was 40 pulses s^{-1}. Figure 2 shows that significantly more spermatozoa were ejaculated in response to stimulation at the higher voltage ($P < 0.05$) and use of the strip electrode gave a greater spermatozoal output than the ring electrode at either voltage ($P < 0.05$).

Comparison of some wave forms of stimulus using the strip electrode

In Experiment 9, four rams were allocated at random to three treatment groups and were each ejaculated twice with an interval of one week between collections. Stimulation was applied for 2 min using a stimulus

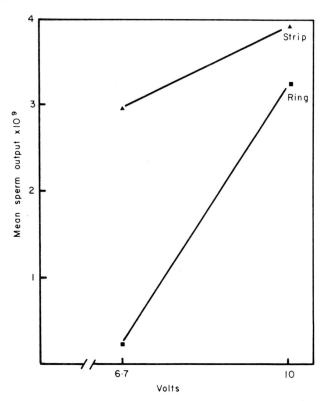

FIG. 2. (Experiment 8). Mean output of spermatozoa using the strip or two-ring rectal probes. Each point is the mean for ejaculates from nine rams. The ejaculating stimulus was square wave DC pulses used at 40 pulses s^{-1} and 0·5 mark space ratio.

cycle of 5 s on in every 10 s and a frequency of 40 pulses s^{-1}. Table VII shows the mean output of spermatozoa in response to a 10 V square wave DC pulse, and to two levels of a differentiated wave form generated from the square wave DC pulse. The peak-to-peak voltages delivered in this form were 12 and 20 V respectively. The differentiated wave form was obviously inferior to the square wave pulse.

A Latin Square design was used in Experiment 10 in which six rams were electrically stimulated with square wave DC pulses, square wave AC pulses and sine wave AC pulses at either 5 or 8 V. Stimuli were applied for 5 s in every 10 s for 2 min and frequency was 40 Hz. Alternating currents produced significantly more spermatozoa in the ejaculates than DC pulses (Fig. 3). There was also a highly significant increase in the output of spermatozoa when the voltage of the DC stimulus was raised from 5 to 8 V. Eight volts DC appeared to be equivalent in effect to 5 V AC (i.e. 10 V peak-to-peak).

TABLE VII

Experiment 9: mean output of spermatozoa per electro-ejaculation using a stimulus of 40 pulses s^{-1} [a]

Wave form (ref. Fig. 1)	Voltage (V)	Mean output of spermatozoa ($\times 10^9$)
F. square	10	8·63
H. differentiated	6	0·51
H. differentiated	10	1·75

[a] Four rams per treatment, each ejaculated twice with an inverval of one week between collections.

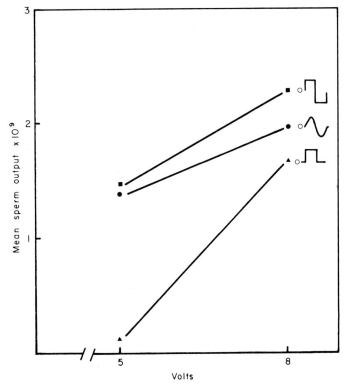

FIG. 3. (Experiment 10). Mean output of spermatozoa per collection for rams electro-ejaculated with stimuli of frequency 40 s^{-1}. ⊙⎍, Square wave AC; ⊙ ⌁ , sine wave AC; ⊙⊓ , square wave DC (both square waves at 0·5 mark space ratio). Means from six rams used in a 6×6 Latin Square design.

In the last experiment, three rams were randomized to each of six treatment groups and, as in Experiment 9, the rams were ejaculated twice at an interval of one week. Frequency and stimulus cycle were the same as were used in Experiment 10. Figure 4 shows the responses to three levels of voltage of square wave and sine wave current. There was a highly significant common linear response to increasing voltage and the square wave was just significantly better than the sine wave ($P < 0.05$).

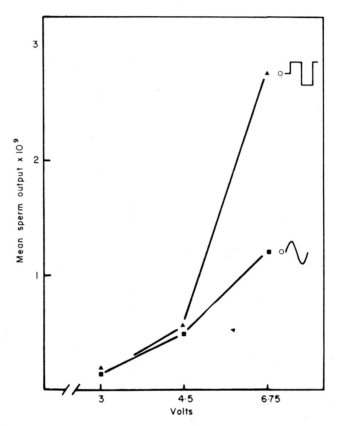

FIG. 4. (Experiment 11). Mean output of spermatozoa from rams electroejaculated with alternating currents of frequency $40\ \mathrm{s}^{-1}$. ⌐⊔, square wave; ∘⋀ , sine wave. Means from two ejaculates from each of three rams randomly allocated to each of the levels of treatment.

Discussion

The ram has proved to be a particularly valuable animal type in which to test techniques of electroejaculation. A high spermatozoal production rate is characteristic of rams and spermatogenesis is not greatly influenced by seasonal changes. They are highly responsive to electro-ejaculation, do not appear to be distressed by stimulation and, accordingly, do not need sedation or anaesthesia. A ram's mating drive is usually also high, so that rams can be trained quickly to serve an artificial vagina. Experiment 7 has shown that the two methods of collection can be employed on the same ram without either technique affecting the other.

It appears from the range of research reviewed earlier in this paper, that techniques of electroejaculation which are satisfactory for the ram and bull can be applied to other species. This generality can be extended over a wide range of mammalian species with, perhaps, the exception of the marsupials (and monotremes) (Rodger and White, this volume, p. 289). Seasonality of spermatogenesis will determine the best time for electroejaculation of wild species. Restraint and sedation of wild animals during electroejaculation are clearly also important. Fortunately, several very useful drugs and mixtures of drugs are already available.

The experiments using square wave DC pulses with mark space ratio 0·5 indicated that frequencies between 32 and 50 pulses s^{-1} gave highest output of spermatozoa, which agreed with the findings of Martin & Rees (1962) who found 40 pulses s^{-1} to be most effective for the electroejaculation of bulls. Direct current pulses of high frequency (1000 pulses s^{-1}) have been used by Freund (1969) to ejaculate guinea-pigs, but most reports of the use of low frequency DC pulses (Warner et al., 1974; Furman et al., 1975) have concluded that such pulses were less satisfactory than sine wave AC. In these cases, however, the pulse widths were small relative to the interval between pulses. Reports of the use of sine wave AC showed that frequencies between 20 and 60 Hz were satisfactory (Dziuk, Graham & Petersen, 1954; Marden, 1954; Dowling, 1961).

In the experiments presented here on the response of rams, it appears that power delivered per pulse was the most important factor. The simplest representation of this is the area contained by the wave form and the zero volt baseline. Power can be increased in the DC wave form by increasing the mark space ratio or by increasing amplitude. However, above a mark space ratio of 0·5, stimuli become progressively less effective until no ejaculatory response is produced by continuous DC voltages. It seems that, with these longer pulses, the animal responds to the "off" time of the cycle. In all of these experiments, amplitude has been restricted to a maximum of 10 V, as higher voltages applied to the circuits used would have given high current flow and a risk of damage to animal tissues. An alternating current with a square wave form and mark space ratio of 0·5 was carrying maximum power, but satisfied the requirement for an intermittent pulse for stimulation.

The power in a sine wave form can only be raised by an increase in amplitude. The root mean square (r.m.s. value) is often quoted in terms of power of a sine wave AC source and is 0·707 of the peak voltage. This means that a continuous DC voltage of 0·707×(AC voltage) would provide the same heating power as the sine wave AC. In the last experiment, the square wave AC at 10 V stimulated the emission of more spermatozoa than were produced by a sine wave having the same amplitude but less power. In both cases, where the DC pulse has been modified and its area reduced, either by use of the differentiated wave form or by reduction of the mark space ratio (Martin & Rees, 1962), spermatozoal output has fallen. It was also observed in Experiment 10 that the response to 8 V DC was equivalent to a 5 V AC stimulus, which, again, is most easily interpreted in terms of power rather than amplitude. During the sine wave cycle, current flow rates rise and fall which could induce a degree of asynchrony in firing of fibres in a nerve trunk according to the excitability of individual nerves. This is possibly an advantage over the square wave stimulus, in which current flow and potential for stimulation is maximal at the moment the pulse is generated.

Warner *et al.* (1974) presented values for current density at the electrode–tissue interface and considered that both in terms of animal safety and the ejaculation response, this value, which incorporates the variables of voltage and electrode–tissue resistance was the most informative expression of the stimulus. They found that ejaculation occurred in chimpanzees and orang-utans when the current density was of the order of 0.25 μA mm^{-2} and that gorillas ejaculated after a higher current density had been reached (mean of 0.42 μA mm^{-2}). The higher current density required was probably related to the use of the same rectal probe in all three species and the greater volume of tissue to be stimulated in the pelvis of the gorilla in comparison with the other apes. Future work with rams and bulls will include measurements of the variables of resistance and voltage to test this hypothesis.

The design of the rectal probe is also important and most workers have found strip electrodes to be more satisfactory than paired, or multiples of pairs, of ring electrodes. Although in the experiments described in this paper, total electrode surface area was kept constant, the distance separating the electrodes differed greatly from the ring to the strip type. Therefore, in position in the rectum, the strip probe had a lower resistance than the ring probe and accordingly, for the same voltage applied, current densities would have been greater. Substantial experimental work is needed to define optimal conditions for the generation of a suitable current flow.

Relatively short periods of stimulation and rest, of the order of 5 s each, have given significantly higher responses than longer periods of stimulation. It is suggested that the stimulus causes emission of semen into the urethra and then, during the rest period, the semen is propelled reflexly along the urethra. In contrast to the response of normal, intact animals, it is interesting to see the reaction of paraplegic men to electroejaculation (Thomas, McLeish & McDonald, 1975). In these cases, particularly in men with injury causing a cord transect at T12, no sciatic motor response was possible and sensation was limited to visceral afferents from the testes. Presumably, some of the sympathetic outflow involved in emission was also still intact, or else the stimuli affected the smooth muscle of the vas deferens and accessory sexual glands directly. However, semen was not collected unless supra-pubic compression (a technique used by paraplegics to empty the bladder) was applied after stimulation. In these cases, some contamination of the semen with urine seems unavoidable.

Although no improvement in sperm output was observed when rams were given a low voltage "conditioning" stimulus, further experiments on stimulus patterns other than simply "on" and "off" appear to be warranted (Ball, 1976). There are many possibilities for the use of integrated circuits to control automatically the application, maintenance and reduction of the stimulus.

Further research is needed on both stimulus patterns and electrode locations so that the nerves of the autonomic system concerned with erection and ejaculation can be selectively stimulated with minimal involvement of somatic nerves and skeletal muscle. To date, it has been demonstrated that, by modifying probe designs, and positioning of electrodes on them, stimuli causing urination can be minimized and erection can be stimulated in a high proportion of animals. However, a technique of electroejaculation which gives reliable synchronization of erection and emission has yet to be developed. Eccles (1955) showed that the sympathetic ganglia responded most efficiently to stimulus frequencies under 40 impulses s^{-1} and concluded that the majority of frequencies transmitted by autonomic ganglia lay between 12 and 30 impulses s^{-1}. In a review of the function of the autonomic nervous system, Koizumi & Brooks (1974) stated that tonic activity in fibres could be as low as 1 impulse $10 s^{-1}$ and were typically of the order of 3 to $5 s^{-1}$. However, when activated, volleys of impulses in both parasympathetic and sympathetic nerve fibres rose to frequencies of 20 to $30 s^{-1}$. The trend in research on electroejaculation has been to select stimuli in this range of frequencies (Furman et al., 1975), but it must be acknowledged that motor neurones to skeletal muscles will also produce a strong response at these frequencies.

ACKNOWLEDGEMENTS

I am indebted to Professor C. W. Emmens for advice during the preparation of this paper.

The experiments reported in this paper were supported by funds from the Australian Meat Research Committee and the Australian Research Grants Committee.

REFERENCES

Aulerich, R. J., Ringer, R. K. & Sloan, C. S. (1972). Electroejaculation of mink (*Mustela vison*). *J. Anim. Sci.* **34**: 230–233.

Austin, J. W., Hupp, E. W. & Murphree, R. L. (1961). Comparison of quality of bull semen collected in the artificial vagina and by electroejaculation. *J. Dairy Sci.* **12**: 2292–2297.

Baker, R. D. & Dziuk, P. J. (1964). Semen composition and ejaculation of the boar after pilocarpine administration. *J. Reprod. Fert.* **8**: 255–256.

Baker, R. D., VanDemark, N. L., Graves, C. N. & Norton, H. W. (1964). Effects of pilocarpine and atropine on copulatory behaviour, ejaculation and semen composition in the bull. *J. Reprod. Fert.* **8**: 297–303.

Ball, L. (1976). Electroejaculation. In *Applied electronics for veterinary medicine and animal physiology*: 394–441. Klemm, W. R. (ed.). Springfield: Charles C. Thomas.

Ball, L. & Furman, J. W. (1972). Electroejaculation of the bull. *Bovine Practitioner* No. 7: 46–48.

Batelli, F. (1922). Une méthode pour obtenir l'émission complète du liquide des vésicules séminales chez le cobaye. *C.r. Soc. Phys. hist. Nat. Genève* **39**: 73–74.

Benham, T. A. & Enders, R. K. (1941). An improved stimulator for obtaining semen from small mammals. *N. Am. Vet.* **22**: 300–301.

Blackshaw, A. W. (1954). A bipolar rectal electrode for the electrical production of ejaculation in sheep. *Aust. vet. J.* **30**: 249–250.

Chang, M. C. (1965). Fertilizing life of ferret sperm in the female ferret. *J. exp. Zool.* **158**: 87–99.

Dalziel, C. F. & Phillips, C. L. (1948). Electric ejaculation. Determination of the optimum electric shock to produce ejaculation in chinchillas and guinea pigs. *Am. J. vet. Res.* **9**: 225–232.

Denny, M. J. S. (1974). Anaesthesia in kangaroos. *Aust. Mammal.* **1**: 294–298.

Dewsbury, D. A. (1972). Patterns of copulatory behavior in male mammals. *Q. Rev. Biol.* **47**: 1–33.

Dowling, D. F. (1961). Electrical stimulation of ejaculation in the bull. *Aust. vet. J.* **37**: 176–181.

Dzuik, P. J., Graham, E. F., Donker, J. D. Marion, G. B. & Petersen, W. E. (1954). Some observations on the collection of semen from bulls, goats, boars and rams by electroejaculation. *Vet. Med.* **49**: 445–458.

Dzuik, P. J., Graham, E. F. & Petersen, W. E. (1954). The technique of electro-ejaculation and its use in dairy bulls. *J. Dairy Sci.* **37**: 1035–1041.

Dzuik, P. J. & Norton, H. W. (1962). Influence of drugs affecting the autonomic nervous system on seminal ejaculation. *J. Reprod. Fert.* **4**: 47–50.

Eccles, R. M. (1955). Intracellular potentials from a mammalian sympathetic ganglion. *J. Physiol.* **130**: 572–584.

Edgar, D. G., Inkster, I. J. & MacDairmid, H. J. (1956). An improved method for the collection and evaluation of ram semen. *N.Z. vet. J.* **4**: 20–24.

Freund, M. (1958). Collection and liquefaction of guinea pig semen. *Proc. Soc. exp. Biol. Med.* **98**: 538–540.

Freund, M. (1969). Interrelationships among the characteristics of guinea pig semen collected by electroejaculation. *J. Reprod. Fert.* **19**: 393–403.

Furman, J. W., Ball, L. & Seidel, G. E. (1975). Electroejaculation of bulls using pulse waves of variable frequency and length. *J. Anim. Sci.* **40**: 665–670.

Gunn, R. M. C. (1936). Fertility in sheep. Artificial production of seminal ejaculation and the characters of the spermatozoa contained therein. *Bull. Coun. scient. ind. Res. Melb.* No. 94: 1–116.

Guttman, L. (1973). *Spinal cord injuries. Comprehensive management and research.* Oxford: Blackwell Scientific.

Healey, P. & Weir, B. J. (1967). A technique for electroejaculation of chinchillas. *J. Reprod. Fert.* **13**: 585–588.

Hoskins, D. D. & Patterson, D. L. (1967). Prevention of coagulum formation with recovery of motile spermatozoa from rhesus monkey semen. *J. Reprod. Fert.* **13**: 337–340.

Hovell, G. J. R., Ardran, G. M., Essenhigh, D. M. & Smith, J. C. (1969). Radiological observations on electrically-induced ejaculation in the ram. *J. Reprod. Fert.* **20**: 383–388.

Hulet, C. V., Foote, W. C. & Blackwell, R. L. (1964). Effect of natural and electrical ejaculation on predicting fertility in the ram. *J. Anim. Sci.* **23**: 418–424.

Jones, R. C. (1973). Collection, motility and storage of spermatozoa from the African elephant *Loxodonta africana. Nature, Lond.* **243**: 38–39.

Kedia, K. R. & Markland, C. (1976). The ejaculatory process. In *Human semen and fertility regulation in men*: 497–503. Hafez, E. S. E. (ed.). St. Louis: C. V. Mosby.

Koizumi, K. & Brooks, C. McC. (1974). The autonomic nervous system and its role in controlling visceral activities. In *Medical physiology* (13th edn): 798–799. Mountcastle, V. B. (ed.). St. Louis: C. V. Mosby.

Kraemer, D. C. & Vera Cruz, N. C. (1969). Collection, gross characteristics and freezing of baboon semen. *J. Reprod. Fert.* **20**: 345–348.

Laplaud, M. & Cassou, R. (1945). Nouveau procède de recolte du sperme par électrode bipolaire rectale unique. *C.r. Acad. agric. Fr.* **31**: 37.

Laplaud, M. & Cassou, R. (1948). Recherches sur l'électro-éjaculation chez le taureau et le verrat. *C.r. Séance Soc. Biol.* **142**: 726–727.

Lapwood, K. R., Martin, I. C. A. & Entwistle, K. W. (1972). The fertility of Merino ewes artificially inseminated with semen diluted in solutions based on skim milk, glucose or ribose. *Aust. J. agric. Res.* **23**: 457–466.

Lundberg, P. O. (1976). Sexual function in men with neurological disorders. In *Human semen and fertility regulation in men*: 504–512. Hafez, E. S. E. (ed.). St. Louis: C. V. Mosby.

Marberger, H. (1974). The mechanisms of ejaculation. In *Basic life sciences*. **4.** *Physiology and genetics of reproduction. Part B*: 99–110. Hollaender, A. E. (ed.). New York, Plenum Press.

Marden, W. G. R. (1954). New advances in electroejaculation of the bull. *J. Dairy Sci* **37**: 556–561.

Martin, I. C. A. & Rees, D. (1962). The use of direct current pulses for the electroejaculation of the bull. *Aust. vet. J.* **38**: 92–98.

Mastroianni, L. & Manson, W. A. (1963). Collection of monkey semen by electroejaculation. *Proc. Soc. exp. Biol. Med.* **112**: 1025–1027.

Mattner, P. E. & Voglmayr, J. K. (1962). A comparison of ram semen collected by the artificial vagina and electro-ejaculation. *Aust. J. exp. Agric. Anim. Husb.* **2**: 78–81.

Moore, C. R. & Gallagher, T. F. (1930). Seminal vesicle and prostatic function as a testis-hormone indicator: the electric ejaculation test. *Am. J. Anat.* **45**: 39–69.

Nichols, G. de la M. & Edgar, D. G. (1964). A transistorised rectal probe for ejaculating rams. *N.Z. vet. J.* **12**: 145–146.

Rock, K. C. (1976). Safe use of electronic equipment. In *Applied electronics for veterinary medicine and animal physiology*: 442–450. Klemm, W. R. (ed.). Springfield: Charles C. Thomas.

Rodger, J. C. & White, I. G. (1975). Electroejaculation of Australian marsupials and analysis of the sugars in the seminal plasma from three macropod species. *J. Reprod. Fert.* **43**: 233–239.

Roussel, J. D. & Austin, C. R. (1967). Enzymatic liquefaction of primate semen. *Int. J. Fert.* **12**: 288–291.

Roussel, J. D. & Austin, C. R. (1968). Improved electroejaculation of primates. *J. Inst. Anim. Techns* **19**: 22–32.

Salamon, S. & Morrant, A. J. (1963). A comparison of two methods of artificial breeding in sheep. *Aust. J. exp. Agric. Anim. Husb.* **3**: 72–77.

Scott, J. V. & Dziuk, P. J. (1959). Evaluation of the electroejaculation technique and the spermatozoa thus obtained from rats, mice and guinea pigs. *Anat. Rec.* **133**: 655–664.

Seager, S. W. J. (1976). Electroejaculation of cats (Domestic and captive wild Felidae). In *Applied electronics for veterinary medicine and animal physiology*: 410–418, Klemm, W. R. (ed.). Springfield: Charles C. Thomas.

Settlage, D. S. F. & Hendrickx, A. G. (1974). Electroejaculation technique in *Macaca mulatta* (rhesus monkey). *Fert. Steril.* **25**: 157–159.

Shump, A. U., Aulerich, R. J. & Ringer, R. K. (1976). Semen volume and sperm concentration in the ferret (*Mustela putorius*). *Lab. Anim. Sci.* **26**: 913–916.

Shump, A. U., Shump, K. A., Aulerich, R. J. & Ringer, R. K. (1977). Effect of electro-ejaculation frequency on semen characteristics in the ferret (*Mustela putorius*). *Theriogenology* **7**: 83–87.

Thomas, R. J. S., McLeish, G. & McDonald, I. A. (1975). Electroejaculation of the paraplegic male followed by pregnancy. *Med. J. Aust.* **1975(ii)**: 798–799.

Valerio, D. A., Ellis, E. B., Clark, M. L. & Thompson, G. E. (1969). Collection of semen from macaques by electroejaculation. *Lab. Anim. Care* **19**: 250–252.

Warner, H., Martin, D. E. & Keeling, M. E. (1974). Electroejaculation of the great apes. *Ann. Biomed. Engin.* **2**: 419–432.

Watson, P. F. (1976). Electroejaculation, semen characteristics and semen preservation of the Brindled gnu. *J. Reprod. Fert.* **47**: 123–126.

Weisbroth, S. & Young, F. A. (1965). The collection of primate semen by electroejaculation. *Fert. Steril.* **16**: 229–235.

Symp. zool. Soc. Lond. (1978) No. 43, 153–173

Semen Preservation in Non-Domestic Mammals

E. F. GRAHAM, M. K. L. SCHMEHL, B. K. EVENSEN and D. S. NELSON

University of Minnesota, St Paul, Minnesota, USA

SYNOPSIS

An extensive review of the literature and some original data are presented on frozen semen of the following animals: bear, camel, cat (domestic and wild, 10 species), chinchilla, deer, dog, elephant, fox, gnu, goat (wild), llama, mouse, mink, moose, oryx, primate (10 species), rabbit, sheep (wild), wolf, yak and zebra. References range from a statement of attempt to freeze semen to complete methodologies with fertility results. Buffer systems, cryo-protectants, methods and procedures, post-freeze recovery and fertility results are presented for original data and from the literature (where available). The application of these procedures to endangered and other species of zoological interest is discussed, together with the importance of viability assays, and the approaches and problems involved in establishing banks of viable frozen semen.

INTRODUCTION

Volumes of literature are available on frozen semen from humans, cattle, sheep and pigs but literature is scarce or non-existent for most species of wild and captive animals. In recent years, research on other domestic animals such as the dog, cat, mink and rabbit has been published which increases the knowledge of procedures and techniques applicable to semen preservation. This paper reviews the literature relating to some of these species and considers the available literature for captive wild animals. In addition, some previously unpublished information on chinchilla, mice and other miscellaneous mammals is presented. Finally, the importance of a bank of frozen semen of endangered species is discussed and some steps towards its establishment and use are considered.

REVIEW OF LITERATURE

Deer and Reindeer Semen (Table I)

Problems of inbreeding and complicated quarantine regulations have created a need for artificial insemination with frozen reindeer semen, but

TABLE I
Experimental extenders, cryoprotectants, freeze methods and fertility results for frozen deer and reindeer semen as compiled from the literature

Reference	Extender	Glycerol (%)	Yolk (%)	Freeze method	Motility post-thaw (%)	Fertility (%)
Dott & Utsi (1973)	Lactose	3	20	P	NA[b]	0
	Reconstituted skim milk	3·5	0			
	Glucose	2·7	30			
	Inositol	2·7	30			
Mkrtchyan & Deryazhentsev (1973)	Lactose-yolk	3·5	NA	P, S	NA	—
Mkrtchyan & Rombe (1973)	Lactose-citrate	3·5	20	S	15–40	36
Jaczewski et al. (1976)	Three extenders	NA	NA	P	35–40	—

P, Pellets; S, straws; NA, information is not available.

little research has been done. Dott & Utsi (1973) reported 16 inseminations of reindeer cows with undiluted, diluted and frozen semen. Semen
frozen in some diluents maintained motility when thawed but only two
cows inseminated with undiluted semen conceived. The poor conception
rate was blamed on the diluent, the presence of glycerol or the smaller
number of spermatozoa reaching the cervix with diluted semen. Plastic
straws yielded better post-thaw motility than pellets when reindeer semen
was frozen by Mkrtchyan & Deryazhentsev (1973). Mkrtchyan & Rombe
(1973) reported that four out of 11 (36%) reindeer cows conceived
following insemination with frozen semen. Most cows were inseminated
twice, the second insemination following the first by eight or 12–14 hours.
Semen collected from red deer stags in autumn and frozen by Jaczewski *et
al.* (1976) maintained good progressive motility when thawed. This trial
indicated that good quality semen can be collected electrically. The semen
was frozen for future trials (see Krzywinski & Jaczewski, this volume, p.
271).

Dog Semen (Table II)

Early studies showed that dog semen could be frozen with subsequent
recovery of motility. Gutierrez Nales (1957) first reported that canine
semen was successfully preserved for eight months in solid carbon dioxide
($-79°C$). Harrop (1961, 1962) found that an equilibration period of two
hours yielded better post-thaw recovery of dog semen than a 24-hour
period. Martin (1963a,b) reported that five-hour equilibration yielded
better results than no equilibration and that centrifugation and resuspension of frozen semen in a glycerol-free medium increased the post-
thaw recovery of motility. Foote (1964) also reported successful freezing
of dog semen. Seager (1969) first reported the birth of a litter of puppies
utilizing frozen semen. Gill *et al.* (1970) inseminated 12 bitches with
frozen dog semen but obtained no pregnancies. Van Gemert (1970)
inseminated one bitch with frozen semen with a litter of nine puppies
born. Andersen (1972a,b) reported no fertility following eight
intravaginal inseminations, but obtained a litter of three pups following
an insemination in the corpus uteri by means of laparotomy. Seager &
Fletcher (1973) found that litter sizes and sex ratios of 21 litters of puppies
from frozen semen did not differ from those of litters from fresh semen.
Andersen (1974) reported nine litters from 10 inseminations and (1975)
10 litters from 11 intrauterine inseminations with frozen dog semen.
Seager, Platz & Fletcher (1975) summarized the results of inseminations
over six years with frozen dog semen (1969–74) showing fertilities of
9·5%, 10·3%, 40·0%, 65·8%, 43·2% and 63·6% for 21, 29, 20, 38, 37 and
11 inseminations, respectively. Seager, Platz & Templeton (1975) found
that insemination with 50 million motile spermatozoa after thawing
resulted in an 85% conception rate when double inseminations were used.
Andersen (1976) reported 80% fertility for 20 bitches inseminated into

TABLE II

Experimental extenders, cryoprotectants, freeze methods and fertility results for frozen dog semen as compiled from the literature

Reference	Extender	Glycerol (%)	Yolk (%)	Freeze method	Motility post-thaw (%)	Fertility (%)
Gutierrez Nales (1957)	Citrate	5·5	37·5	DI	NA	Yes
Harrop (1961)	NA	NA	NA	NA	50	NA
Martin (1963a)	Yolk-Ringer	2, 4, 8	25	A–F	4–30	NA
	Yolk-citrate	2, 4, 8	25			
	Ringer-fructose-PO_4	8	0			
Martin (1963b)	Skim milk powder	7·5–10	0	A–F	36·7	NA
	Ringer-fructose-PO_4		0		17·7	
	citrate-fructose-PO_4		0		6·2	
Foote (1964)	Tris-yolk-glucose	11	20	A–DI	41	NA
	Yolk-citrate-glucose	8	20		27	
Seager (1969)	11% lactose	4	20	P	NA	NA
Gill et al. (1970)	Tris-citric acid-fructose	8·8	20	NA	40–50	0
Van Gemert (1970)	NA	NA	NA	P	25	100
Andersen (1972b)	Lactose-yolk	4·7	20	S	40–60	0
	Tris-citric acid-yolk	7	20			33
Andersen (1975)	Tris-fructose-citrate	8	20	S	50–70	91
Seager, Platz & Fletcher (1975)	11% lactose	4	20	P	NA	9·5
						10·3
						40·0
						65·8
						43·2
						63·6
Andersen (1976)	Tris-fructose-citric acid	8	20	S	50–70	80
Takeisi et al. (1976)	Milk-glucose	4	0	S	35–50	75

A, Ampule; DI, dry ice (solid carbon dioxide); F, controlled freezing apparatus; P, pellet form; S, straws in liquid nitrogen vapor; NA, information is not available.

the uterus with one to nine puppies per litter. Takeisi *et al.* (1976) reported that three of four bitches whelped after intracervical insemination with frozen semen stored for between two and 12 months. Seager (1976) reported that the 500 puppies born using frozen dog semen in his program compared closely in infant mortality, congenital defects, birth weight and sex ratios with naturally bred litters. In summary, the technology of dog semen preservation is practical and usable for intracontinental semen shipment, genetic semen banking or other applications.

Fox Semen (Table III)

Fox semen has been frozen only recently. Aamdal, Andersen & Fougner (1972) obtained no pregnancies following 14 intravaginal inseminations with good frozen semen (Trial 1), and they found that spermatozoa were not reaching the oviducts and uteri of the two females examined (Trial 2). No fertilization had taken place in seven inseminated females. Seven of 13 females conceived (54%) with surgical inseminations (Trial 3). Three females were killed and fertile eggs or fetuses found; four others gave birth to normal pups (mean 6·7 per litter). This trial demonstrated the need for intrauterine semen deposition. Fougner, Aamdal & Andersen (1973) developed a method of non-surgical intrauterine insemination, and reported 82% fertility: nine of 11 females conceived, seven gave birth to an average of eight pups, two aborted. In 1974, Aamdal described an apparatus for non-surgical intrauterine insemination. Thirty females were inseminated with frozen semen and 20 had litters. A single, rather than a double, insemination of 150 million spermatozoa reduced the percentage of fertile eggs recovered from females killed 144 hours after insemination. This was followed by a field trial (Aamdal, Nyberg & Fougner, 1976) with fresh and frozen semen inseminated by trained technicians. Conception rates varied from farm to farm, but with two inseminations of 100 or 150 million frozen spermatozoa, 77 of 107 females conceived (72%). The females had average-sized litters. Inseminations with fresh semen produced 79% fertility. Problems included the detection of heat and the timing of insemination.

Mink Semen (Table IV)

Mink semen has not been frozen successfully but the use of fewer males with high genetic potential would be advantageous. Ahmad, Kitts & Krishnamurti (1975a) froze mink semen in tris (tris, citric acid, glycine, fructose, glycerol and egg yolk), PVP (tris extender with polyvinyl pyrrolidone and caproic acid), milk (boiled and filtered with glycerol) and sodium citrate extender. Semen frozen in tris for two hours maintained 3–10% motility after thawing. Milk and sodium citrate buffers were unsuitable freezing media and PVP was unsuitable at any temperature. After storage for seven days at −196°C, motility decreased to <7% with

TABLE III
Experimental extenders, cryoprotectants, freeze methods and fertility results for frozen fox semen as compiled from the literature

Reference	Extender	Glycerol (%)	Yolk (%)	Freeze method	Motility post-thaw (%)	Fertility (%)
Aamdal, Andersen & Fougner (1972)	Lactose-yolk	4·7	20	S	60	0–54
	Tris-citric acid-fructose	7	20	S		
Fougner et al. (1973)	Tris-citric acid-fructose	8	20	S	55–65	82
Aamdal (1974)	Tris-citric acid-fructose	7	20	S	60–70	67
Aamdal, Nyberg & Fougner (1976)	Tris-citric acid-fructose	7	20	S	60	72

S, Straws.

TABLE IV
Experimental extenders, cryoprotectants and freeze methods for frozen mink semen as compiled from the literature

Reference	Extender	Glycerol (%)	Yolk (%)	Freeze method	Motility Post-thaw (%)
Ahmad et al. (1975a)	Tris-citric acid-fructose	3	10	S	3–10
	Tris-PVP-caproic acid	0	0	S	0
	Milk extender	3	0	S	0
Ahmad et al. (1975b)	Combinations of tris-citrate-milk	0–12·5	0–25	S	0–50

S, Straws.

tris buffer. In a second paper, Ahmad, Kitts & Krishnamurti (1975b) compared four extenders. The base buffer was tris, sodium citrate, citric acid, bactopeptone, glycine and fructose with yolk, glycerol and/or milk added for the remaining three extenders. Tris base–yolk–glycerol and tris base–yolk–glycerol–milk were unsuitable for storage at −196°C. With short pre-freeze storage and slow cooling, motilities of 12–19% were obtained after two hours at −196°C. High initial motility (>50%) led to better motilities after thawing (15–26%). Epididymal spermatozoa (initial motility 81%) yielded post-thaw motilities of 50% and 58% in the two freeze buffers. Epididymal spermatozoa were also more resistant to thermal shock than semen obtained from the vagina.

Non-Human Primate Semen (Table V)

Sadleir (1966) first reported successful freezing of non-human primate semen. Semen from a chimpanzee male was frozen with some motility after one to three days' storage at −196°C. Roussel & Austin (1967) reported 22–30% post-thaw motility for rhesus monkey semen and 25–30% for chimpanzee semen after three days' storage at −196°C. Results for the semen of other monkeys were: stumptail macaque, 15–30%; patas, 15–26%; African green, 20–31%. However, Ackerman & Roussel (1971) used semen from the above trial 6–12 months later for metabolic studies and reported less than 10% motile spermatozoa. Kraemer & Vera Cruz (1969) reported successful freezing of baboon semen with no difference in freezing ability between spontaneously ejaculated or electroejaculated semen. Leverage *et al.* (1972) were the first to record a pregnancy with frozen semen when one of 48 rhesus monkeys inseminated became pregnant but aborted. Cho & Honjo (1973) found 60% survival rates of frozen macaque spermatozoa after 20 weeks storage at −196°C. Cho, Honjo & Makita (1974) reported pregnancies from two of 13 inseminations of eight females with frozen macaque semen but both aborted six to eight weeks after insemination. Denis *et al.* (1976) froze squirrel monkey semen for use in timed fertilization studies with post-thaw motilities of 53·8 ± 17·1 (mean ± s.d.) and 51·8 ± 22·9 for two males. Fertility data was not given. In summary, offspring from frozen non-human primate semen seem possible in the near future but are not yet a reality.

Rabbit Semen (Table VI)

In 1950, Emmens & Blackshaw reported that rabbit spermatozoa survived freezing, and Smith & Polge (1950) were the first to report fertility with the use of frozen semen, one out of 53 eggs being fertilized. Smirnov (1951) reported 174 offspring from 61 inseminations with frozen semen. In another study, 13 offspring resulted from three does

TABLE V
Experimental extenders, cryoprotectants, freeze methods and fertility results for frozen non-human primate semen as compiled from the literature

Reference	Extender	Cryoprotectant	Final cryoprotectant concentration (%)	Yolk (%)	Freeze method	Motility post-thaw (%)	Fertility (%)
Sadleir (1966)	Sodium citrate	G	4, 7	20	NA	NA	NA
	Glucose	G	or 10	20			
	Skim milk powder	D	7	5			
Roussel & Austin (1967)	Sodium glutamate	G	14	20	A	30	—
Kraemer & Vera Cruz (1969)	Sodium glutamate	G	14	20	A	48	—
Leverage et al. (1972)	Bicarbonate-citrate-K$_2$HPO$_4$	G	7	20	A	47	2
Cho & Honjo (1973)	Lactose-glucose-raffinose	G	5	9	P	60	—
Cho et al. (1974)	Lactose-glucose-raffinose	G	5	9	P	NA	25
Denis et al. (1976)	11% lactose	G	4	20	P	53	—

D, Dimethyl sulfoxide; G, glycerol; A, ampule; P, pellet; NA, information is not available.

which conceived out of six inseminated with frozen semen (Fox, 1961). Fox & Burdick (1963) and Fox & Plummer (1964) investigated the effects of various extenders, cryophylactics and freeze methods on semen recovery after thawing. Sawada & Chang (1964) reported four out of nine rabbits pregnant with intravaginal insemination and five out of eight pregnant with intrauterine inseminations with frozen semen. Wales & O'Shea (1968) recorded a fertility of 25% with tris buffered Tyrode's extender and 50% fertility with skim milk powder extender, each with eight does inseminated. O'Shea & Wales (1969) showed that washing spermatozoa after thawing, and storage of frozen rabbit semen for six months, did not affect fertility. Stranzinger, Maurer & Paufler (1971) studied four methods of freezing rabbit semen and found no significant difference in fertility at artificial insemination between pellet-frozen and liquid semen. Higher fertility was recorded when inseminations were made five hours after injection of ovulating hormone rather than at the time of the injection. Maurer & Foote (1971), however, reported contradictory findings; fresh semen yielded significantly higher fertility than frozen semen (98·7% versus 92·9%) and the fertility for inseminations made at the time of the injection of luteinizing hormone was 81·1% whereas it had fallen to 75·3% for inseminations made five hours later. Freezing and storage of rabbit spermatozoa did not affect either their transport in the female tract or the subsequent cleavage of fertilized ova in culture when compared with fresh spermatozoa (Murdoch & O'Shea, 1973). Weitze, Hellemann & Krause (1976) inseminated a total of 124 does in five experiments with frozen semen and reported a fertility of 86·7% for pelleted semen (5·3 per litter), 94·7% for semen frozen in straws (6·4 per liter) and 75% for fresh semen controls. R. Andrieu & M. Courot (unpublished) found that 69–84% of does inseminated with frozen semen gave birth to young. In summary, there are apparently several extenders and freeze methods available for successful preservation of rabbit spermatozoa.

Semen of Miscellaneous Mammals

Semen has also been successfully frozen in other species (Table VII). Semen from the Arabian oryx (Rowlands, 1964), Grevy's zebra (Sadleir, 1966) and the brindled gnu (Watson, 1976) has been frozen with post-thaw recovery of motile spermatozoa. Frozen semen from Norfolk horn sheep (Jones & Hime, 1972) was used in an attempt to breed ewes, but none conceived. Jones (1973) and Jones, Bailey & Skinner (1975) reported procedures for the collection and freezing of elephant semen with recovery of motile spermatozoa after thawing. Nevill *et al.* (1976) used the frozen elephant semen for artificial insemination but the attempt was unsuccessful (P. F. Watson, pers. comm.). Seager, Platz & Hodge (1975) successfully froze wolf semen, and inseminations subsequently resulted in the birth of two litters of wolf cubs. Platz *et al.* (1976) reported the birth of

TABLE VI
Experimental extenders, cryoprotectants, freeze methods and fertility results for frozen rabbit semen as compiled from the literature

Reference	Extender	Cryoprotectant	Final cryoprotectant concentration (%)	Yolk (%)	Freeze method	Motility post-thaw (%)	Fertility (%)
Smith & Polge (1950)	Distilled water	G	10	0	NA	0	
	0·25% gum arabic	G	20	0		0	2
	0·1% methyl cellulose	G	30	0		0	
Emmens & Blackshaw (1950)	Alcohols and sugars (26)	E, P	7·5	0	A, T	25–30	NA
Smirnov (1951)	NA	NA	NA	NA	NA	5–30	NA
Fox (1961)	Na citrate, dextrose glycine, Na bicarbonate, KCl, citric acid	G	8	20	V-DI	NA	50
Fox & Burdick (1963)	4 variations of Fox (1961)	E, G	0–16	20	V-DI	Yes	NA
Fox & Plummer (1964)	Na citrate	NA	NA	Yes	V	NA	NA
Sawada & Chang (1964)	Tyrode's plus 20% tris	D	10–22·5	0–20	T-F	7–50	44
	Tyrode's plus 20% tris plus fructose and glycine						62
	Glycine-yolk-citrate	G	8	20		0–6	0

Wales & O'Shea (1968)	Tris-buffered Tyrode	D	1·4-8	NA	V-DI	3-4	25
	Ca-free Ringer-phosphate	E	NA				NA
	Skim milk powder	G	NA				50
O'Shea & Wales (1969)	Skim milk powder plus fructose	D	4-18	0	V-DI	Index 47·0-93·6	63
							46
							69
Stranzinger et al. (1971)	Tris-yolk-12·5% DMSO	G	3	NA	A	25-50	29·8
					PT		41·7
					S		25·0
					P		52·6
Rohloff & Laiblin (1976)	Tris-yolk	D	5	20	P	49	0
		G	1·3				
Weitze et al. (1976)	Tris-citric acid-glucose	D	4·5	20	P	60	86·7
		G	1		S	Loss	94·7
R. Andrieu & M. Courot (unpublished)	Buffer-glucose-yolk	P	NA	NA	S	NA	69-84
	Lactose-yolk	G	NA	NA	S	NA	

D, Dimethyl sulfoxide; E, ethylene glycol; G, glycerol; P, propylene glycol; A, ampule; P, pellet on dry ice; PT, PVC tubing; S, straws; T–F, test tubes in controlled freezer; V–DI, vials in dry ice-alcohol bath; NA, information is not available.

TABLE VII
Experimental extenders, cryoprotectants, freeze methods and selected fertility results for frozen semen compiled from the literature

Reference	Animal	Available information
Rowlands (1964)	Arabian oryx	Semen has been frozen
Sadleir (1966)	Grevy's zebra	Epididymal semen *post mortem* was frozen in citrate-yolk, glucose-yolk and skim milk-yolk extender with 5, 10 or 15% glycerol and 5, 10 or 15% DMSO. Semen extended in 5% DMSO—glucose extender showed >50% motility after 5 months storage
Jones & Hime (1972)	Norfolk horn sheep	Semen was frozen and used to inseminate Scottish half-bred ewes with no pregnancies
Jones (1973)	African elephant	Semen from 7 males was frozen in buffered citrate-yolk extender in straws with 2, 4, 8 or 16% glycerol or DMSO. 7% DMSO with 1% glycerol maintained the best post-thaw motility (10–60%, 44·3% ave)
Seager, Platz & Hodge (1975)	Timber wolf	Semen was frozen in lactose-yolk-glycerol diluent in pellet form on dry ice. Post-thaw motility ranged from 40 to 60%. Two litters of cubs have been born
Nevill et al. (1976)	African elephant	One female was inseminated with the semen frozen by Jones (1973). Pregnancy was not achieved.
Watson (1976)	Brindled gnu	Semen was frozen in glucose-phosphate and citrate extenders with 7·5% glycerol and 6·25–25·0% egg yolk and frozen in French straws. Semen frozen in citrate 6·25% yolk maintained the highest (60%) motility after 6 months storage
Platz et al. (1976)	Domestic cat	Semen was extended in egg-yolk-lactose glycerol buffer and pellet frozen on dry ice. One kitten has been born following AI with frozen semen
Seager & Platz (1976)	Wild cats (10 species) bear (2 species) camel primate (5 species) red fox and wolf	Semen was frozen in lactose-yolk-glycerol extender by pellet method on dry ice

the first kitten from frozen domestic cat semen. Seager (1974, 1977) and Seager & Platz (1976) reported collection and successful freezing of semen from many species of large cats, bears, wolves and non-human primates. To date, S. W. J. Seager & C. C. Platz (unpublished data) have electroejaculated 17 feline species, 12 carnivores, 10 non-human primates, 12 herbivores, three reptiles, one rodent and one bird species and had good recovery of spermatozoa after thawing (25–50% motile) for the following animals: African cheetah, African lion, baboon, clouded leopard, degu, Dorcas gazelle, Eld's deer, jaguar, leopard cat (70%), margay, meerkat, mongoose, North Chinese leopard, sloth bear, Speke's gazelle, spotted leopard, tamarin and timber wolf. They also obtained some post-thaw recovery of spermatozoa (1–20% motile) from the following species: Angola python, Bengal tiger, brown-headed spider monkey, Canadian lynx, cougar, Geoffroy's cat, moloch gibbon, ocelot, orang-utan, red fox, South American tapir, white-cheek gibbon and yellow-backed duiker. Post-thaw motility was zero for the following species: African elephant, Bactrian camel, bushbaby, golden cat, lesser panda and polar bear. No semen was obtained from the following animals: American black bear, Père David's deer, fallow deer, Siberian tiger, turtle and uakari monkey. Post-freeze data were not available on the following species: binturong, bobcat, hog-nose snake, lowland gorilla, palm civet and tree kangaroo.

UNPUBLISHED DATA

Chinchilla Semen

D. W. Schilling (unpublished) has frozen chinchilla semen with TEST buffer (Graham, Crabo & Brown, 1972) plus glucose, sodium citrate, vitamin B complexes, steroids, antibiotics and 25% egg yolk. Semen was collected directly into the extender, placed in 4°C cooler for 20 min, pellet-frozen on dry ice and stored at −196°C. Motility estimates after thawing were 15–50% live spermatozoa with 28–66% live spermatozoa using the filter assay of Graham, Vazquez *et al.* (1976). Glycerol was found to be detrimental to the recovery of motile spermatozoa.

Mouse Semen

G. L. Rapatz and K. J. Zimmerman (unpublished) have developed a method for the collection, dilution, freezing, storage, thawing and artificial insemination of mouse epididymal spermatozoa. Diluents were evaluated for their effect on pre-freeze storage of epididymal spermatozoa, for their capability of exerting a cryoprotective effect during freezing and storage of spermatozoal suspensions and for their effect on fertilizing capacity of collected spermatozoa. Epididymal spermatozoa suspended in diluents which did not inhibit their progressive motility were frozen and thawed over a wide range of cooling and rewarming rates to determine optimal conditions. The fertilizing capacity of the frozen-thawed spermatozoa from samples with the highest recovery of motile cells was tested by artificial insemination with receptive females.

Two groups of mice inseminated with frozen-thawed spermatozoa produced litters. When spermatozoa were suspended in a skim milk (10·5% solids) diluent containing 0·75% fructose, three animals out of a group of 106 gave birth to a total of nine young (3 per litter). In contrast 34 out of 143 animals inseminated with unfrozen samples of the same spermatozoa gave birth to a total of 324 young (9·5 per litter).

In a second group, spermatozoa were suspended in skim milk (10·5% solids) containing 0·75% fructose, 1% bovine serum albumin and 4 mM calcium chloride. Nine out of 84 gave birth to 61 young (6·8 per litter). In contrast 23 out of 102 animals inseminated with unfrozen samples of the same spermatozoa gave birth to 178 young (7·7 per litter).

The results demonstrated that frozen–thawed mouse epididymal spermatozoa retained some fertilizing capacity but conception rates were lower than when fresh spermatozoa were used. In some cases the litter size was also reduced.

Semen of Miscellaneous Mammals

Semen from many species of wild animals has been collected and frozen at the University of Minnesota (Table VIII). In many cases, animals were injured or dead so electroejaculation or flushing of the epididymis or vas deferens was used to collect the semen. Samples were small and only one buffer and freeze method was used for each animal. The samples were thawed, motilities estimated, and semen filtered using the method of Graham, Vazquez et al. (1976). Even without detailed studies to determine optimal conditions, semen quality was excellent for most species.

Future of Frozen Semen of Endangered Species

Yearly, more animals are added to the list of endangered species while others are dropped from the list because they have become extinct. We have been killing off our wild animals while making little progress on preserving existing animal species. Use of artificial insemination and frozen semen banking could help alleviate this problem. Some steps towards accomplishing this goal are elaborated below.

1. An interest in the preservation of species should be developed together with funding, both public and private, to support research. The general public should be made more aware both of species on the endangered and critically endangered lists and of the interests and capabilities of research workers. Private and governmental agencies should also be continuously informed of these interests and capabilities. In addition, information should be

made available on the actual and potential developments of a world effort for improving habitats and also for preserving germ plasm of all species for future recall. A worldwide cooperative effort is necessary because animals throughout the entire planet are in danger of becoming extinct.

2. Cooperative efforts with zoos should be encouraged in which reproductive tracts of dead, sick or injured males and females could be shipped to research workers for obtaining gametes and performing anatomical studies. Spermatozoa from an epididymis cooled to near 0°C can be removed up to 48 hours later and processed for freezing with excellent recovery. This method of banking spermatozoa would allow storage where the material is usually lost. Likewise, female ovaries could be classified as to ovarian development to establish breeding seasons, and ova could be recovered from fully developed follicles for preservation for future fertilization *in vitro* with preserved spermatozoa.

3. A systematic method should be developed for testing variables of the preservation process, such as buffer solutions, cryoprotective agents and cooling rates. A simple system would employ storage of prepared basic buffer systems including sodium citrate, organic buffers and milk, each containing either glycerol, ethylene glycol or dimethylsulfoxide in small known quantities. When gametes are available, vials of buffer could be quickly thawed for immediate gamete processing. It is well known that gametes from different species require different preservative systems. Different cooling rates may then be employed, thus accumulating comparative data by species. The gametes processed with the most optimal system can be stored for later insemination trials.

4. Standard methods could be developed and used for evaluating frozen spermatozoa. A simple method for post-thaw evaluation of preserved spermatozoa is suggested which gives opportunity to determine objectively: the initial concentration of spermatozoa, the total concentration of spermatozoa per insemination dose and the total numbers of motile spermatozoa per insemination dose. Many other evaluation methods are also reported in the literature. A standard and consistent system of evaluation techniques should be adopted by all researchers to facilitate comparison of results.

5. Artificial insemination tools and techniques need to be developed. Utilizing anatomical information from reproductive tracts of disposed animals, artificial insemination techniques and equipment could be developed for each animal species.

6. There is a need to obtain fertility data for spermatozoal numbers and the volume and site of deposition. Much of the information obtained so far is of little value because of the lack of this information. Each species probably requires different numbers of live spermatozoa per insemination dose and different volumes of

TABLE VIII
Methods of collection, preservation and post-thaw data of semen from wild animals

| Species | Collection | | Dilution Ratio | Sperm ml^{-1a} ($\times 10^6$) | Freeze | | Motility (%) | | Filter data[b] | |
	Month	Method			Buffer[c]	Method[d]	pre-freeze	post-freeze	Live (%)	Years
Llama (*Lama glama L.*)	May	E[e]	1:8	460	T-G-Y	P	50	45	30	8
Camel—dromedary (*Camelus dromedarius L.*)	Jan	E	1:10	345	R-G-Y	P	60	50	41	13
Yak (*Bos grunniens L.*)	June	E	1:5	573	T-G-Y	P	45	40	20	8
White-tailed deer (*Odocoileus virginianus*)	Dec	E	1:5	429	R-G-Y	S	60	40	38	12
Moose (*Alces alces L.*)	Oct	E	1:4	772	L-G-Y	P	35	20	8	7
Bison—American buffalo	Sept	E	1:10	394	T-G-Y	P	50	45	44	5
(*Bison bison L.*)		E	1:10	421	T-G-Y	P	50	40	32	5
Rocky mountain goat (*Oreamnos americanus*)	Dec	E	1:5	1026	T-G-Y	S	50	40	3	6
Markhor	Mar	E	1:4	1920	T-G-Y	S	55	40	34	6
(*Capra falconeri*)		E	1:4	3515	T-G-Y	S	50	40	30	6

Species	Month	Collection	Dilution	No.[a]	Buffer[c]	Freeze[d]				
Bighorn sheep (*Ovis canadensis*)	Dec	E	1:4	1362	T-G-Y	S	50	50	30	6
Mouflon (*Ovis musimon*)	Mar	E	1:4	3405	T-G-Y	S	50	35	22	6
		E		3615	T-G-Y	P	25	15	3	6
Dall sheep (*Ovis dalli dalli*)	Mar	E	1:4	1910	T-G-Y	S	50	35	27	6
		E	1:4	3295	T-G-Y	P	40	20	4	6
		Epididymal	1:1	8440	T-G-Y	P	40	20	19	6
Black bear (*Euarctos americanus*)	May	E	1:1	90	L-G-Y	P	35	10	—	—
Polar bear (*Thalarctos maritimus*)	May	E	1:1	120	L-G-Y	P	20	10	—	—
Leopard (*Panthera pardus*)	Dec	Vas flush	~1:5	203	T-G-Y	P	70	50	59	8
Chinchilla (*Chinchilla laniger*)	May	Massage	~1:5[f]	438	T-G-Y	P	85	10	32	1 day
		massage	~1:5[f]	203	S	P	80	35	41	1 day
		massage	~1:35	101	S	S	75	30	51	1 day
Little brown bat (*Myotis lucifugus*)	Feb	Female flush	~1:5	35	C-G-Y	P	75	60	—	—
		epididymal flush	~1:5	2020	C-G-Y	P	70	60	—	—

[a] Determined by Coulter counter, after dilution.
[b] Method by Graham, *Vazques et al.* (1976), years indicate time in storage before analysis.
[c] Buffer code: C, citrate; G, glycerol; L, lactose; R, raffinose; S, Schilling's extender; T, test; Y, yolk.
[d] Freeze method code: P, pellet; S, straw.
[e] E, Electroejaculation.
[f] $\bar{x}$ of 2 samples.

inseminate which also vary according to the site of semen deposition. Moreover, each animal has an optimum insemination time in regard to the onset of estrus. This information is necessary for the successful application of artificial insemination. Careful observations and records of observed natural breedings as well as attempts at artificial insemination could be invaluable towards accomplishing this goal.

A well-designed program can result in success and add to basic knowledge in regard to the preservation of germ plasm from mammals chosen for propagation.

REFERENCES

Aamdal, J. (1974). Insemination of the fox with deep frozen semen. *Proc. Nordic Vet. Congr. Reykjavik* **12**: 155–156.

Aamdal, J., Andersen, K. & Fougner, J. A. 1972. Insemination with frozen semen in the blue fox. *VII Int. Congr. Anim. Reprod. Artif. Insem.* [Munich] **2**: 1713–1716.

Aamdal, J., Nyberg, K. & Fougner, J. 1976. Artificial insemination in foxes. *VIII Int. Congr. Anim. Reprod. Artif. Insem.* (Krakow) **4**: 956–959.

Ackerman, D. R. & Roussel, J. D. (1971). Citric acid, lactic acid and oxygen metabolism of frozen-thawed semen from four subhuman primate species. *J. Reprod. Fert.* **27**: 441–443.

Ahmad, M. S., Kitts, W. D. & Krishnamurti, C. R. (1975a). Mink semen studies I. liquid preservation and prospect of freezing spermatozoa. *Theriogenology* **4**: 15–22.

Ahmad, M. S., Kitts, W. D. & Krishnamurti, C. R. (1975b). Mink semen studies II. freeze-preservation of spermatozoa. *Theriogenology* **4**: 77–98.

Andersen, K. (1972a). Fertility of frozen dog semen. *Acta vet. scand.* **13**: 128–130.

Andersen, K. [1972b]. Fertility of frozen dog semen. *VII Int. Congr. Anim. Reprod. Artif. Insem.* [Munich] **2**: 1703–1706.

Andersen, K. (1974). Intrauterine insemination with frozen semen in dogs. *Proc. Nordic Vet. Congr. Reykjavik* **12**: 153–154.

Andersen, K. (1975). Insemination with frozen dog semen based on a new insemination technique. *Zuchthyg.* **10**: 1–4.

Andersen, K. [1976]. Artificial uterine insemination in dogs. *VIII Int. Congr. Anim. Reprod. Artif. Insem.* (Krakow) **4**: 960–963.

Cho, F. & Honjo, S. (1973). A simplified method for collecting and preserving cynomolgus macaque semen. *Japan. J. med. Sci. Biol.* **26**: 261–268.

Cho, F., Honjo, S. & Makita, T. (1974). Fertility of frozen-preserved spermatozoa of cynomolgus monkeys. In *Contemporary Primatology, Int. Congr. Primat.* **5**: 125–133.

Denis, L. T., Poindexter, A. N., Ritter, M. B., Seager, S. W. J. & Deter, R. L. (1976). Freeze preservation of squirrel monkey sperm for use in timed fertilization studies. *Fert. Steril.* **27**: 723–729.

Dott, H. M. & Utsi, M. N. P. (1973). Artificial insemination of reindeer, *Rangifer tarandus. J. Zool., Lond.* **170**: 505–508.

Emmens, C. W. & Blackshaw, A. W. (1950). The low temperature storage of ram, bull and rabbit spermatozoa. *Aust. Vet. J.* **26**: 226–228.

Foote, R. H. (1964). Extenders for freezing dog semen. *Am. J. vet. Res.* **25**: 37–40.

Fougner, J. A., Aamdal, J. & Andersen, K. (1973). Intrauterine insemination with frozen semen in the blue fox. *Nord Vet. Med.* **25**: 144–149.

Fox, R. R. (1961). Preservation of rabbit spermatozoa: fertility results from frozen semen. *Proc. Soc. exp. Biol. Med.* **108**: 663–665.

Fox, R. R. & Burdick, J. F. (1963). Preservation of rabbit spermatozoa: ethylene glycol vs glycerol for frozen semen. *Proc. Soc. exp. Biol. Med.* **113**: 853–856.

Fox, R. R. & Plummer, B. H. (1964). One-half-dram glass vial as a container for frozen semen. *J. Anim. Sci.* **23**: 1234.

Gill, H. P., Kaufman, C. F., Foote, R. H. & Kirk, R. W. (1970). Artificial insemination of Beagle bitches with freshly collected, liquid-stored and frozen-stored semen. *Am. J. vet. Res.* **31**: 1807–1813.

Graham, E. F., Crabo, B. G. & Brown, K. I. (1972). Effect of some zwitter-ion buffers on the freezing and storage of spermatozoa. 1. Bull. *J. Dairy Sci.* **55**: 372–378.

Graham, E. F., Vazquez, I. A., Schmehl, M. K. L. & Evensen, B. K. [1976]. An assay of semen quality by use of sephadex filtration. *VIII Int. Congr. Anim. Reprod. Artif. Insem.* [Krakow] **4**: 896–899.

Gutierrez Nales, N. (1957). [Dilution and storage of canine semen.] *Revta Patron. Biol. anim.* **3**: 189–236. (*Anim. Breed. Abstr.* **26**: No. 197.) (In Spanish.)

Harrop, A. E. [1961]. Semen preservation and canine artificial insemination. *IV Int. Congr. Anim. Reprod. Artif. Insem.* [The Hague] **4**: 898–901.

Harrop, A. E. (1962). Artificial insemination in the dog. In *The semen of animals and artificial insemination*: 304–315. Maule, J. P. (ed.). Farnham Royal, England: Comm. Agric. Bur.

Jaczewski, Z., Morstin, J., Kossakowski, J. & Krzywiński, A. [1976]. Freezing the semen of red deer stags. *VIII Int. Congr. Anim. Reprod. Artif. Insem.* [Krakow] **4**: 994–997.

Jones, R. C. (1973). Collection, motility and storage of spermatozoa from the African elephant *Loxodonta africana*. *Nature, Lond.* **243**: 38–39.

Jones, R. C., Bailey, D. W. & Skinner, J. D. (1975). Studies on the collection and storage of semen from the African elephant *Loxodonta africana*. *Koedoe* **18**: 147–164.

Jones, R. C. & Hime, J. M. (1972). In Scientific Report 1969–1971 of the Zoological Society of London. Mammalian spermatozoa: preservation. *J. Zool., Lond.* **166**: 571–572.

Kraemer, D. C. & Vera Cruz, N. C. (1969). Collection, gross characteristics and freezing of baboon semen. *J. Reprod. Fert.* **20**: 345–348.

Leverage, W. E., Valerio, D. A., Schultz, A. P., Kingsbury, E. & Dorey, C. (1972). Comparative study on the freeze preservation of spermatozoa. Primate, bovine and human. *Lab. Anim. Sci.* **22**: 882–889.

Martin, I. C. A. (1963a). The freezing of dog semen to −79°C. *Res. vet. Sci.* **4**: 304–314.

Martin, I. C. A. (1963b). The deep-freezing of dog spermatozoa in diluents containing skim-milk. *Res. vet. Sci.* **4**: 315–325.

Maurer, R. R. & Foote, R. H. (1972). Effects of frozen semen and insemination time on early embryonic development in the rabbit. *Biol. Reprod.* **7**: 103.

Mkrtchyan, M. E. & Deryazhentsev, V. I. (1973). (An experiment on deep freezing of reindeer semen.) *Sb. nauch. Rab. murmansk. olenevodch. opyt. Stn.* **2**: 34–36. (*Anim. Breed. Abstr.* **41**: No. 630.) (In Russian.)

Mkrtchyan, M. E. & Rombe, S. M. (1973). (Artificial insemination of reindeer with deep frozen semen.) *Zhivotnovodstvo* **6**: 72–74. (In Russian.)

Murdoch, B. E. & O'Shea, T. (1973). Effect of storage of rabbit spermatozoa at −79°C on their subsequent transport and fertility in the rabbit doe. *Aust. J. biol. Sci.* **26**: 645–651.

Nevill, G. F., Crompton, W. G., Hennessy, M. A. & Watson, P. F. (1976). Instrumentation for artificial insemination in the African elephant. *Int. Zoo Yb.* **16**: 166–171.

O'Shea, T. & Wales, R. G. (1969). Further studies of the deep freezing of rabbit spermatozoa in reconstituted skim milk powder. *Aust. J. biol. Sci.* **22**: 709–719.

Platz, C., Follis, T., Demorest, N. & Seager, S. [1976]. Semen collection, freezing and insemination in the domestic cat. *VIII Int. Congr. Anim. Reprod. Artif. Insem.* [Krakow] **4**: 1053–1056.

Rohloff, D. & Laiblin, C. (1976). Deep freezing of rabbit spermatozoa. *Berl. Münch. tierärztl. Wschr.* **89**: 181–183.

Roussel, J. D. & Austin, C. R. (1967). Preservation of primate spermatozoa by freezing. *J. Reprod. Fert.* **13**: 333–335.

Rowlands, I. W. (1964). Artificial insemination of mammals in captivity. *Int. Zoo Yb.* **5**: 105–106.

Sadleir, R. M. F. S. (1966). The preservation of mammalian spermatozoa by freezing. *Lab. Pract.* **15**: 413–417.

Sawada, Y. & Chang, M. C. (1964). Motility and fertilizing capacity of rabbit spermatozoa after freezing in a medium containing dimethyl sulfoxide. *Fert. Steril.* **15**: 222–229.

Seager, S. W. J. (1969). Successful pregnancies utilizing frozen dog semen. *A. I. Digest* **17**: 6, 16.

Seager, S. W. J. (1974). Semen collection and artificial insemination in captive wild cats, wolves and bears. *Ann. Proc. Am. Assn Zoo vets* **1974**: 29–33.

Seager, S. W. J. [1976]. Freezing and transportation of dog semen. *VIII Int. Congr. Anim. Reprod. Artif. Insem.* [Krakow] **5**: 1251–1252.

Seager, S. W. J. (1977). A program of semen collection and freezing in captive wild Felidae. In *The world's cats* **3**: 112–119. Eaton, R. L. (ed.). Chicago: Zool. Soc.

Seager, S. W. J. & Fletcher, W. S. (1973). Progress on the use of frozen semen in the dog. *Vet. Rec.* **92**: 6–10.

Seager, S. W. J. & Platz, C. C. [1976]. Semen collection and freezing in captive wild mammals. *VIII Int. Congr. Anim. Reprod. Artif. Insem.* [Krakow] **4**: 1075–1078.

Seager, S. W. J., Platz, C. C. & Fletcher, W. S. (1975). Conception rates and related data using frozen dog semen. *J. Reprod. Fert.* **45**: 189–192.

Seager, S. W. J., Platz, C. C. & Hodge, W. (1975). Successful pregnancy using frozen semen in the wolf. *Int. Zoo Yb.* **15**: 140–143.

Seager, S. W. J., Platz, C. C. & Templeton, J. W. (1975). Canine genetics and frozen semen. *Transplantation Proc.* **7**: 571–573.

Smirnov, I. V. (1951). (The storage of livestock semen at a temperature of −78°–183°.) *Socialist Zivotn.* **1**: 94–95. (*Anim. Breed. Abstr.* **19**: No. 155). (In Russian.)

Smith, A. U. & Polge, C. (1950). Survival of spermatozoa at low temperatures. *Nature, Lond.* **166**: 668–669.

Stranzinger, G. F., Maurer, R. R. & Paufler, S. K. (1971). Fertility of frozen rabbit semen. *J. Reprod. Fert.* **24**: 111–113.

Takeisi, M., Mikami, T., Kodama, Y., Tsunekane, T. & Iwaki, T. (1976). Studies on reproduction in the dog. VIII Artificial insemination using frozen semen. *Jap. J. Anim. Reprod.* **22**: 28–33.

Van Gemert, W. (1970). Pups from deep-frozen semen. *Tijdschr. Diergeneesk.* (*Anim. Breed. Abstr.* **95**: Nos. 697–699).

Wales, R. G. & O'Shea, T. (1968). The deep freezing of rabbit spermatozoa. *Aust. J. biol. Sci.* **21**: 831–833.

Watson, P. F. (1976). Electroejaculation, semen characteristics and semen preservation of the Brindled gnu. *J. Reprod. Fert.* **47**: 123–126.

Weitze, K. F., Hellemann, C. & Krause, D. [1976]. Insemination with rabbit semen frozen in plastic straws. *VIII Int. Congr. Anim. Reprod. Artif. Insem.* [Krakow] **4**: 1100–1103.

Symp. zool. Soc. Lond. (1978) No. 43, 175–193

The Detection of Oestrus

F. D'SOUZA

Zoological Society of London, Regent's Park, London, England

SYNOPSIS

Several methods exist for the detection of either oestrus or ovulation, each having varying degrees of accuracy, practicality and predictive value. The most common methods are discussed in the context of their use with non-domestic rather than domestic or laboratory species. It is emphasized that both individual and interspecific female reproductive differences are of the greatest importance and that the initial use of several techniques and the subsequent construction of a reproductive cycle profile will, in the end, prove more informative than results derived predominantly from concentration on a single method.

INTRODUCTION

Artificial breeding of non-domestic animals has not, in the past decade, been as successful as the investment in research effort has warranted. The reasons for this relative lack of success are, of course, numerous and undoubtedly due to variables in both males and females. However, it could be argued that research techniques developed for the collection and storage of semen are more tried and tested than the methods dealing with prediction and monitoring of events in the reproductive cycle of the female. Thus for example in the male, erection and emission can generally be achieved by training the individual and/or by mechanical means; the quality of semen can be checked rapidly with the light microscope; abnormalities can often be rectified by methods other than hormonal therapy, e.g. a higher percentage of motile spermatozoa may be obtained by more frequent ejaculation; spermatozoal storage has a long and respectable history of success. There is, therefore, a sophisticated framework for experimentation within the field of male reproductive biology. At the risk of seriously offending the majority of contributors to this volume, the male side of artificial breeding presents problems to these workers but theirs would seem to be the easier task in that the problems are more amenable to rapid identification and solution. If, indeed, this is the case then the greater part of the comparative lack of success must be the fault of the female or even, perhaps, the fault of the female investigator!

This paper attempts a review of some causes of these female difficulties and suggests possible approaches and solutions, particularly when dealing with relatively unfamiliar non-domestic species.

Animals present a bewildering range of female reproductive responses when compared to the more restricted male function of spermatozoal production and insemination (Nalbandov, 1973): the females may be induced or reflex ovulators or they may ovulate spontaneously; they can show various degrees of behavioural and physiological receptivity at different stages in relation to release of the egg, which in itself implies a very complex interaction of sex hormones; the fertilized egg may implant within a few days of mating or be subject to long obligatory delays; placentation can range from the human invasive type to a relatively non-invasive means of embryonic and foetal sustenance. Generally however, oestrus, by whatever means it is achieved, is a stage in the reproductive cycle when the female is both physiologically and behaviourally receptive to the male. For the purposes of fertilization, the crucial event is ovulation and this is where major detection problems arise. While oestrus itself is, depending on the species concerned, usually heralded and accompanied by a variety of morphological, physiological and behavioural changes, the actual occurrence of ovulation is far more obscure. Indeed the ovulatory event is mysterious enough for one investigator to have said that it must occur "at night and always in the dark" (Keefer, 1965).

In spite of the vast body of research literature which has accumulated over the decades, there is as yet no single, rapid and unequivocal ovulatory test, let alone a method which reliably predicts ovulation. Since this is true of the two main experimental "animal models"—the laboratory rodent and the human female—it is even more applicable to those species which fall into the "non-domestic" category. The causes of this state of affairs would appear to be two-fold

1. not enough attention has been given to individual variation between animals of the same species and between the cycles of one individual (Bosu, Johansson & Gemzell, 1973).

2. interspecific variation, though acknowledged, is in fact obscured by the common practice of extrapolating not only methods but their results from one species to another.

The battery of tests and procedures for estimating receptive periods and ovulation were developed predominantly in rats, monkeys and humans. Furthermore, the results of these investigations have, more often than not, been tailored to give the so-called "normal" picture. Students therefore learn that the female rat has an oestrous cycle of 4·5 days and that there are well-defined stages within this cycle, easily detectable, for example, by changes in vaginal cytology and/or fluctuations in levels of gonadotrophins in the blood plasma. In the case of the human female it is accepted that ovulation occurs spontaneously at the end of the follicular phase and that this event is part of a delicately balanced, yet predictable and controlled physiological process. In fact, cycle length in the rat can vary from three to nine days and individuals may not show some of the classic signs of oestrus and yet be normally fertile (Mandl, 1951). More surprising are the findings suggesting that the human female can conceive at any point in the cycle since copulation itself can be a sufficient trigger mechanism for ovulation at the most unlikely times (Jöchle, 1970).

Nevertheless laboratory tests developed for certain species have been applied indiscriminately to non-rodent, non-primate species. The absence of clearcut results has been taken as an indication of abnormality. Unfortunately however, non-domestic species do not often behave in a textbook fashion and some of the disappointing results of co-ordinating insemination with ovulation must be attributable to the fact that many mammalian species are not, on the whole, uniform.

Deliberate reference has been made in the title of this paper to detection of oestrus rather than simply detection of ovulation since, although the end point of any investigation into the female side of artificial insemination is the determination of ovulation, the events which precede and follow it may be of immense importance. The mere fact of ovulation need not be sufficient for fertilization to take place; the quality of oestrus is probably highly significant. For this reason some baseline knowledge on what may be called a reproductive cycle profile has to be established for individual animals and eventually for the species. Essentially, a reproductive cycle profile is concerned with evaluating the relative changes which occur during the various stages of reproductive activity in the mature non-pregnant non-lactating animal. The establishment of these profiles can, and often does, yield scientific information which justifies the time and effort involved.

METHODS

The undoubted indications of ovulation having taken place are recovery of ova, direct observation of a ruptured follicle and pregnancy. However, establishing ovulation by these means can be difficult, costly, hazardous, time-consuming or quite simply too late for artificial insemination. For these reasons indirect detection methods are more often used. Table I lists some of the more common events which accompany ovulation in mammals and are thus useful in establishing that ovulation has occurred, although possibly less useful in predicting that it will take place and at what precise time. Thus the majority of events listed are valuable predictors only if the individual animal's cycle is well known and this in turn depends on knowledge built up over several previous cycles. For example, analysis of vaginal smears often reveals a progressive picture as do the female reproductive hormones, notably follicle stimulating hormone (FSH) and the oestrogens, as ovulation approaches. The thermal shift, however (in human females), occurs at, or immediately after, ovulation and thus one is recording the *post hoc* rather than the preceding stages. The progressive picture, however, may be highly idiosyncratic and not

TABLE I

Oestrus phenomena: events which can occur immediately prior to, during and following ovulation in female mammals

	Present/elevated	Absent/depressed
Reproductive tract		
Cornified epithelial cells	+	
Erythrocytes	+	
Leucocytes		+
Spinnbarkeit	+	
Ferning	+	
pH	+	
Glucose	+	
Chloride	+	
Plasma/urine		
Follicle stimulating hormone	+	
Luteinizing hormone	+	
Oestrogens	+	
Progesterone	+	
Visual cues		
Vaginal opening	+	
Perineal swelling	+	
Perineal colour changes	+	
Lordosis	+	
Level of activity	+	
Basal body temperature		+

conform to the expectations of what should happen—once again this problem of a discrepancy between what actually occurs and what should occur can be resolved by awareness of both species and individual variation.

Hormonal Changes

The female reproductive system depends entirely on certain hormones which promote the initial development of reproductive organs and regulate reproductive cycles in the adult female. Measurement of the gonadotrophins (FSH and luteinizing hormone, LH) and the steroid hormones (oestrogens and progesterone) should, therefore, indicate the waxing and waning of oestrous periods.

Hormone assay work, however, in unfamiliar species has both practical and theoretical problems. Blood or urine has to be collected regularly and often it is the latter which is more difficult to obtain. Colleagues at the Wellcome Laboratories (Zoological Society of London), however, have had some success in detecting oestrous cycles by assaying urine collected from the cage floors of various species (Martin, Seaton & Lusty, 1976;

Martin, 1976). If an animal is to be sedated, blood and urine (by means of catheterization) can be collected and hormone assay results subsequently matched for accuracy. A further major problem lies in designing suitable assay methods, especially with the protein hormones, for individual species.

Theoretical difficulties in establishing the hormone profile revolve around the extreme diversity shown by different species in the nature, amount and timing of hormones produced and their relationship to ovulation. Although the same pituitary and ovarian hormones are common to most species, there is, hormonally, no "typical" female (Short, 1972). Salient differences in the timing of the production of FSH, LH, oestrogens and progesterone, in the rat, sheep and human female are well known. One should, therefore, expect significant deviations in other species.

Providing that a regular pattern of hormone secretion and a suitable assay method can be established, further factors should be considered. For example, there is good evidence to suggest that the oestrogens and pituitary hormones fluctuate markedly during a 24-hour period in some, if not all, species (Baranczuk & Greenwald, 1973; Weitzman, 1976). Sampling must take this diurnal rhythm into account and in animals having short cycles twice daily sampling may be necessary to determine accurately the point of ovulation.

Hormone assay techniques have become greatly refined within the last decade and there are now several methods which are both rapid and accurate (Seaton, Lusty & Watson, 1976; Batra, 1976). As a predictive method, however, these methods are less useful than direct observation of the ovary. Analyses of the fluctuation of hormone levels in urine and peripheral plasma may not show a constant relationship to the same hormones in the ovarian artery. Careful matching of results from all three sources over a relatively long time-period in potentially or known fertile animals can, of course, provide an excellent basis for future use of only one less time-consuming method. Once again it is clear that no single method avoids the need for constructing a reproductive cycle profile when confronted with an unfamiliar species.

Changes in the Vaginal Contents

The vagina or urogenital tract as a target organ for the steroid hormones is an extremely rich source of information on the reproductive status, normal and pathological, of an animal. The vagina or cervical epithelium itself, the sloughed-off cells and the mucus, ideally all show characteristic changes under the influence of oestrogens and progesterone. Theoretically, at least, tests designed to evaluate changes in the reproductive tract are immensely valuable: since ovarian activity and reproductive tract phenomena appear to be intimately related, one is not only monitoring the ovaries indirectly but therefore also, most importantly, utilizing a method which has some value in predicting ovulation. And yet the results

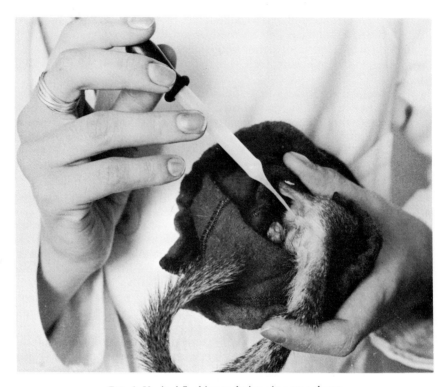

FIG. 1. Vaginal flushing technique in a tree-shrew.

of these various techniques when applied to unfamiliar species have not revealed as much information as one would have hoped, and the underlying reasons are worthy of more detailed consideration.

Techniques for obtaining samples from the reproductive tract can be modified and applied to most mammals varying in size from tree-shrews to elephants, larger animals obviously requiring sedation or a co-operative disposition. Apart from biopsy, sampling methods include swabbing, scraping, flushing and aspiration of fluids. In order to get consistent results, however, it is important that the tract be sampled uniformly, e.g. if

FIG. 2. Ideal relationship between changes in the vaginal epithelium (left) and exfoliated cells (right). a. Pro-oestrus and pre-ovulation; the vaginal epithelium is not yet fully cornified and the sloughed-off cells are of the immature, intermediate type with some leucocytes occurring. b. Oestrus or ovulation; the vaginal epithelium shows a deep layer of fully cornified anucleated cells and the shed cells show a characteristic squarish shape with pyknotic nuclei. Few or no leucocytes. c. Post-oestrus or post-ovulation; the vaginal epithelium is considerably reduced in height and the cornified cells become degenerate, folded and clump together. Leucocytes begin to reappear.

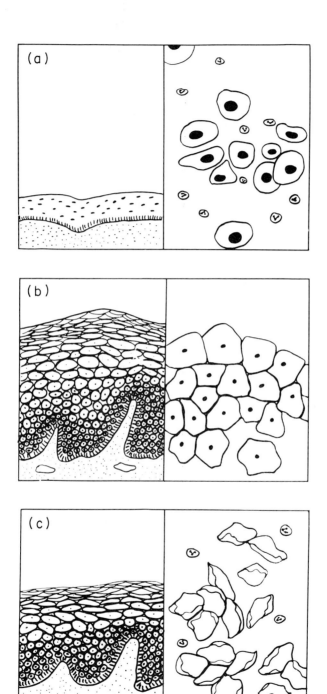

flushing is the chosen method precisely the same amount of saline should be used on each occasion; if a probe is used it should always be inserted to the same depth and in the same direction (see Fig. 1). Standardized means of fixation and staining are obvious counterparts for analysing and measuring "real" changes as opposed to artifacts of the technique. Some species, including the human female, show the cytological changes which have become so closely associated with oestrus and ovulation in a specific area of the vagina only. Hafez (1973) reports that the human cervical epithelium does not reflect the cyclical ovarian activity as accurately as the endometrium or the vaginal epithelium. Wied & Bibbo (1970) refined this further in indicating that samples should not be taken from the lower portion of the vagina since this part of the genital tract does not reflect the pronounced hormonal changes which are found in the upper genital tract. These seemingly small differences in sampling techniques are important and, in the author's experience with exotic animals, vital if vaginal cytology is to be a useful indicator. Care in sampling the vagina itself is obviously important when dealing with those species having a urogenital sinus (Watson & D'Souza, 1975; Nevill et al., 1976). In these species the lower part of the tract is not only less, if at all, responsive to hormonal changes, but is also often contaminated with urinary tract cells and urine, which may make an assessment of hormonal status impossible.

Cellular changes

The vaginal epithelium, under the influence of ovarian oestrogens, proliferates and forms a cornified layer of cells which is then shed and passed into the lumen (Fig. 2). Exfoliative cytology, then, is concerned with observing and assessing progressive cornification and eventual degeneration of vaginal epithelial cells. In the "ideal" human or rat these gradual yet perceptible changes can be charted and categorized into the follicular or pro-oestrous phase, the ovulatory or oestrous phase and the luteal or dioestrous phase. However to rely simply on the cornification criterion is to ignore the many associated cell fluctuations which are also indicative of ovarian activity and potentially as valuable. Thus the traditional analytical techniques such as the karyopyknotic index and the eosinophilic index (Wied & Bibbo, 1970) are dependent on the presence and microscopic appearance of mature cells and their nuclei. The maturation index (Wied & Bibbo, 1970) however is an expression of the number of cornified cells in relation to immature or intermediate cells and is thus more useful when dealing with relatively unfamiliar species.

Not all species show these classic stages of vaginal cornification and even if this phenomenon is a species characteristic, not all individuals may show these changes to the same degree. Furthermore, the somewhat transitory peak of cornification may well be obscured by even slight infection or minor changes in staining techniques. For example, recent observations on domestic cats showed several individuals which never manifested complete cornification but only the presence of epithelial and intermediate cells and yet became pregnant quite normally (F. D'Souza & J. Rivers, unpublished). Extended investigations of vaginal flushings of tree-shrews, *Tupaia belangeri* and *T. minor*, almost never revealed any regular pattern of cornification. The reasons probably include the anatomical fact of a urogenital sinus, and thus, because of the small size of the animals, difficulty in sampling the upper part of the tract. However, a genuine periodicity in leucocytosis was apparent in most individuals and this leucocytic invasion was linked to the presence of erythrocytes in the vaginal flushing immediately prior to the leucocytic phase, suggesting ovarian or endometrial activity (F. D'Souza, unpublished). The animals also showed a characteristic clumping of epithelial cells having an eccentric position of the nuclei at regular times in relation to the presence of both erythrocytes and leucocytes. This is a post-ovulatory pattern first noted by Papanicolaou (1933) in the human female. These linked criteria were indicative of a "low level" oestrous cycle and thus of a periodicity in ovarian activity which was later confirmed by mating and fertilization tests. Hamilton (1951) reported that even the classic reflex ovulator, the rabbit, also showed a cycle of vaginal changes in the absence of males. This is surprising since in those species which are dependent on induced ovulation one would expect (in the non-pregnant, non-lactating mature female) a constant behavioural oestrus with the attendant high oestrogen levels together with, perhaps, persistent vaginal cornification. It is therefore more than likely that artificial insemination of a reflex ovulator may require more than mechanical stimulation in order to achieve ovulation— presumably stimulation of the cervix must be undertaken at the right point in the cycle. Similarly hormonal induction of ovulation may prove more successful at certain times in the cycle.

In animals such as the African elephant, *Loxodonta africana* (Watson & D'Souza, 1975), and some of the large Felidae, *Acinonyx jubatus, Felis concolor, Lynx lynx* (D'Souza, Nevill & Jones, in preparation), on the other hand, leucocytes were found to be an extremely rare phenomenon and analysis of vaginal or urogenital smears was based largely on intermediate and cornified cells (fully mature and degenerate). So, in the field of exfoliative cytology, cornification of cells reflects the ovarian oestrogen output but regular fluctuation of almost any cell type can provide very useful information on the oestrous cycle, ovulation times and thus the breeding potential of the animal.

Histological and histochemical studies of reproductive tract tissue are valuable as additional methods. However, when dealing with nondomestic, non-laboratory species, these methods have a lesser priority since the aim is to keep the sometimes rare and valuable animals alive and well. Material taken at biopsy or *post mortem* is important in establishing species differences in the basic structure, ultrastructure and histophysiology of the cervix and vagina. Detailed histochemical studies can provide vital background information on what happens to the tract at receptive times and how, in particular, it affects the viability, motility and migration of spermatozoa. For example, recent studies of the mouse vagina using scanning electron microscopy have shown characteristic patterns of corrugation of squamous cells at the peak of oestrus suggesting that profound changes do occur and are related to spermatozoal viability and migration (Rubio, 1976).

Electrical changes

Repeated series of tests have been carried out to measure the increase in electro-potential between vaginal and suprapubic electrodes in laboratory rodents and the human female. Until recently the results of these tests have been difficult to categorize; there appears to be a cutaneous vascular change during the oestrous or menstrual cycle but how closely related this change is to ovulation is still a matter for debate. A more recent study (Edwards & Levin, 1974) utilized electrodes to estimate the "resistance" of the vestibular epithelium and genital secretions in cows. These authors demonstrated that there is a significant change in the resistance of the mucosal surface of the vestibular epithelium when cervical mucus is produced during oestrus. During oestrus the resistance is considerably lower than at other phases of the cycle and the lowest resistance coincides with the optimal insemination time resulting in a 75% breeding success rate at the first insemination. The optimal insemination time, as these authors emphasize, may not necessarily be at the exact time of ovulation but towards the end of oestrus. Thus the electrical changes measured in the vagina of cows, sheep and pigs reflect a reproductive tract environment which is conducive to spermatozoal motility and survival.

Changes in cervical mucus

Gross characteristics: volume, colour, elasticity. Changes in the amount, content and appearance of cervical mucus have been correlated with ovarian changes in numerous mammalian species. These changes are under the influence of ovarian oestrogens and, undoubtedly, are related to the ease of spermatozoal migration in the female reproductive tract at or near the time of ovulation (Linford, 1974). Characteristically the amount of mucus increases as ovulation approaches, it becomes transparent or is only faintly opaque and shows a tendency towards elasticity. This last feature is due to the capacity of the mucus micelles to become arranged longitudinally to form strands (Elstein, Moghissi & Borth, 1973).

The advantages of assessing the reproductive status of an animal by means of noting the state of cervical mucus are many; only very small quantities are needed and these are usually easy to obtain, the state of the mucus is either immediately apparent or needs only simple and rapid testing. Furthermore, since cervical mucus changes gradually in response to rising levels of circulating oestrogens, it is possible to construct a picture of impending ovulation.

However, tests devised to determine cervical mucus characteristics depend on whether or not mucus is secreted, and this appears to be species specific and also subject to individual and intra-individual variation. The rate of mucus secretion is a function of both "the number of mucus-secretory units in the cervical canal . . . and the percentage of mucus-secretory cells per unit" (Hafez, 1973: 30). A transitory vaginal or cervical infection can induce copious mucus secretion throughout the cycle and obscure any transparency changes at the time of ovulation. The author's investigations of domestic and some large cats suggest that cervical secretions are generally scanty, however they appear to be extremely variable in tree-shrews, *T. minor*, *T. belangeri*, *Lyonogale tana*. For example two equally reproductively successful tree shrews, *T. belangeri*, showed respectively no measurable quantity of cervical mucus and excessive amounts throughout non-pregnancy, pregnancy and lactation, with no evidence of infection.

Finally, there is as yet no standard method for evaluating gross characteristics of cervical mucus (D'Souza, in preparation). This in itself implies that relative changes are the only useful indicators and this in turn implies regular, probably daily, sampling. This is particularly important since individual animals may show significant variations from one cycle to another and thus the presence, for example, of a transparent mucus may be less useful an observation than the degree of transparency.

Spinnbarkeit. Spinnbarkeit, a rheological change which occurs markedly at ovulation, is due to the large amount of water bound within the cervical mucins. The structure of the mucus at this stage is such that spermatozoa can migrate more easily. After ovulation, in monkeys and human females, cervical mucus becomes thick and sticky and the density precludes passage of spermatozoa (Elstein *et al.*, 1973). The phenomenon of spinnbarkeit or elasticity of cervical mucus is well known in cattle (Studer, 1975) and sheep, and is well documented. However, it is not known whether spinnbarkeit itself is a necessary condition for spermatozoal survival, and therefore fertilization, in all mammals. Tests *in vitro* indicate that in women, spermatozoal penetration and velocity are highest in pre-ovulatory mucus (Elstein *et al.*, 1973). If it could be shown that this is also the case in other mammals, further research into mucus changes together with the development of a standardized index would be invaluable for the purposes of artificial insemination.

Ferning. Ferning or crystallization of cervical mucus refers to the fern-like pattern seen in dried mucus at oestrus and especially at ovulation. The ferning phenomenon together with other rheological changes such as spinnbarkeit and consistency of mucus, is largely determined by hormonal factors and is manifested in the periovulatory phase (Elstein *et al.*, 1973). Tests *in vitro* suggest that crystallization is again related to spermatozoal penetration (Elstein *et al.*, 1973). The chief advantage of using ferning as an index of impending ovulation is that assessment can be extremely rapid; all that is required is a small sample of mucus and a microscope. Regular periodic ferning of mucus taken from the urogenital tract of an African elephant coincided with the appearance of fully cornified superficial cells (Watson & D'Souza, 1975). It was interesting to note that the ferning phenomenon had a slightly longer duration than that of mature cell peaks and that the build-up picture was gradual and therefore less informative of the time of ovulation. In the human female ferning apparently occurs only at the time of ovulation (Garcia, 1967).

Chemical changes: chloride, glucose, peroxidase. Other methods of dealing with cervical mucus include assessment of both glucose and chloride concentrations (McSweeney & Sbarra, 1965; Mahoney, 1970). Investigators of human menstrual cycles have had some success in predicting ovulation by using a relatively simple method of inserting chemically impregnated applicators into the vagina and constructing a colour graph (Birnberg, Wexler & Gross, 1963). However, in all cases the colour graphs appear to have validity only when used in conjunction with other indicators such as basal body temperature and/or analyses of plasma samples for oestrogens and progesterone. It was also emphasized that continuous daily observations were essential to account for even minor fluctuations in an individual cycle. Initial studies showed a peak in the pH of the vagina at the time of ovulation (Zuck & Duncan, 1939). However this method has become somewhat discredited owing to the large amount of individual variation manifested, and therefore great difficulty in establishing a pattern for the human female, let alone any other species.

Linford (1974) discussed the role of peroxidase in the cervical mucus of heifers. This enzyme is well defined and its presence is easily detectable. In her study of maiden heifers, Linford found that the peroxidase content of the mucus fluctuated throughout the cycle but dropped to very low levels for two days before ovulation. The single animal in which no decrease in the enzyme was apparent showed no oestrous behaviour and later failed to conceive when inseminated.

These methods are worth more research time and effort, chiefly because of their practicability. However the same precautions necessary in vaginal smearing also apply to chemical testing; false positive results are common and this may be due in part to inconsistent sampling together with contamination by secretions from the lower reproductive tract.

Laparotomy, Laparoscopy and Palpation

If the vagina is directly affected by ovarian hormones causing a change in cell type then one can usually also witness corresponding changes in other cervical products such as ferning of the mucus or an increase in both chloride and glucose levels. These associated changes, although valuable if eventually only one rapid detection method is to be used in future investigations, do not solve the problem of determining the exact time of ovulation. It is important to be able to relate temporally the events in the ovary with those in the vagina (for example, the stage of the ovulation process coinciding with full cornification of vaginal cells).

Direct observation of the ovaries by means of laparoscopy or laparotomy is a seemingly obvious technique for obtaining unequivocal information on the appropriate time for insemination. These methods are also invaluable as a preliminary procedure for examining the pelvic organs generally, noting any anatomical abnormalities, congenital or acquired, and thus assessing the reproductive potential of the individual concerned and possibly saving a great deal of time and effort. Regular and successful use of laparoscopy has been demonstrated in several species (see, for example Dukelow, this volume, p.195). This technique, however, is more costly in terms of staff time, equipment and the greater risk of damage to the animal than are the indirect methods. Although it has been used repeatedly in monkey species (Jewett & Dukelow, 1972; Bosu, 1973) with undoubted success, it may not always be possible with other species for two main reaons

1. regular sedation of large animals can and does incur problems such as tolerance to the drug and longer recovery periods;
2. it has been suggested (Bosu *et al.*, 1973) that the surgical stress involved can result in a reduction of the ovarian blood supply and thereby seriously interfere with normal ovarian activity.

Furthermore, few zoos concerned with the dual problem of providing a display of animals and encouraging breeding have either the facilities, expertise or time to undertake regular laparoscopic examinations. These restrictions are particularly pertinent since, if it is to be maximally informative, laparoscopy as a method for oestrus detection should probably be carried out at least three times per week whatever the species concerned. As a technique it is possibly generally most useful as an initial exploratory exercise and also as a means of early detection of pregnancy.

Monitoring ovarian changes utilizing palpation has obvious advantages and just as obvious drawbacks. The advantages lie in the fact that no surgical stress is involved although a mild degree of sedation will, again, depending on the species, be necessary. Studer (1975) commented that heat detection in cows is primarily a management problem which is easily overcome by rectal palpation. He devised a scheme for classification and coding of the reproductive condition of the animal by examining the structures of the ovary. Although no success rate is given, it is clear that an experienced practitioner may well be able to predict the time of the next oestrus and ovulation with accuracy. Studer's (1975) work, however, in common with the majority of detection procedures, also relies on the evidence of the state of the vaginal mucus, and the supportive histological evidence of the ovaries after slaughter. One of the most successful studies utilizing palpation as one method in a relatively small mammal is that of

Mahoney (1970). This study combined three methods to detect ovulation in the rhesus monkey, *Macaca irus*: cyclical changes in the karyopyknotic index; the apparent variation in the chloride concentration of the vaginal mucus and the palpable physical changes in the urinogenital tract. Briefly reviewing the accuracy of the two former methods, Mahoney (1970) points out that the chloride spot test devised by McSweeney & Sbarra (1965) reflects not the concentration of sodium chloride but the amount of mucus in the tract. This test in Mahoney's study occasionally gave a false indication of ovulation in some anovulatory cycles but generally was useful only as a warning of impending ovulation and not an accurate index of ovulation.

Cytologically there appeared to be no constant relationship between the peak of the karyopyknotic index and the day of ovulation. For example, lower amplitude peaks in cornified cells were apparent during the luteal phase. Moreover in the majority of cycles Mahoney (1970) noted that a well-defined peak developed during the menstrual phase on Day 3 or 4. This contradictory result has led him to suggest that oestrogens are not necessarily responsible for the nuclear pyknosis of the vaginal epithelium, at least in rhesus monkeys.

Finally, however, Mahoney (1970) concluded that rectal palpation of the genital organs was the only method which differentiated with certainty between ovulatory and non-ovulatory cycles and therefore could determine precisely the moment of ovulation.

Changes in Sites Other than the Reproductive Organs

Almost all areas of the female body have been used as testing sites for detecting ovarian activity on the premise "that a female is female all over".

Analyses of buccal smears of several species carried out by the author have on the whole provided confusing rather than useful corroborative evidence of oestrus cycles. The rationale behind these investigations was that the buccal epithelium may reflect to some extent rising and falling levels of circulating oestrogens. If this was the case one might expect a periodicity in fully cornified cells, although not necessarily exactly coincident with vaginal cornification, at least phased in relation to it. Buccal smears of two elephants, *L. africana* and *Elephas maximus*, over a period of 18 months showed little regular periodicity or indeed any real correlation with cells from the urogenital tract. Similar series of tests on some cat species, *A. jubatus* and *L. lynx*, also proved disappointing. However, more detailed observations on the various cell types and more sophisticated measurement techniques (for example, the use of an image analyser) may, in future, reveal a more consistent pattern.

Various other tests ranging from assessment of alveolar carbon dioxide tension and enzyme assay of nasal mucosa (Garcia, 1967) to observing changes in uterine tissue transplanted onto the surface of the eye (David, Czernobilsky & Kaplur, 1974) have been attempted. Such exotic methods, however, have not proved very useful even in a controlled laboratory context; it is clear that they would be even less so in animals which are difficult and costly to restrain and keep under continuous observation.

Behavioural Changes

A large body of literature exists on the behavioural parameters of oestrus and ovulation. At the outset one must acknowledge the great value of the behavioural approach by saying simply that, on the whole, male animals of a given species are adaptively excellent at responding to a receptive female and that females of the same species are usually equally adept in their response to the male. Careful observations of the interactions between a pair of animals before, during and after oestrus help to elucidate the sort of stimuli apparently necessary and sufficient for successful mating. Behaviour can range from overt solicitation on the part of the female (seen for example in several monkey species) to fleeing from the male at the height of oestrus and thereby inducing chasing behaviour and aggressive sexual interactions as in some rodents (Weir, 1973) and in the tree shrews, *T. belangeri* (Martin, 1968), *T. minor* and *L. tana* (D'Souza, in preparation).

Behaviour is, however, so complex that analysis of the individual units is a time-consuming task and often only applicable in a laboratory context. Nevertheless, interesting and valid observations have been made either in the field or in a captive context. Herbert (1972), for example, noted that rodents wriggle their ears and monkeys adopt a characteristic mating posture at oestrus. Female cats, depending to some extent on the species, howl at oestrus and numerous species, including cats and rats, show lordosis (Komisaruk & Diakow, 1973). These types of behaviour and visual cues are hormone dependent; however, an absence of oestrous behaviour, the so-called "silent oestrus," is not uncommon. Furthermore, unequivocal oestrous behaviour need not necessarily be accompanied by a normal heat period which includes ovulation. Observations on a female puma, *F. concolor*, in which oestrus had been induced hormonally and ovulation confirmed by laparoscopy, showed no behavioural signs of the oestrus (D'Souza, Nevill & Jones, in preparation).

Nevertheless some species show such specific behaviour at different phases of the oestrus, or, in the case of primates, menstrual cycles that it is possible to use their behaviours as taxonomic indicators (see, for example, Dixson, 1977).

CONCLUSIONS

It is clear that no single test of oestrus and ovulation exists, chiefly because of intra- and inter-individual differences which, in turn, are superimposed on considerable species differences. Nalbandov (1973), quoting Sir John Hammond, writes "... each of the species is an entity unto itself and ... it is highly improbable that information on one species would be necessarily applicable to another". The reproductive cycle of any species is the product of an evolutionary process whereby each component of that cycle has proved adaptive and has been selected for. However, extrapolation of methods between species can be a valuable exercise. There are broad underlying similarities which can be elucidated by various methods if all possible parameters are sought and tested. The goal is to narrow the detection field and eventually find a single method which is reliable. To do this, however, the relative changes which occur in the individual animals during the oestrous cycle must be recorded.

It could be argued that the problem of oestrus detection can be solved quite simply by inducing heat and ovulation. There are many advantages, not the least being that the most propitious time for insemination can be predicted and manipulated to occur at convenient times. As with some of the more sophisticated techniques, however, hormonal induction of oestrus in unfamiliar species is a matter of trial and error; understimulation of the ovaries is wasteful of time and effort, while overstimulation can result in multiple ovulation with unfavourable results. Frequent hormonal manipulation of the oestrus cycle could seriously impair the reproductive potential of the animal. Advances in the study of reproductive physiology of uncommon species have been based on numerous careful observations of untreated individuals. The overall picture resulting from such painstaking observations and analyses has provided valuable insights into species and individual differences and their causes. Hormonal induction of oestrus, furthermore, does not indicate necessarily the precise time for insemination. This information results from a combination of chemical, surgical, mechanical and behavioural tests and observations. If the complete cycle is ignored and investigators are content to examine and manipulate only parts of it, detection problems may be increased rather than solved. Once a reproductive cycle profile has been established and the broad reproductive mechanisms of a species ascertained, oestrus and ovulation induction is more likely to be successful.

Hormonal induction of oestrus has, however, a further perspective. If it is successful in that ovulation is confirmed, other tests can be carried out and their usefulness assessed. For example, vaginal cornification, ferning, fluctuations in the enzymes of nasal or cervical mucosa and behavioural changes can be confirmed or not, as the case may be. This approach was used with the female puma, *F. concolor*, and although the observations are ongoing and not all the parameters have been fully analysed, it would seem that vaginal cornification provides the clearest indication of oestrus (D'Souza, Nevill & Jones, in preparation).

Some of the most successful studies on the detection of oestrus in various species have relied on two or three methods each of which measures the different ovarian effects on target organs (Mahoney, 1970; Morrow, Swanson & Hafs, 1976). The keynote of these studies is flexibility in approaching the problem and a willingness to undertake short periods of intensive study of all the available parameters. While not fully in agreement with Wood (1976) who describes the oestrus cycle "in terms of an interacting family of mathematical models", I consider that it is, nevertheless, a phenomenon which is detectable providing that a concentration on what should happen is not at the expense of noting those sometimes obscure relative changes which actually happen.

ACKNOWLEDGEMENTS

Thanks are due to Dr R. D. Martin for his helpful comments on the manuscript and to Rachael Simper for typing the manuscript.

REFERENCES

Baranczuk, R. & Greenwald, G. S. (1973). Peripheral levels of estrogen in the cyclic hamster. *Endocrinology* **92**: 805–812.

Batra, S. (1976). New simplified procedures for the determination of progesterone by competitive protein binding and radioimmunoassay. *J. Steroid Biochem.* **7**:131–134.

Birnberg, C. H., Wexler, D. J. & Gross, M. (1963). Estimation of ovulation phase by serial tests of cervical and vaginal glucose. *Obstet. Gynec. N.Y.* **21**:194–200.

Bosu, W. T. K. (1973). Laparoscopic technique for the examination of the ovaries in the rhesus monkey. *J. med. Primatol.* **2**:124–129.

Bosu, W. T. K., Johansson, E. D. B. & Gemzell, C. (1973). Ovarian steroid patterns in peripheral plasma during the menstrual cycle in the rhesus monkey. *Folia primat.* **19**: 218–234.

David, A., Czernobilsky, B. & Kaplur, S. (1974). A new technical approach for the study of endometrial regeneration in the rabbit. *Lab. Anim. Sci.* **24**: 552–558.

Dixson, A. F. (1977). Observations on the displays, menstrual cycles and sexual behaviour of the 'Black ape' of Celebes (*Macaca nigra*). *J. Zool., Lond.* **182**: 63–84.

D'Souza, F. (In preparation). *Reproductive cycles in the tree-shrews* Tupaia belangeri, T. minor *and* Lyonogale tana.

D'Souza, F., Nevill, G. F. & Jones, D. M. (In preparation). *Detection of oestrus in some large cat species.*

Edwards, D. F. & Levin, R. J. (1974). An electrical method of determining optimum time to inseminate cattle, sheep and pigs. *Vet. Rec.* **15**: 416–420.

Elstein, M., Moghissi, K. S. & Borth, R. (eds) (1973). *Cervical mucus in human reproduction*: 11–22. Cervical mucus: present state of knowledge. Copenhagen: Scriptor.

Garcia, C. R. (1967). Detection and diagnosis of ovulation. *Clin. Obstet. Gynec.* **10**: 380–389.

Hafez, E. S. E. (1973). Histology and microstructure of the cervical epithelial secretory system. In *Cervical mucus in human reproduction*: 23–32. Elstein, M., Moghissi, K. S. & Borth, R. (eds). Copenhagen: Scriptor.

Hamilton, C. E. (1951). Evidence of cyclic reproductive phenomena in the rabbit. *Anat. Rec.* **10**: 557–568.

Herbert, J. (1972). Behavioural patterns. In *Reproduction in mammals*: **4**. *Reproductive patterns*: 34–68. Austin, C. R. & Short, R. V. (eds). Cambridge: Cambridge University Press.

Jewett, D. A. & Dukelow, W. R. (1972). Infra-red photolaparographic techniques for ovulation studies in Primates. *J. med. Primatol.* **1**: 193–195.

Jöchle, W. (1970). Predicting ovulation—a reply. *Science, N.Y.* **169**: 717.

Keefer, C. V. (ed.) (1965). *Human ovulation*. London: Churchill.

Komisaruk, B. & Diakow, C. (1973). Lordosis reflex intensity in rats in relation to the estrous cycle, ovariectomy, estrogen administration and mating behavior. *Endocrinology* **93**: 548–557.

Linford, E. (1974). Cervical mucus: an agent or barrier to conception? *J. Reprod. Fert.* **37**: 239–250.

Mahoney, C. J. (1970). Study of the menstrual cycle in *Macaca irus* with special reference to the detection of ovulation. *J. Reprod. Fert.* **21**: 153–163.

Mandl, A. M. (1951). The phases of the oestrous cycle in the adult white rat. *J. exp. Biol.* **28**: 576–584.

Martin, R. D. (1968). Reproduction and ontogeny in the tree shrews (*T. belangeri*) with reference to their general behaviour and taxonomic relationships. *Z. Tierpsychol.* **25**: 409–532.

Martin, R. D. (1976). Breeding great apes in captivity. *New Scient.* **72**: 100–102.

Martin, R. D., Seaton, B. & Lusty, J. A. (1976). Application of urinary hormone determinations in the management of gorillas. *Rep. Jersey Wildl. Pres. Trust* **12**: 61–70.

McSweeney, D. J. & Sbarra, A. J. (1965). Rapid ovarian hormone and ovulation test. *Obstet. Gynec. N.Y.* **26**: 201–206.

Morrow, D. A., Swanson, E. V. & Hafs, H. D. (1976). Estrous behavior and ovarian activity in peripubertal heifers. *Theriogenology* **6**: 427–435.

Nalbandov, A. V. (1973). Puzzles of reproductive physiology. The Fourth Hammond Memorial Lecture. *J. Reprod. Fert.* **34**: 1–8.

Nevill, G. F., Crompton, W. G., Hennessy, M. A. & Watson, P. F. (1976). Instrumentation for artificial insemination in the African elephant *Loxodonta africana*. *Int. Zoo Yb.* **16**: 166–171.

Papanicolaou, G. N. (1933). The sexual cycle in the human female as revealed by vaginal smears. *Am. J. Anat.* **52**: 519–616.

Rubio, C. A. (1976). Exfoliating cervico-vaginal surface, 2. Scanning electron microscopical studies during the estrous cycle in mice. *Anat. Rec.* **185**: 359–372.

Seaton, B., Lusty, J. A. & Watson, J. (1976). A practical model for steroid hormone radioimmunoassays. *J. Steroid Biochem.* **7**: 511–516.

Short, R. V. (1972). Role of hormones in sex cycles. In *Reproduction in mammals:* **3**. *Hormones in reproduction*: 42–72. Austin, C. R. & Short, R. V. (eds). Cambridge: Cambridge University Press.

Studer, E. (1975). Palpation of the genital tract for prediction of estrus in the cow. *Vet. Med. Small Anim. Clin.* **70**: 1337–1341.

Watson, P. F. & D'Souza, F. (1975). Detection of oestrus in the African elephant (*Loxodonta africana*). *Theriogenology* **4**: 203–209.

Weir, B. J. (1973). The role of the male in the evocation of oestrus in the cuis (*Galea musteloides*) [Rodentia: Hystricomorpha]. *J. Reprod. Fert.* (Suppl.) No. 19: 421–432.

Weitzman, E. (1976). Biologic rhythms and hormone secretion patterns. *Hosp. Pract.* **11**: 79–90.

Wied, G. L. & Bibbo, M. (1970). The effects of sex steroids on the vaginal epithelium. In *Advances in steroid biochemistry and pharmacology*: 188–213. Briggs, M. H. (ed.). New York and London: Academic Press.

Wood, P. D. P. (1976). A note on detection of oestrus in cattle bred by artificial insemination, and the measurement of embryonic mortality. *Anim. Prod.* **22**: 275–278.

Zuck, T. T. & Duncan, D. R. L. (1939). Time of ovulation in the human female as determined by the measurement of the hydrogen ion concentration of vaginal secretion. *Am. J. Obstet. Gynec.* **38**: 310–313.

Symp. zool. Soc. Lond. (1978) No. 43, 195–206

Ovulation Detection and Control Relative to Optimal Time of Mating in Non-Human Primates

W. R. DUKELOW

Michigan State University, East Lansing, Michigan, USA

SYNOPSIS

Adaptation of feral non-human primates to captive breeding conditions requires an appreciation of the animal's reproductive biology as well as knowledge of its response to captivity. Adaptation to captivity, as evidenced by normal cyclicity, requires from nine to 24 months in non-human primates and patterns of seasonality, although attenuated and often shifting temporally, will persist for many years. In the squirrel monkey, the seasonal response is very pronounced but ovulation and pregnancy can occur throughout the year by the use of exogenous hormones. Detection of ovulation and prediction of probable ovulation are important in breeding programs. Laparoscopy provides a quick and easy method of detecting and predicting ovulation, but other methods, including hormonal assay, can be utilized. Varying systems of mating can be utilized, on an experimental basis, to determine the optimal time for mating to yield a pregnancy. Traditionally zoo animals have been housed continuously in pairs or harem groups to provide maximum opportunity for mating and pregnancy while at the same time providing a "family group" display. The same approach applies to large troops of primates in compounds. In the caged environment females are housed separately or in small groups and exposed to the male for limited periods of time. This strategy, termed "timed mating", provides valuable data for teratological studies and is also applicable to the breeding of primates in captivity. The present study concerns analysis of 144 "timed" pregnancies from two species of macaques where male–female exposure ranged from 20 min to 24 hours. Using extremely short periods of exposure (20–30 min) a conception rate of about 19% (i.e. 5·26 matings per conception) can be expected. With longer exposure times the conception rate increases. In determining the optimal time for mating, the day of breeding to cycle length (DB/CL) ratio was calculated. Optimal DB/CL ratios for successful pregnancies were $0·41 \pm 0·05$ (mean ± s.e.), and $0·45 \pm 0·04$ for the stumptailed macaque and the cynomolgus macaque respectively. Ovulation occurs, on a corresponding scale, at $0·48 \pm 0·08$ in the latter species.

INTRODUCTION

The captive breeding of non-human primates, whether in the zoo or in the laboratory, imposes several unique reproductive problems not encountered with common domestic and laboratory animal species. These problems generally relate to adaptation, physiologically and behaviorally, to captivity and to patterns of seasonality of reproduction found in the wild which are carried over in captivity. These aspects have been reviewed in recent years for a wide variety of non-human primates (Vandenbergh, 1973; Butler, 1974; Dukelow, 1974; Wolf, Harrison & Martin, 1975).

Ovulation has been induced in non-human primates, chiefly in *Macaca mulatta*. As early as 1935, Hisaw, Greep & Fevold, using crude ovine anterior pituitary extracts, reported induction of ovulation in three out of four macaques. Hartman (1938) reported a total of seven ovulations out of 104 cycles in anovulatory adult rhesus monkeys. In 1942, he reported an additional six ovulations out of 46 cycles. Pfeiffer (1950) attempted to prevent ovulation in rhesus monkeys with 0·5 mg of progesterone daily from the tenth to the fourteenth day of the cycle. Four ovulations were obtained from 11 monkeys after the end of treatment.

Most of the work on the induction of ovulation in non-human primates has been carried out by van Wagenen and her colleagues and has been summarized (van Wagenen, 1968). Beginning in 1935, she unsuccessfully attempted to induce ovulation in macaques with pregnant mare's serum gonadotrophin (PMSG) (van Wagenen & Cole, 1938). Later work with purified follicle stimulating hormone (FSH) and interstitial cell stimulating hormone (ICSH) of ovine origin was ineffective in inducing ovulation. Knobil, Morse & Greep (1956) reported the importance of species specificity in some primate pituitary hormones. This led to the important observation that in a significant number of cases, multiple ovulation was obtained with gonadotrophins of primate origin (van Wagenen & Simpson, 1957a,b; Simpson & van Wagenen, 1957). Subsequently, Knobil, Kostyo & Greep (1959) induced ovulation in hypophysectomized macaques by treatment with porcine FSH and human chorionic gonadotrophin (HCG).

In the intervening years, various ovulation-inducing agents have been studies for their effect on ovulation in rhesus monkeys. Dede & Plentl (1966) used Pergonal (human menopausal gonadotrophin, HMG) injections for eight to ten days followed by two-day injections of HMG and HCG. The animals were then mated or inseminated artificially and pregnancies were obtained.

Wan & Balin (1969), using HMG/HCG, clomiphene citrate and DL-18-methyl oestriol to induce ovulation in macaques, were successful in 60%, 59% and 32% of the treated cycles, respectively. They were also successful in obtaining high incidences of single ovulations in contrast to Simpson & van Wagenen (1962), who reported multiple ovulations on each ovary.

Bennett (1967a,b) induced multiple ovulation in the squirrel monkey with various regimens of PMSG and HCG. He demonstrated that ova can be recovered from the oviducts and suggested that, probably because of the elevated level of oestrogen from the ovary, tubal transport could be speeded up by the high level of PMSG employed. Bennett began injections without reference to the stage of the cycle.

The development of sensitive radioimmunoassay techniques has led to rapid advances in determining the endocrinological relationships in the reproductive cycle of non-human primates (Knobil, 1972, 1973; Hodgen, 1975). These hormonal assays have been applied to breeding programs (Parkin & Hendrickx, 1975; Hobson *et al.*, 1976).

Other measures of detecting ovulation have been used with success in breeding programs. These include detumescence of the perineum (sex skin) in species exhibiting this characteristic (Hendrickx & Kraemer, 1969), laparotomy (Betteridge, Kelly & Marston, 1970), laparoscopy (Dukelow *et al.*, 1971; Jewett & Dukelow, 1971; Graham *et al.*, 1973; Dukelow, 1975a; Dierschke & Clark, 1976; Dukelow & Ariga, 1976) and body temperature changes, cervical and vaginal criteria (Fujiwara & Imamichi, 1966; Jarosz, Kuehl & Dukelow, 1977). In addition, many types of "calendar-method" breeding systems have been used, with varying periods of male–female exposure, to insure mating at a time approximating ovulation (Honjo, Fujiwara & Cho, 1975; Valerio & Dalgard, 1975; Anderson & Erwin, 1975).

MATERIALS AND METHODS

Data are included from three colonies of non-human primates. The conditions of housing and management for the colony of 104 squirrel monkeys, *Saimiri sciureus*, and the colony of 24 cynomolgus macaques, *Macaca fascicularis*, have previously been described (Dukelow, 1975a; Dukelow, 1976). Both colonies were initiated in 1969. The *Saimiri* were housed indoors continuously until 1974. Since that time they have been housed in outdoor cages from May to November (Jarosz & Dukelow, 1976). Ovulation data are also included from a colony of 20 *Saimiri* at the University of Georgia which were used in induction studies for an eight-month period in 1968–69.

For comparative purposes, data are included to compare breeding results in the above colony of *M. fascicularis* with a colony of 95 *M. arctoides* maintained at Sandoz, Ltd in Basel, Switzerland (Brüggemann & Grauwiler, 1972). The *M. fascicularis* and *M. arctoides* data cover a five-year period of breeding and observation, and timed matings were carried out with 20- to 30-min male–female exposure once each cycle.

Induction of ovulation in squirrel monkeys was by a regimen consisting of five days of progesterone pretreatment (5 mg day^{-1}, intramuscularly) followed by four days of follicle stimulating hormone (FSH-P, Armour-Baldwin Co., Omaha; 1 mg day^{-1}, subcutaneously) and a single injection of HCG (500 i.u., intramuscularly) on the evening of the final day of FSH. This regimen has been shown (Dukelow, 1970) to result in single or double ovulations at a level of 60% during the breeding months. This effect is seasonal (Harrison & Dukelow, 1973) and during the summer months it is necessary to increase the FSH injection schedule to five days (Kuehl & Dukelow, 1975a). In the present studies on seasonality, only animals receiving the breeding season regimen (i.e. four days of

FSH) are included. In all other studies, animals receiving both the breeding season and summer regimens are combined. Ovulations were not induced in the *M. fascicularis* until 1977; the colony was previously used to study follicular maturation and natural ovulation (Jewett & Dukelow, 1972; Rawson & Dukelow, 1973). In the 1977 studies, six animals received an ovulation induction regime consisting of four days of FSH (1 mg) followed by a single injection of HCG (500 i.u.) on the last day of FSH. The injections were begun on Day 3 or 4 of the normal menstrual cycle and ovulation occurred on Day 10 or 11 in three of the six animals.

All ovarian observations and pregnancy diagnoses in the *Saimiri* and *M. fascicularis* colonies were carried out laparoscopically (Dukelow *et al.*, 1971; Dukelow & Ariga, 1976). Timed matings in the *M. fascicularis* colony were based on laparoscopic assessment of the probable day of ovulation. Timed matings in the *M. arctoides* colony were based on calendar methods with matings two to three days before mid-cycle but with individual adjustments based on each animal's previous history. Timed matings were not carried out with *Saimiri* but successful matings occurred after induction of ovulation both in the summer and regular seasons. This work has previously been reported (Jarosz *et al.*, 1977).

To allow for individual differences in cycle length, all macaque data were adjusted to a day of breeding to cycle length ratio (DB/CL ratio). This ratio, when multiplied by 100, represents the percentage of the total cycle which had elapsed when breeding occurred. Where ovulation was known to have occurred (by laparoscopic observation) a similar day of ovulation to cycle length ratio (DO/CL ratio) was calculated. All data were subjected to analysis of variance on the Michigan State University CDC 6500 computer.

RESULTS AND DISCUSSION

The results of 623 inductions of ovulation in squirrel monkeys, by month of injection, are shown in Fig. 1. A significant seasonal effect ($P < 0.01$) was observed for the percentage of animals ovulating, with a lower percentage ovulating in July, August and September. Depressions of the percentage of animals ovulating in March and December reflect the variability in response of newly arrived animals. Harrison & Dukelow (1973) reported on an 18-month study of animals acclimatized to the laboratory for varying times up to two years. These animals (whose data are included in the present eight-year study) exhibited greater variation with a trend to decreasing ovulatory responsiveness in January–February and September–October. Animals acclimatized for two years, on the other hand, showed an orderly response with minimal values in late summer (August–September) and maximal values in late winter (February–March). Acclimatization appears to require about nine months before a standard response to the ovulation regime occurs (Harrison &

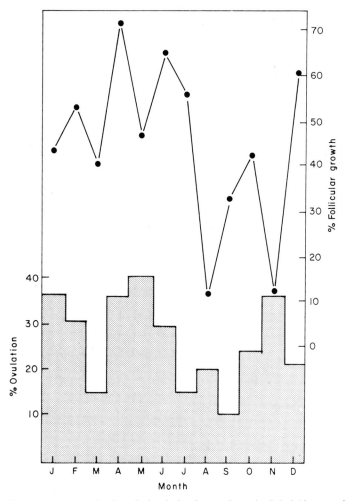

FIG. 1. Response to a standard ovulation induction regimen in *Saimiri* by month.

Dukelow, 1973). The percentage of animals showing a follicular growth response to the induction regimen is also shown in Fig. 1 and, again, the lowest response is observed in the late summer months. Based on present and previous data the ovulation-inducing regimen can be expected to result in ovulation in about 60% of the cases when used with acclimatized *Saimiri* during the breeding season compared with about 40% of all *Saimiri* regardless of degree of acclimatization. This response is diminished during the summer season. One hypothesis for the decreased sensitivity during the summer season was that the standard ovulation-inducing regimen used minimal gonadotrophic stimulation and that the

season of low response represents a period of low endogenous gonadotrophin levels. Thus the regimen becomes subminimal. Substantiation of
this hypothesis came with the observation that an increase in the dose or
extension of the treatment period with FSH, resulted in increased ovulation during the summer months, whereas alteration of HCG levels did not
(Kuehl & Dukelow, 1975a). Further evidence in support of the hypothesis, from behavioral and vaginal cytology studies, has been reported
(Jarosz *et al.*, 1977).

In the macaques, seasonality can easily be measured by the occurrence
of amenorrhea and by variation in the length of the menstrual cycle.
Several years ago Riesen, Meyer & Wolf (1971) indicated a seasonal
anovulatory period in many *M. mulatta*. Some animals show ovulatory
cycles during the entire year, but others enter an anovulatory period
during the summer months. This results in a lower ovulatory response in
M. mulatta colonies when calculated on a yearly basis than is found with
other macaques (Wallach, Virutamasen & Wright, 1973; Dukelow,
1975b).

In the present studies of *M. fascicularis* and *M. arctoides*, ovulations and
conceptions occurred throughout the year. Based on laparoscopic observations, 89·9% of all *M. fascicularis* cycles were ovulatory. The seasonal
variation in cycle length for the two species is shown in Fig. 2. The data
were analysed over 13 periods of 28 days (lunar months). A slight seasonal
trend was noted in *M. fascicularis* ($P < 0·05$) with longer cycles appearing
in the spring of the year (lunar months 4, 5 and 6) and with slightly shorter

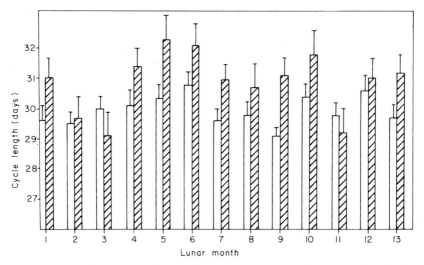

FIG. 2. Mean (±s.d.) seasonal variation of menstrual cycle length in *M. fascicularis* (hatched
bars) and *M. arctoides* (open bars) by lunar month.

cycles occurring in the winter months. No significant effects were found with *M. arctoides* nor were seasonal effects on duration of menstrual bleeding observed for either species. These results are based on a total of 1960 menstrual cycles between the two species, over a five-year period of analysis. Of 1253 timed matings (20- to 30-min male–female exposure time) in *M. arctoides*, 16·3% resulted in pregnancy with no significant seasonal effects. Of 80 timed matings in *M. fascicularis*, 13·3% resulted in pregnancy with no seasonal effect.

Short cycles (i.e. less than 20 days) occurred in 2·9% of all cycles in *M. fascicularis* and in 1·5% of all *M. arctoides* cycles. Again, there was no significant seasonal variation, but there was a tendency in *M. fascicularis* for a higher percentage of short cycles in the second and third lunar months (8·5 and 8·6% respectively). In *M. fascicularis*, where the laparoscopic observation of ovulation made calculation of luteal phase length possible, the luteal phase was 16·1 ± 0·5 days (mean ± s.e.) whereas the follicular phase was 14·4 ± 0·3 days. Traditional concepts have stated that the greater variability in cycle length of human primates is due to variation in the follicular phase of the cycle. The reverse occurred in animals of the present study.

Thus it appears that the seasonality exhibited by *Saimiri* and the anovulatory season exhibited by *M. mulatta*, even after captivity, distinguish them from *M. fascicularis* and *M. arctoides*. The latter two species show a fast adaptation to captive conditions. Evidence of seasonality present in the feral condition is not present in captivity. It must be emphasized that the captive condition of the present animals includes indoor housing in individual cages. Animals maintained in outdoor compounds exhibit a degree of their native seasonality as evidenced by the presence of a "birth season" (I. Bernstein, pers. comm.).

The establishment of a precisely timed pregnancy is important for both teratological testing of pharmaceutical compounds and for basic reproductive studies of capacitation, fertilization and implantation. Such studies also provide background information of the exact time of fertilization which can then be applied to breeding programs of captive primates.

Accordingly, a study was carried out of 144 pregnancies in *M. fascicularis* and *M. arctoides* using short periods of male–female exposure (20 to 30 min). The DB/CL ratios for successful pregnancies for each species are shown in Table 1.

In *M. fascicularis* the DO/CL ratio was 0·48 ± 0·08 (mean ± s.e.) compared to a DB/CL ratio for the same species of 0·45 ± 0·04. The difference between these two, which in an animal with a 30-day cycle represents about 21 hours, would represent the time required for sperm capacitation and the degree of error in estimating the time of ovulation based on laparoscopic observation of follicular development (Jewett & Dukelow, 1972). The need for capacitation in non-human primates was suggested by the early work of Hartman (1933) who reported that the

TABLE I

Day of breeding to cycle length ratios for successful pregnancies with limited mating exposure[a]

Species	DB/CL ratio	DO/CL ratio
Macaca fascicularis	$0·45 \pm 0·04$[b]	$0·48 \pm 0·08$
Macaca arctoides	$0·41 \pm 0.05$	

[a] 20–30 min exposure.
[b] Mean ± s.e.

mean and model ovulation date for *M. mulatta* was Day 13 of the cycle; and the work of van Wagenen (1945) who stated that the optimal time for insemination was from noon of Day 11 to noon of Day 12 for this species. These data provide for the requirement of a period of uterine incubation before fertilization of the ovum occurs. Marston & Kelly (1968) have suggested that three to four hours is required for the capacitation of *M. mulatta* spermatozoa. Recent studies on *in vitro* fertilization in both non-human and human primates have supported the need for capacitation in these species (Kuehl & Dukelow, 1975b; Soupart & Morgenstern, 1973; Edwards, 1973). For a recent discussion of capacitation, readers are referred to Dukelow & Fujimoto (1975).

The conception rates in *M. fascicularis* and *M. arctoides* at differing DB/CL ratios between 0·36 and 0·47 are shown in Fig. 3. Maximum fertility occurred (with 20- to 30-min matings) at a DB/CL ratio of 0·40 to 0·41. In an animal with a 30-day cycle this would represent breeding in the morning of Day 12 of the cycle, where Day 1 represents the first day of

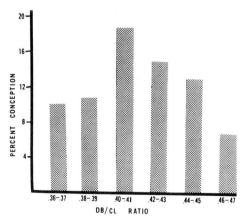

FIG. 3. Percentage conception in macaques (*M. fascicularis* and *M. arctoides*) by DB/CL ratio.

menstrual bleeding. This represents a slightly earlier mating time than was suggested by van Wagenen (1945) (whose $\dot{M}$. *mulatta* normally exhibited 28-day cycles) and is also earlier than the "midcycle minus two days" breeding formula used in many teratological breeding programs. The conception rate had dropped to 6·5% at a DB/CL ratio of 0·47 which translated to a 30-day cycle, would represent breeding on Day 13 (midcycle minus one day).

Weick *et al.* (1973) have reported that, in *M. mulatta*, the luteinizing hormone (LH).peak is preceded by the oestrogen peak at an interval of nine to 15 hours. Ovulation was never observed earlier than 28 hours after the LH peak. If one assumes that ovulation in *M. arctoides* (the species with the majority of pregnancies in our study) occurs 0·06 DB/CL ratio units after the optimal breeding time (as is evident in *M. fascicularis*) then it occurs at a DB/CL ratio of 0·47. If one further assumes that the temporal relationships reported by Weick *et al.* (1973) can be applied from *M. mulatta* to *M. arctoides*, then the LH peak must occur at a DB/CL ratio of 0·43 or lower with the oestrogen peak occurring at a DB/CL ratio of 0·41–0·42. These values correspond closely with the optimal time for mating as illustrated in Fig. 3. This timing also corresponds to the report by Parkin & Hendrickx (1975) of two *M. mulatta* pregnancies occurring when the animals were mated 12·5 hours after the oestrogen peak and four *M. radiata* pregnancies where mating occurred eight hours before, and 7·5, 12·0 and 12·5 hours after the oestrogen peak, respectively. Further support of this conceptual strategy for timed mating comes from the recent publication by Wilks (1977) who reported an LH peak occurring in *M. arctoides* at 11·9±0·5 (mean ±s.e.) days of cycles which averaged 29·1±0·6 days. Translated into terms of the DB/CL ratio, this places the LH peak at a value of 0·41, nearly identical to the calculated value above based on the report of Weick *et al.* (1973). Wilks (1977) reported that the highest oestradiol levels were found on the day preceding the LH peak, i.e. at a DB/CL ratio approximating 0·38, a value corresponding to the optimal mating time in the present study.

These results confirm earlier suggestions that the optimal time for breeding is near the time of the oestrogen surge or soon thereafter. Ideally, quick hormone assay systems can be used to enhance breeding efficiency but if this is not possible, breeding at a time corresponding to between 40 and 41% of the cycle length of the animal in question, measured from Day 1 of the cycle, should yield optimal results.

ACKNOWLEDGEMENT

I wish to acknowledge the assistance of Drs T. J. Kuehl and W. T. Magee in the statistical analyses of the data and to Mr Chris Theodoran and Miss Julie Howe for invaluable assistance with the data handling. Special thanks go to Drs Jules Grauwiler and Sigrid Brüggemann (Sandoz Ltd,

Basel) for providing basic breeding data to be used in the analysis. Finally, thanks are expressed to former postdoctoral and graduate students of the Endocrine Research Unit who provided valuable data to the present study in the course of their own investigations. These include Drs R. M. Harrison, D. A. Jewett, S. J. Jarosz, J. M. R. Rawson, T. J. Kuehl, S. Fujimoto, J. Werbinski and S. Ariga. Approved by the Director, Michigan State University Agricultural Experiment Station, Journal Series No. 8247.

REFERENCES

Anderson, B. & Erwin, J. (1975). A comparison of two timed-mating strategies for breeding pigtailed monkeys (*Macaca nemestrina*). *Theriogenol.* **4**: 153–156.

Bennett, J. P. (1967a). The induction of ovulation in the squirrel monkey (*Saimiri sciureus*) with pregnant mares serum (PMS) and human chorionic gonadotrophin (HCG). *J. Reprod. Fert.* **13**: 357–359.

Bennett, J. P. (1967b). Artificial insemination of the squirrel monkey. *J. Endocr.* **37**: 473–474.

Betteridge, K. J., Kelly, W. A. & Marston, J. H. (1970). Morphology of the rhesus monkey ovary near the time of ovulation. *J. Reprod. Fert.* **22**: 453–459.

Brüggemann, S. & Grauwiler, J. (1972). Breeding results from an experimental colony of *Macaca arctoides*. *Med. Primatol.* **1**: 216–226.

Butler, H. (1974). Evolutionary trends in primate sexual cycles. *Contrib. Primat.* **3**: 2–35.

Dede, J. A. & Plentl, A. A. (1966). Induced ovulation and artificial insemination in a rhesus colony. *Fert. Steril.* **17**: 757–764.

Dierschke, D. J. & Clark, J. R. (1976). Laparoscopy in *Macaca mulatta*: specialized equipment employed and initial observations. *J. med. Primatol.* **5**: 100–110.

Dukelow, W. R. (1970). Induction and timing of single and multiple ovulations in the squirrel monkey (*Saimiri sciureus*). *J. Reprod. Fert.* **22**: 303–309.

Dukelow, W. R. (1974). Captive breeding of nonhuman primates. *Proc. Am. Assoc. Zoo Veter.* **1974**: 52–67.

Dukelow, W. R. (1975a). The morphology of follicular development and ovulation in nonhuman primates. *J. Reprod. Fert.* Suppl. **22**: 23–51.

Dukelow, W. R. (1975b). Similarities and dissimilarities of the reproductive physiology of *Macaca mulatta* & *M. fascicularis*. *Lab. Primate Newsl.* **14**: 1–4.

Dukelow, W. R. (1976). Ovulatory cycle characteristics in *Macaca fascicularis*. *J. med. Primatol.* **6**: 33–42.

Dukelow, W. R. & Ariga, S. (1976). Laparoscopic techniques for biomedical research. *J. med. Primatol.* **5**: 82–99.

Dukelow, W. R. & Fujimoto, S. (1975). Capacitation of sperm. In *Progress in infertility*: 745–763. Behrman, S. J. & Kistner, R. W. (eds). Boston: Little, Brown and Company.

Dukelow, W. R., Jarosz, S. J., Jewett, D. A. & Harrison, R. M. (1971). Laparoscopic examination of the ovaries in goats and primates. *Lab. Anim. Sci.* **21**: 594–597.

Edwards, R. G. (1973). Physiological aspects of human ovulation, fertilization and cleavage. *J. Reprod. Fert.* Suppl. No. 18: 87–101.

Fujiwara, T. & Imamichi, T. (1966). Breeding of cynomolgus monkeys as an experimental animal. *Jap. J. med. Sci. Biol.* **19**: 225–226.

Graham, C. E., Keeling, M., Chapman, C., Cummins, L. B. & Haynie, J. (1973).

Method of endoscopy in the chimpanzee: relations of ovarian anatomy, endometrial histology and sexual swelling. *Am. J. phys. Anthrop.* **38**: 211–216.

Harrison, R. M. & Dukelow, W. R. (1973). Seasonal adaption of laboratory-maintained squirrel monkeys (*Saimiri sciureus*). *J. med. Primatol.* **2**: 277–283.

Hodgen, G. D. (1975). Comparative primate reproductive endocrinology: advancements important in domestic breeding programs. *Lab. Anim. Sci.* **25**: 793–797.

Honjo, S., Fujiwara, T. & Cho, F. (1975). A comparison of breeding performance of individual cage and indoor gang cage systems in cynomolgus monkeys. *Int. Congr. Primat.* **5**: 98–105.

Hartman, C. (1933). Pelvic (rectal) palpation of the female monkey with special reference to the ascertainment of ovulation time. *Am. J. Obstet. Gynec.* **26**: 600–608.

Hartman, C. G. (1938). The use of gonadotropic hormones in the adult rhesus monkey. *Bull. John Hopkins Hosp.* **63**: 351–360.

Hartman, C. G. (1942). Further attempts to cause ovulation by means of gonadotropes in the adult rhesus monkey. *Contr. Embryol.* **30**: 111–113.

Hendrickx, A. G. & Kraemer, D. C. (1969). Observation on the menstrual cycle, optimal mating time and pre-implantation embryos of the baboon, *Papio anubis* and *Papio cynocephalus*. *J. Reprod. Fert.* Suppl. No. 6: 119–128.

Hisaw, F. L., Greep, R. O. & Fevold, H. O. (1935). Experimental ovulation of Macacus rhesus monkeys. *Anat. Rec.* Suppl. **61**: 24.

Hobson, W., Dougherty, W., Lowry, J., Fuller, G. & Coulston, F. (1976). Increased fertility in rhesus monkeys by breeding after the preovulatory LH surge. *Lab. Anim. Sci.* **26**: 63–65.

Jarosz, S. J. & Dukelow, W. R. (1976). Temperate season outdoor housing of *Saimiri sciureus* in the northern United States. *J. med. Primatol.* **5**: 176–185.

Jarosz, S. J., Kuehl, T. J. & Dukelow, W. R. (1977). Vaginal cytology, induced ovulation and gestation in the squirrel monkey (*Saimiri sciureus*). *Biol. Reprod.* **16**: 97–103.

Jewett, D. A. & Dukelow, W. R. (1971). Laparoscopy and precise mating techniques to determine gestation length in *Macaca fascicularis*. *Lab. Primate Newsl.* **10**: 16–17.

Jewett, D. A. & Dukelow, W. R. (1972). Serial observations of follicular morphology near ovulation in *Macaca fascicularis*. *J. Reprod. Fert.* **31**: 287–290.

Knobil, E. (1972). Hormonal control of the menstrual cycle and ovulation in the rhesus monkey. *Acta Endocr.* (Suppl. 166) **71**: 137–144.

Knobil, E. (1973). On the regulation of the primate corpus luteum. *Biol. Reprod.* **8**: 246–258.

Knobil, E., Kostyo, J. L. & Greep, R. O. (1959). Production of ovulation in the hypophysectomized rhesus monkey. *Endocrinology* **65**: 487–493.

Knobil, E., Morse, A. & Greep, R. O. (1956). The effects of beef and monkey pituitary growth hormone on the costochondral junction in the hypophysectomized rhesus monkey. *Anat. Rec.* **124**: 320–324.

Kuehl, T. J. & Dukelow, W. R. (1975a). Ovulation induction during the anovulatory season in *Saimiri sciureus*. *J. med. Primatol.* **4**: 23–31.

Kuehl, T. J. & Dukelow, W. R. (1975b). Fertilization *in vitro* of *Saimiri sciureus* follicular oocytes. *J. med. Primatol.* **4**: 209–216.

Marston, J. H. & Kelly, W. A. (1968). Time relationships of spermatozoan penetration into the egg of the rhesus monkey. *Nature, Lond.* **217**: 1073–1074.

Parkin, R. F. & Hendrickx, A. G. (1975). The temporal relationship between the preovulatory estrogen peak and the optimal mating period in rhesus and bonnet monkeys. *Biol. Reprod.* **13**: 610–616.

Pfeiffer, C. A. (1950). Effects of progesterone upon ovulation in the rhesus monkey. *Proc. Soc. exp. Biol. Med.* **75**: 455–458.

Rawson, J. M. R. & Dukelow, W. R. (1973). Observation of ovulation in *Macaca fascicularis. J. Reprod. Fert.* **34**: 187–190.

Riesen, J. W., Meyer, R. K. & Wolf, R. C. (1971). The effect of season on occurrence of ovulation in the rhesus monkey. *Biol. Reprod.* **5**: 111–114.

Simpson, M. E. & van Wagenen, G. (1957). Experimental induction of ovulation in the Macaque monkey. *Fert. Steril.* **9**: 386–399.

Simpson, M. E. & van Wagenen, G. (1962). Induction of ovulation with human urinary gonadotrophins in the monkey. *Fert. Steril.* **13**: 140–152.

Soupart, P. & Morgenstern, L. L. (1973). Human sperm capacitation and *in vitro* fertilization. *Fert. Steril.* **24**: 462–478.

Valerio, D. A. & Dalgard, D. W. (1975). Experience in the laboratory breeding of nonhuman primates. *Lab. Anim. Handb.* **6**: 49–62.

Vandenbergh, J. G. (1973). Environmental influence on breeding in rhesus monkeys. *Int. Congr. Primat.* **4(2)**: 1–19.

van Wagenen, G. W. (1945). Mating and pregnancy in the monkey. *Anat. Rec.* **91**: 304–306.

van Wagenen, G. W. (1968). Induction of ovulation in *Macaca mulatta. Fert. Steril.* **19**: 15–28.

van Wagenen, G. W. & Cole, H. H. (1938). Failure of continued injections of gonadotrophic hormones to prevent menstrual cycles and pregnancy in *Macaca mulatta. Am. J. Physiol.* **123**: 208–212.

van Wagenen, G. W. & Simpson, M. E. (1957a). Induction of multiple ovulation in the rhesus monkey (*Macaca mulatta*). *Endocrinology* **61**: 316–318.

van Wagenen, G. W. & Simpson, M. E. (1957b). Experimentally induced ovulation in the rhesus monkey (*Macaca mulatta*). *Revue suisse Zool.* **64**: 807–819.

Wallach, E. E., Virutamasen, P. & Wright, K. H. (1973). Menstrual cycle characteristics and side of ovulation in the rhesus monkey. *Fert. Steril.* **24**: 715–721.

Wan, L. S. & Balin, H. (1969). Induction of ovulation in rhesus monkeys: a comparative study. *Fert. Steril.* **20**: 111–126.

Weick, R. F., Dierschke, D. J., Karsch, F. J., Butler, W. R., Hotchkiss, J. & Knobil, E. (1973). Periovulatory time courses of circulating gonadotrophic and ovarian hormones in the rhesus monkey. *Endocrinology* **93**: 1140–1147.

Wilks, J. W. (1977). Endocrine characterization of the menstrual cycle of the stumptailed monkey (*Macaca arctoides*). *Biol. Reprod.* **16**: 474–478.

Wolf, R. H., Harrison, R. M. & Martin, T. W. (1975). A review of reproductive patterns in new world monkeys. *Lab. Anim. Sci.* **25**: 814–821.

Symp. zool. Soc. Lond. (1978) No. 43, 207–218

Artificial Breeding of Non-Primates

S. SEAGER, D. WILDT and C. PLATZ

Baylor College of Medicine/ Texas A and M University, Houston, Texas, USA

SYNOPSIS

It is encouraging that more of the world's zoos are awakening to the fact that they can no longer be places only for the maintenance and observation of animals. Many zoological societies have become aware that zoos are not only areas for display of animals but also for research and perpetuation of endangered species; there are a number of species maintained in zoos or private collections that are no longer found in the wild. The attrition rate in many species that seem to be incapable of breeding, or at best breed poorly, in captivity has been so great that many are becoming extinct or so rare that the exportation from the country of origin is prohibited. Unfortunately, it is recognized that such restrictions imposed by the importing country (whether the importation be for the fur trade, zoos or private collections), do not necessarily mean the continuation of these species in their natural habitat. Habitats of many species are being rapidly encroached upon by man, and man may therefore have to resort to artificial insemination to maintain the populations of certain species even in their own natural surroundings. Breeding by artificial insemination, however, is not the ideal method of species perpetuation but unless such radical steps as this are taken, extinction of the species is almost certain.

Development of artificial means of semen collection and insemination in captive wild mammal populations has been, for the most part, negligible. This is in contrast to the high degree of efficiency seen in many animals of agricultural importance.

The reasons for establishing an artificial insemination program are listed.

1. To eliminate the risk and expense of shipping captive wild animals for breeding.
2. To inseminate the female without the male being present, or to eliminate the problem of incompatible pairs.
3. To prevent possible disease transmission from the introduction of new animals for breeding.
4. To introduce new bloodlines into the gene pool from the wild animal game parks, reserves and zoos.
5. To help eliminate complicated legal procedures and animal shipment involved with breeding loans.
6. To improve bloodlines, with possible reduction or elimination of undesirable genetic traits, by means of superior males of the species.
7. To make available the possibility of progeny testing males.

Artificial insemination *per se* is only one of many major components involved in a successful artificial breeding program.

In 1972, a study involving investigation of semen collection, storage and artificial insemination of captive wild animals was initiated in this laboratory. It was first necessary to establish certain satisfactory methods including: semen collection and evaluation, detection of estrus, time and frequency of breeding during the estrous cycle, successful methods of cryopreservation and thawing (if frozen semen was to be used), restraint or anesthesia of the animals, artificial insemination and pregnancy diagnosis.

To date, our studies have been successful with respect to semen collection and preservation in a number of species. Artificial insemination, estrus detection and ovulation induction with exogenous gonadotropin therapy is being investigated in several of the large species of Felidae and other captive wild animals.

SEMEN COLLECTION

Semen collection is an essential part of any artificial reproduction program. In the non-primate this is normally performed using an artificial vagina, electroejaculation or surgical methods. There have been reports of semen collection by artificial vagina in several species (see Watson, this volume, p. 97), and we have used this technique for the Bengal tiger, clouded leopard, cheetah and wolf (Seager & Platz, 1974; Seager, Platz & Hodge, 1974). However, the majority of non-primate animals require electroejaculation under anesthesia. There have been a number of reports of specific aspects of electroejaculation techniques for wild Felidae (Sadleir, 1966; Mayo, 1967; Seager, 1976) and Martin (this volume, p. 127) has considered the basic principles. The list of species electroejaculated in our program is shown in Table I.

TABLE I

Species that have been electroejaculated in this study

Non-human primates	
Lowland gorilla	*Gorilla gorilla gorilla*
Ring-tailed lemur	*Lemur catta*
Baboon	*Papio cyanocephalus*
Squirrel monkey	*Saimiri sciureus*
Red uakari	*Cacajao rubicundus*
Orang-utan	*Pongo pygmaeus*
Formosan rock macaque	*Macaca cyclopis*
Geoffroy's spider monkey	*Ateles geoffroyi*
Lar gibbon	*Hylobates lar*
Moloch gibbon	*Hylobates moloch*
Geoffroy's tamarin	*Saguinus geoffroyi*
Black and white colobus monkey	*Colobus polykomes*
Hamlyn's monkey	*Cercopithecus hamlyni*
Cotton-headed tamarin	*Saguinus oedipus*
White-lipped tamarin	*Saguinus fuscicollis*

Herbivores	
Greater kudu	*Tragelaphus strepsiceros*
Persian onager	*Equus heminonus onager*
Fallow deer	*Dama dama*
Axis deer	*Axis axis*
Black rhinoceros	*Diceros bicornis*
Sitatunga	*Tragelaphus spekei*
Bactrian camel	*Camelus bactrianus*
Arabian oryx	*Oryx leucoryx*
Père David's deer	*Elaphurus davidianus*

TABLE I—*continued*

Herbivores	
Red sheep	*Ovis orientalis*
Eld's deer	*Cervus eldii eldii*
Red lechwe	*Kobus leche leche*
Speke's gazelle	*Gazella spekei*
Slender horned gazelle	*Gazella leptoceros*
Brazilian tapir	*Tapirus terrestris*
Yellow-backed duiker	*Cephalophus silvicultor*
Dorcas gazelle	*Gazella dorcas*

Felines	
Cougar	*Felis concolor*
Canadian lynx	*Lynx canadensis*
Bobcat	*Lynx rufus*
Golden cat	*Felis temmincki*
Ocelot	*Felis pardalis*
Geoffroy's cat	*Felis geoffroyi*
Clouded leopard	*Neofelis nebulosa*
Lion	*Panthera leo*
Bengal tiger	*Panthera tigris tigris*
Siberian tiger	*Panthera tigris altaica*
Leopard	*Panthera pardus*
Jaguar	*Panthera onca*
Cheetah	*Acinonyx jubatus*
Domestic cat	*Felis catus*
Leopard cat	*Felis bengalensis*
Margay	*Felis wiedi*
N. Chinese leopard	*Panthera pardus orientalis*

Carnivores and omnivores	
Binturong	*Arctictis binturong*
Lesser panda	*Ailurus fulgens*
Meerkat	*Suricata suricatta*
Matschie's tree kangaroo	*Dendrolagus matschiei*
African mongoose	*Herpestes ichneumon*
West African brush-tailed porcupine	*Atherurus africanus*
Palm civet	*Paradoxurus hermaphroditus*
Sugar glider	*Petaurus australis*
Spotted hyena	*Crocuta crocuta*
West African water mongoose	*Atilax paludinosus*
Polar bear	*Thalarctos maritimus*
American black bear	*Euarctos americanus*
Kodiak bear	*Ursus arctos middendorffi*

TABLE I—*continued*

Carnivores and omnivores	
Sloth bear	*Melursus ursinus*
Bush dog	*Speothos venaticus*
Red fox	*Vulpes fulva*
Crab-eating fox	*Cerdocyon thous*
Canadian timber wolf	*Canis lupus canadensis*
Domestic dog	*Canis familiaris*

Rodents	
Degu	*Octodon degus*
Laboratory rat	*Rattus norvegicus*

Aves	
Pigeon	*Columba livia*

Reptiles	
Angola python	*Python anchietae*
Bull snake	*Pituophis melanoleucus sayi*
Hog nose snake	*Heterodon contortrix*

SEMEN PRESERVATION

Once the ejaculate is obtained it is examined for the following criteria: volume, color, turbidity, viscosity, pH, percentage of motile cells, progressive motility (based on a scale from 0 to 5, 5 being maximal), spermatozoal count and morphology (Platz & Seager, 1977). Semen from the many animals that have been electroejaculated has been frozen by a variety of methods with varying extenders and dilution rates. There are three principal methods of cryopreservation of semen: glass vials, plastic straws of varying size from 0·25 to 2·0 ml and "pellets", the latter obtained by placing drops of extended semen into a solid block of carbon dioxide and then subsequent immersion into liquid nitrogen. The post-thawing motility results of semen frozen in an egg yolk-lactose-glycerol extender by the "pellet" method are given in Table II.

INDUCTION OF ESTRUS AND OVULATION

There have been few published reports on estrus behavior and ovulation in wild non-primates (Rowlands & Sadleir, 1968; Rowlands & Hime,

1970). In our research we have adapted techniques of semen collection in the domestic cat for use in wild species of Felidae. Colby (1970) and Sojka, Jennings & Hamner (1970) have reported on induced estrus and timed pregnancies in domestic cats, and artificial insemination. Wildt, Kinney & Seager (1978) and Wildt & Seager (1978) also have reported successful artificial induction of estrus and ovulation in the domestic cat and have had preliminary success with similar hormone regimens in the jaguar. We have induced ovulation (preceded by artificial insemination with frozen semen) in the jaguar (D. Wildt, C. Platz & S. Seager, unpublished) as well as successfully induced estrus in the ocelot followed by natural breeding and the birth of five kittens (S. Seager, W. Hodge & C. Platz, unpublished).

LAPAROSCOPY

Laparoscopy has been a valuable technique for monitoring reproductive cyclicity in both domestic and wild mammals. It has been used successfully to monitor the effects of exogenous hormones on follicle development and ovulation in a number of species in this laboratory. To date we have performed 925 (63 animals) laparoscopic examinations in the domestic cat and 512 (56 animals) examinations in the dog for determination of ovarian morphology and cyclicity (Wildt, Kinney & Seager, 1977, 1978; Wildt, Levinson & Seager, 1977; Wildt & Seager, in press). Labaroscopy has also been used extensively for determining the time of ovulation in our wild animal studies (Table III). One jaguar underwent 14 laparoscopic examinations over a 12-month interval with no adverse effects.

ARTIFICIAL INSEMINATION

We have obtained over 1000 puppies from frozen semen in the domestic dog (Seager, Platz & Fletcher, 1975; Seager & Platz, 1978). This information has provided much insight into semen collection, evaluation and freezing in the domestic dog. This information was utilized to bring about the first report of pregnancy from the use of frozen semen in the wolf (Seager, Platz & Hodge, 1974). A South American bush dog has also been inseminated but this did not result in a conception. Aamdal, Nyberg & Fougner (1976) have reported successful artificial insemination and conceptions using frozen and fresh semen in foxes.

In the domestic cat pregnancies have been obtained following hormonal induction of estrus and ovulation followed by artificial insemination of previously frozen semen (Platz, Follis et al., 1976; Seager, 1977; Platz, Wildt & Seager, 1978). Using similar techniques developed first in the domestic cat, a jaguar and three ocelots have been artificially inseminated but no conceptions have resulted.

TABLE II

Representative data on semen collected by electroejaculation together with post-thaw motility[a]

Species	Total sperm count ($\times 10^6$) (hemacytometer)	Pre-freeze		Post-thaw	
		Percentage motility	Progressive motility	Percentage motility	Progressive motility
Père David's deer (*Elaphurus davidianus*)	65	0	—	—	—
Eld's deer (*Cervus eldii eldii*)	1·2	40	3·5	30	5
Speke's gazelle (*Gazella spekei*)	361	35	4	35	4·5
Brazilian tapir (*Tapirus terrestris*)	11814	50	3·5	1	2
Angola python (*Python anchietae*)	206	90	5	10	2
Hog nose snake (*Heterodon contortrix*)	26	60	4	—	—
Degu (*Octodon degus*)	39·3	80	5	50	4·7
African mongoose (*Herpestes ichneumon*)	70	28	4·7	30	4·5
Palm civet (*Paradoxurus hermaphroditus*)	1·25	80	5	—	—

Binturong (*Arctictis binturong*)	4	50	4	—	—
Sloth bear (*Melursus ursinus*)	220	70	4	40	3
Polar bear (*Thalarctos maritimus*)	12	20	1	0	0
American black bear (*Euarctos americanus*)	0	0	0	0	0
Matschie's tree kangaroo (*Dendrolagus matschiei*)	10	80	5	—	—
Lesser panda (*Ailurus fulgens*)	0·57	10	5	0	0
Bush dog (*Speothos venaticus*)	10	10	4·0	0	0
Red fox (*Vulpes fulva*)	8	30	3	20	4
Clouded leopard (*Neofelis nebulosa*)	30	65	4·8	50	5
Cougar (*Felis concolor*)	10	14	4·2	5	2·5
Canadian lynx (*Lynx canadensis*)	10	60	5	20	4
Ocelot (*Felis pardalis*)	150	55	5	20	4·5
Bengal tiger (*Panthera tigris tigris*)	108	53	4·5	16	4·0

TABLE II—*continued*

Species	Total sperm count ($\times 10^6$) (hemacytometer)	Pre-freeze		Post-thaw	
		Percentage motility	Progressive motility	Percentage motility	Progressive motility
Leopard (*Panthera pardus*)	40	50	4·5	35	4·5
Siberian tiger (*Panthera tigris altaica*)	0	—	—	—	—
Fallow deer (*Dama dama*)	0	—	—	—	—
Bactrian camel (*Camelus bactrianus*)	82	37	3·5	0	0
Timber wolf (*Canis lupus*)	308	59	4·6	35	4·7
Geoffroy's cat (*Felis geoffroyi*)	194	80	5	20	1
Golden cat (*Felis temmincki*)	0·15	80	2·5	0	—
Jaguar (*Panthera onca*)	489	71	5	40	4·5
Leopard cat (*Felis bengalensis*)	49	90	5	70	5
Lion (*Panthera leo*)	180	71	5	41	4

Margay (*Felis wiedi*)	44	90	5	30	5
North Chinese leopard (*Panthera pardus orientalis*)	101	65	5	30	4·5
Bobcat (*Lynx rufus*)	0·6	5	3	—	—
Cheetah (*Acinonyx jubatus*)	110	65	4·7	38	5
Baboon (*Papio cyanocephalus*)	38	50	4·5	25	3·5
Moloch gibbon (*Hylobates moloch*)	90	5	4·5	5	2
Brown-headed spider monkey (*Ateles fusciceps*)	13	70	5·0	20	5
Lar gibbon (*Hylobates lar*)	2·2	20	4·5	2	5
Lowland gorilla (*Gorilla gorilla gorilla*)	148	5	2·5	—	—
Orang-utan (*Pongo pygmaeus*)	41	70	5	5	4·5
Geoffroy's tamarin (*Saguinus geoffroyi*)	25	90	5	40	4·0
Red uakari (*Cacajao rubicundus*)	0	—	—	—	—
Meerkat (*Suricata suricatta*)	6·7	52	4·7	50	1·8

[a] Freezing in pellets, extender—egg yolk, lactose and glycerol.

TABLE III

Zoo animal species in which laparoscopy has been used

Matschie's tree kangaroo	*Dendrolagus matschiei*
Red-crowned mangabey	*Cercocebus torquatus*
Gibbon	*Hylobates* sp.
Moor macaque	*Macaca maurus*
Kodiak bear	*Ursus arctos middendorffi*
Polar bear	*Thalarctos maritimus*
Palm civet	*Paradoxurus hermaphroditus*
Lion	*Panthera leo*
Bengal tiger	*Panthera tigris tigris*
Jaguar	*Panthera onca*
Cheetah	*Acinonyx jubatus*
Clouded leopard	*Neofelis nebulosa*
Bush dog	*Speothos venaticus*

CONCLUSION

Possibly one of the most important achievements of this symposium was to bring together people from many parts of the world who are involved in aspects of artificial breeding of captive wild species. While much has been done in this area in relation to domestic species, we have only "scratched" the surface of what can and will be done in wild species. We advocate the use of an artificial breeding program only as a part or an aid in overall breeding programs. We believe that animals given ideal situations and habitat need little help from man to reproduce. When man has put such constraints on the "ideal" situation, whether it be by disease, caging, diet or other stress, we suggest that artificial insemination can play a vital role in wild mammal reproduction. The success that has been obtained by our research and by other investigators leads us to believe that the capability of breeding many captive wild animals, including some of the endangered species, can be brought about by using artificial methods.

ACKNOWLEDGEMENTS

We wish to express gratitude to a number of private owners and to the following zoos and game parks for their cooperation and assistance in this endeavor: The National Zoo, Washington, DC; Olympic Game Farm, Sequim, Washington; Woodland Park Zoo, Seattle, Washington; Marine World, Los Angeles, California; Gladys Porter Zoo, Brownsville, Texas; San Diego Zoo, San Diego, California; San Diego Wild Animal Park,

Escondido, California; Japanese Deer Park, San Francisco, California; Hermann Park Zoo, Houston, Texas; Memphis Zoo, Memphis, Tennessee; St. Louis Zoo, St. Louis, Missouri; Henry Doorly Zoo, Omaha, Nebraska; Kings Dominion, Doswell, Virginia.

REFERENCES

Aamdal, J., Nyberg, K. & Fougner, J. [1976]. Artificial insemination in foxes. *VIII Int. Congr. Anim. Reprod. Artif. Insem.* [Krakow] **4**: 956–959.

Colby, E. (1970). Induced estrus and timed pregnancies in cats. *Lab. Anim. Care* **20**: 1075–1080.

Mayo, J. (1967). Tranquilization of a male snow leopard for semen extraction. *Int. Zoo Yb.* **7**: 148.

Platz, C., Follis, T., Demorest, N. & Seager, S. [1976]. Semen collection, freezing, and insemination in the domestic cat. *VIII Int. Congr. Anim. Reprod. Artif. Insem.* [Krakow] **4**: 1053–1056.

Platz, C. & Seager, S. (1977). Successful pregnancies using concentrated frozen beagle semen. *Lab. Anim. Sci.* **27**: 1013–1016.

Platz, C., Wildt, D. & Seager, S. (1978). Pregnancies in the domestic cat using artificial insemination with previously frozen spermatozoa. *J. Reprod. Fert.* **52**: 279–282.

Rowlands, I. & Hime, J. (1970). Induction of ovulation and oestrus in a jaguar and a Saiga antelope. *Scient. Rep. zool. Soc. Lond.* **1967–69**: 62.

Rowlands, I. & Sadleir, R. (1968). Induction of ovulation in the lion, *Panthera leo. J. Reprod. Fert.* **16**: 105–111.

Sadleir, R. (1966). The diagnosis of fertility in a male leopard *Panthera pardus. Vet. Rec.* **79**: 397–398.

Seager, S. (1976). Electroejaculation of cats. In *Electronics and veterinary medicine*: 410–418. Klein, W. R. (ed.). Springfield, Ill.: Charles C. Thomas Co.

Seager, S. (1977). Semen collection, evaluation and artificial insemination of the domestic cat. In *Current verterinary therapy* **6**: 1252–1254. Kirk, R. W. (ed.). Philadelphia: W. B. Saunders Co.

Seager, S. & Platz, C. (1974). Semen collection and artificial insemination in captive wild cats, wolves, and bears. *Proc. Am. Assoc. Zoo Vet.* **1974**: 29–35.

Seager, S. & Platz, C. (1978). One thousand puppies from frozen semen. *American Kennel Club Gazette*:

Seager, S., Platz, C. & Fletcher, W. (1975). Conception rates and related data using frozen dog semen. *J. Reprod. Fert.* **45**: 189–192.

Seager, S., Platz, C. & Hodge, W. (1974). Successful pregnancy using frozen semen in the wolf. *Int. Zoo Yb.* **15**: 140–143.

Sojka, N., Jennings, L. & Hamner, C. (1970). Artificial insemination in the cat. *Lab. Anim. Care* **20**: 198–204.

Wildt, D., Kinney, G. & Seager, S. (1977). Laparoscopy for direct observation in the internal organs of the domestic cat and dog. *Am. J. Vet. Res.* **38**: 1429–1432.

Wildt, D., Kinney, G. & Seager, S. (1978). Gonadotropin induced reproductive cyclicity in the domestic cat. *Lab. Anim. Sci.* **28**: 301–307.

Wildt, D., Levinson, C. & Seager, S. (1977). Laparoscopic exposure and sequential observation of the ovary of the cycling bitch. *Anat. Rec.* **189**: 443–450.

Wildt, D. & Seager, S. (1978). Ovarian response to the estrual cat receiving varying dosages of HCG. *Hormone Research.*

Wildt, D. & Seager, S. (In press). Ovarian and uterine morphology during various stages of the domestic cat reproductive cycle as determined by laparoscopy. In *Current therapy and theriogenology.* Morrow, D. (ed.). Philadelphia: W. B. Saunders Co.

Symp. zool. Soc. Lond. (1978) No. 43, 219–240

Artificial Insemination and a Note on Pregnancy Detection in the Non-Human Primate

A. G. HENDRICKX, R. S. THOMPSON, D. L. HESS AND S. PRAHALADA

University of California, Davis, California, USA

SYNOPSIS

Semen intended for artificial insemination (AI) of non-human primates is normally collected by electroejaculation with either penile or rectal stimulation (e.g. macaques) or by masturbation (e.g. chimpanzee). Although detailed comparative experiments have not been done, the penile method is considered to produce the best quality semen when electroejaculation is used and therefore may be preferred when dilution of semen is required. Vaginal, intracervical, intrauterine and intraperitoneal insemination are four methods previously used in non-human primates. While substantial data are not yet available, the conception rate for each of the four methods appears to be equal to, or greater than, that with natural mating. An important aspect of all AI trials is an accurate means of determining optimal mating time. Although simian semen may be successfully frozen with recovery of near pre-treatment motility, no term pregnancies have been reported using frozen semen, in contradistinction to the well documented success for similar techniques in domestic animals and man. The literature indicates that failure to achieve successful pregnancy beyond the first trimester is probably due to acrosomal damage of spermatozoa. Consequently, to maximize benefits from AI, techniques for semen preservation in non-human primates must be further developed.

A valuable adjunct to a simian AI program would be a rapid and reliable index of early pregnancy. Various methods of pregnancy detection are briefly reviewed and preliminary data are presented which suggest that previously described elevations in serum progesterone and/or estrogen may prove useful in detecting pregnancy immediately following implantation (i.e. 12–17 days post-mating).

INTRODUCTION

The scientific basis for artificial insemination (AI) was established in the late 18th century by an Italian priest, Spallanzani, who successfully achieved pregnancy in the bitch after deposition of semen in the vagina. The animal subsequently delivered three puppies, which further strengthened the foundation of this procedure (Perry, 1960; Jones, 1971). The greatest emphasis given to AI came in the early part of the 20th century when a Russian scientist began using this procedure in mares (Perry, 1960; Jones, 1971); however, widespread acceptance came after the

Second World War when it became an essential part of agricultural production technology. The tremendous impact that this simple procedure has had on world animal production can be appreciated when it is realized that in the 1960s, 59 million cows, 47 million ewes, one million sows, 125 thousand mares and 56 thousand goats were artificially inseminated (Jones, 1971). To date, AI has been successfully utilized in cattle, sheep, swine, horses, goats, poultry, rabbits, dogs (Perry, 1960; Jones, 1971; Adams, 1976), and, more recently, humans (Perloff, Steinberger & Sherman, 1964; Behrman & Sawada, 1966; Behrman & Ackerman, 1969; Matheson, Carlborg & Gemzell, 1969; Friberg & Gemzell, 1973) and non-human primates (Dede & Plentl, 1966; Bennett, 1967a; Van Pelt, 1970; Valerio, Leverage, Bensenhaver et al., 1971; Settlage, Swan & Hendrickx, 1973; Czaja, Eisele & Goy, 1975; Martin, Graham & Gould, 1977).

The procedure of AI and the use of frozen semen have proven to be very beneficial to medical science in that they provide a means for treatment of certain forms of infertility and also allow banking of semen for future insemination and/or experimental evaluation. However, it has only been within the past five to ten years that those individuals engaged in human-related research, where the non-human primate was the animal model of choice, have taken advantage of some of the unique features of AI and semen preservation. These techniques are not only utilized for experimental purposes, but also can provide benefits for domestic breeding of non-human primates.

Artificial insemination encompasses four basic components: semen collection, evaluation, preservation and insemination.

SEMEN COLLECTION

Masturbation/Artificial Vagina

The simplest means of obtaining semen in the human primate is by manual masturbation (Matheson et al., 1969; Jones, 1971). However, only a few non-human primate species, such as the chimpanzee, respond consistently to this method (Fussell, Franklin & Frantz, 1973; Martin et al., 1977). Usually, a water-jacketed artificial vagina is used to masturbate the animal which has been trained to present for ejaculation. When chimpanzee semen is collected in this manner, almost the entire ejaculate is in the form of coagulum; however, this will normally partially liquefy within 30 min after ejaculation (Fussell, Franklin et al., 1973; Martin et al., 1977).

Electroejaculation

The most common method of eliciting ejaculation in macaques and other non-human primates utilizes electrical stimulation either with a rectal probe designed to stimulate the area around the prostate (Weisbroth & Young, 1965; Bennett, 1967b; Fussell, Roussel & Austin, 1967; Roussel & Austin, 1968; Kraemer & Vera Cruz, 1969; Van Pelt & Keyser, 1970) or penile electrodes (Mastroianni & Manson, 1963; Dede & Plentl, 1966; Valerio, Ellis *et al.*, 1969; Van Pelt & Keyser, 1970; Settlage & Hendrickx, 1974a). Semen is normally collected in pre-warmed glassware (37–38°C) to alleviate cold shock and then allowed to come to room temperature gradually.

Rectal

This particular method of semen collection is based on the principle that passage of a suitable electrical current across the rectal wall in the vicinity of the prostate gland will stimulate this and other organs to contract and thereby produce an ejaculate. The rectal probe is usually manufactured from plastic or lucite in the form of a rod in which several longitudinal metal bands or cross-sectional rings are embedded (Weisbroth & Young, 1965; Healey & Sadleir, 1966; Fussell, Roussel *et al.*, 1967). The power supply can be either a reduced power variation of commercially available equipment for cattle or sheep or a custom-made unit (Weisbroth & Young, 1965; Fussell, Roussel *et al.*, 1967). The animal is normally anesthetized, and alternate periods of rest and stimulation are administered until ejaculation occurs. This usually occurs within 10 min at about 4–7 V for most macaques (Weisbroth & Young, 1965; Fussell, Roussel *et al.*, 1967) and 8–10 V for baboons (Kraemer & Vera Cruz, 1969). The advantages of this method are that no prior training of an animal is needed in order to obtain semen and that it appears to be more reliable than the penile method (Fussell, Roussel *et al.*, 1967; Van Pelt & Keyser, 1970). The fact that semen quality has been reported to be reduced (Van Pelt & Keyser, 1970) and urine contamination can occur are, however, definite disadvantages.

Penile

The penile method of electroejaculation is perhaps a more difficult method than the rectal probe procedure in that the animal is fully conscious and so must be in a calm state to respond appropriately to stimulation. The foundation for the routine use of penile stimulation in non-human primates lies in the method of Mastroianni & Manson (1963). This has been the method most widely utilized or modified (Valerio, Ellis *et al.*, 1969; Van Pelt & Keyser, 1970; Settlage & Hendrickx, 1974a). The animals are normally restrained either on their backs or in a seated position, and two metal electrodes are applied to the penis, one at the base and one just below the glans. A square wave direct current (DC) stimulus is applied usually in one of three ways. The first involves a surge maneuver wherein the voltage is turned up and down repeatedly. The second method involves an initial phase of a small series of fixed voltage stimulations followed by the surge maneuver of the first method. The third procedure employs the gradual increase of voltage to approximately 40–60 V over a 10- to 15-s interval. If ejaculation does not occur, the procedure is repeated.

In the past, attempts have been made to compare the relative merits of rectal versus penile electroejaculation, especially with regard to the reliability of the methodology and the quality of semen collected in non-human primates (Fussell, Roussel *et al.*, 1967; Van Pelt & Keyser, 1970). The existing data on the two methods of collection are limited and show considerable variation, which makes clear and objective discrimination between them difficult. From the arguments of the proponents of the two methods, it would appear that the rectal probe method is more reliable, while the penile method provides a better quality ejaculate. However, it seems to us that these differences require substantially more experimental verification and are, for all practical purposes, of no real significance at the present time.

Consecutive Ejaculates

The number of consecutive ejaculations obtainable by penile electrode stimulation varies considerably among animals (Settlage & Hendrickx, 1974a). In a study by Settlage (1971), three to 12 sequential ejaculates were obtained from five rhesus monkeys during a total collection period of 45 to 80 min. The number of ejaculations to exhaustion from this group of males was $6·8 ± 1·0$ (mean ± s.e.). Analysis of the ejaculate and spermatozoal characteristics of the first four ejaculates indicated that there was no notable difference in the total volume of each successive ejaculate (Table I). However, there was a significant decrease in spermatozoal concentration from the first to the second ejaculate ($P < 0·05$)

TABLE I

Sperm characteristics of consecutive ejaculates in rhesus monkeys[a]

	Ejaculate			Spermatozoa		
Number	Total volume (ml)[b]	Sperm concentration ($\times 10^6$ ml^{-1})[b]	Number/ ejaculate ($\times 10^6$)	Motile forms (%)	Dead (%)[b]	Abnormal (%)[b]
1	$1·52 ± 0·16$	$1054 ± 53$	1623	100	$1·04 ± 0·15$	$2·84 ± 0·18$
2	$1·58 ± 0·12$	$793 ± 26$	1253	99	$1·24 ± 0·08$	$3·05 ± 0·17$
3	$1·44 ± 0·11$	$729 ± 48$	1050	97	$1·25 ± 0·07$	$2·65 ± 0·18$
4	$1·34 ± 0·12$	$425 ± 104$	570	98	$1·25 ± 0·09$	$3·44 ± 0·62$

[a] Mean number of consecutive ejaculates to exhaustion = $6·8 ± 1·0$ ($n = 5$); mean time period required (minutes) = $59·0 ± 3·56$.
[b] Values expressed as mean ± s.e.
Adapted from Settlage (1971) and Settlage & Hendrickx (1974a).

and from the third to the fourth ejaculate ($P < 0.05$). The overall decline in spermatozoal density from the first to the fourth ejaculate is highly significant ($P < 0.01$). No marked difference in spermatozoal concentration was observed between the second and third ejaculates ($P < 0.10$). The total number of spermatozoa (calculated from the ejaculate volume and spermatozoal density data) decreased in each consecutive ejaculate, with the most notable reduction occurring between the third and fourth ejaculates. There was no significant difference in the percentage of motile, dead, or abnormal spermatozoa in the sequential ejaculates.

SEMEN EVALUATION

The most common parameters used in evaluating a given semen sample are: volume, coagulum, concentration, viability, motility, morphology and the percentage of abnormal forms. The literature provides a wealth of varied information describing the components of semen obtained from numerous primate species (Weisbroth & Young, 1965; Roussel & Austin, 1968; Valerio, Leverage & Munster, 1970; Van Pelt & Keyser, 1970). However, only a few of the important parameters are reported in any single publication. There is also a wide range in values for any given semen characteristic due to variations in the method of collection, handling, processing and the elapsed time between collection and evaluation. Consequently, although an attempt has been made to identify and collate common factors, the data summarized in Table II must be of necessity incomplete.

Volume

Semen volume, usually measured in pipettes or calibrated capillary tubes, is commonly reported as the summation of the fluid and coagulum volumes. Since these two components are seldom reported individually, the proportions of spermatozoa in either fraction are unknown. The coagulum usually forms during or immediately after ejaculation and is the larger volume component (as measured by saline displacement). Fluid volume, however, is one of the key parameters in determining the maximum number of inseminations that can be performed with a given ejaculate.

Coagulum

The coagulum retains varying numbers of spermatozoa as it contracts, which prevents complete utilization of the semen sample. During the 15 to 30-min period after ejaculation, the coagulum contracts and exudes both fluid and spermatozoa (Van Pelt & Keyser, 1970; Settlage &

TABLE II

Semen characteristics of non-human primates

Method of electroejaculation	No. of animals	No. of specimens	Total semen volume[a] (ml)		Sperm conc. (×10⁶ ml⁻¹)		Motile sperm (%)	
			Mean	Range	Mean	Range	Mean	Range
Penile								
Rhesus monkey[b]	3	9	0·12	0·02–0·29	2383	914–4472	85	60–95
Rhesus monkey	80	128	3·36	0·2–5·0	447	100–1500	—	—
Rhesus monkey	6	6	—	—	400	93–807	76	30–99
Crab-eating monkey	14	40	0·83	0·05–6·0	347	110–1120	—	—
Pigtail monkey	1	1	—	—	218	—	90	—
Stumptail monkey	1	1	—	—	264	—	35	—
Rectal								
Chimpanzee	7	19	1·9	0·5–6·2	609	231–1269	30	10–60
Gibbon	3	13	1·3	0·5–4·0	152	51–350	9	0–20
Gelada baboon	2	5	1·0	0·5–2·0	503	351–651	21	5–30
Baboon	4	42	3·6	—	216	—	124	—
Mangabey	1	2	1·3	1·1–1·5	576	542–609	60	55–65
Rhesus monkey	5	33	1·1	0·2–4·5	1069	100–3600	58	10–85
Stumptail monkey	7	22	1·6	0·4–4·0	468	214–1268	49	10–80
Crab-eating monkey	5	17	1·2	0·6–3·0	458	161–830	57	25–75
African green monkey	4	23	0·9	0·3–2·0	440	166–811	39	15–70
Patas monkey	3	21	0·6	0·4–1·0	1153	250–3600	45	10–70
Capuchin	2	15	0·6	0·3–1·0	161	56–740	24	10–50
Squirrel monkey	4	15	0·4	0·2–1·5	206	81–311	52	40–80
Tree-shrew	2	4	0·1	0·1–0·1	103	90–117	52	50–60

[a] Includes both fluid and coagulated fractions.
[b] Values obtained from liquid fractions only.
Adapted from: Mastroianni & Manson (1963); Roussel & Austin (1968); Kraemer & Vera Cruz (1969); Van Pelt & Keyser (1970); Valerio, Leverage & Munster (1970).

Hendrickx, 1974b). If the coagulum is allowed to remain in contact with the fluid portion beyond this period, it may act as a sponge or wick and actually remove spermatozoa from the fluid fraction (Settlage & Hendrickx, 1974b). In some instances, part of the coagulum has been digested by enzymatic action in an attempt to free spermatozoa from this mass (Hoskins & Patterson, 1967; Roussel & Austin, 1967a; Kraemer & Vera Cruz, 1969). This procedure is of particular value when samples are composed almost entirely of coagulum with very little fluid fraction. There are apparently no significant effects of enzyme treatment on the liberated spermatozoa after coagulum dissolution when trypsin is used, but a marked reduction in motility does occur when chymotrypsin is used (Roussel & Austin, 1967a). To avoid the problem of coagulum formation entirely, surgical isolation of the prostate gland has been attempted with some modicum of success (Greer, Roussel & Austin, 1968).

Concentration

One of the most important parameters in semen analysis is the concentration of spermatozoa per unit volume. This is normally determined by direct microscopic count, but a Coulter counter can also be calibrated to provide this information (Brotherton, 1965). The most useful measure of concentration is the number of spermatozoa per ml of fluid semen. Frequently, the spermatozoal concentration has been reported on the basis of total semen volume (including coagulum) which is not a particularly useful measure as the volume includes a large mass which contributes few spermatozoa even when digested. The concentration of living, motile spermatozoa per ml of fluid provides the best basis for determining the maximum number of spermatozoa available for insemination or dilution.

Viability

It is essential to know the number of viable spermatozoa as well as the total number of spermatozoa in an ejaculate. Most methods employ vital staining to differentiate live and dead spermatozoa, and the most commonly used stain combination is an eosin/nigrosin mixture (Swanson & Bearden, 1951). Successful AI requires that competent spermatozoa be available in the oviduct for fertilization of the ovum. The use of vital staining affords the opportunity of determining the relative proportion of living spermatozoa in the fluid portion of an ejaculate. This information is then utilized to adjust the concentration of spermatozoa within an inseminate to provide adequate numbers of living spermatozoa at the site of administration.

The proportion of viable spermatozoa varies considerably among species and even between ejaculates from a single individual.

Motility

Another measurement associated with viable spermatozoa is their motile activity. Motility is usually divided into two categories and each is evaluated by subjective means. Gross motility usually refers to the swirls and eddies observed in a drop of semen, either with the unassisted eye or through the microscope, and includes all motion, even the passive movement of dead spermatozoa induced by surrounding live spermatozoa. Progressive motility is a more accurate indication of true spermatozoal activity as it refers to the extent and quality of forward progression of individual spermatozoa. Those spermatozoa displaying vigorous progressive motility have the greatest probability of surviving long enough to achieve successful fertilization within the oviduct. Therefore, a semen sample with a high percentage of spermatozoa demonstrating vigorous progressive motility is more desirable for AI than one with a lesser rating even though the latter might have greater spermatozoal numbers.

Abnormal Forms

All of the microscopic procedures previously alluded to can also provide information regarding the number and type of abnormal forms present in the sample. As the percentage of abnormal cells increases, the quality of the semen sample decreases and, consequently, its potential fertilizing capacity is reduced. The presence of a cytoplasmic droplet around the spermatozoal neck or mid-piece is usually indicative of immaturity and is not regarded as an abnormality. However, misshapen acrosomes, tail-less or headless spermatozoa and multiple-headed or -tailed spermatozoa are regarded as severe structural defects which probably render the affected spermatozoa incapable of functioning normally. Abnormal spermatozoa are thought to result from disease processes, immaturity or senility of the donor, or may be associated with ageing of the spermatozoa within the male reproductive tract (i.e. epididymis and ductus deferens). The apparent wide divergence in values quoted for the proportions of abnormal forms is partially attributable to the criteria used for identifying abnormalities as well as other influences alluded to previously.

DILUTION

In general, if semen is to be introduced into a site other than the vagina, it is usually diluted beforehand in order to increase the number of animals inseminated with a single ejaculate. It is reasonable to argue that fewer spermatozoa are required to produce adequate conception rates if they are placed in the uterus, as opposed to the cervix and/or vagina, since there would be a shorter distance to travel and therefore a lower loss during transit. The actual number of living, motile spermatozoa required to produce an acceptable conception rate is partially based on empirical data and varies with the method of insemination and whether the sample is to be frozen before insemination. Czaja *et al.* (1975) routinely extended rhesus semen to contain 200 million live spermatozoa ml^{-1} and inseminated 20 million live spermatozoa per insemination in their studies on intrauterine AI, and achieved a conception rate of 39·9%, which was higher than all other methods of insemination. However, additional studies are required to determine the minimum number of spermatozoa necessary to maintain the conception rate when deposited at sites closer to the actual point of fertilization.

The fluid used to extend semen is a very important factor and must be carefully evaluated. Although most commercially available culture media can maintain varying degrees of spermatozoal motility, their relative merits as semen diluents in an AI program have yet to be determined. Two of the more common extension media that have proven successful for simian AI are: a mixture of 25% egg yolk and 75% 0·2 M Tris-buffered glucose without glycerin, potassium or antibiotics (Czaja *et al.*, 1975) and a more complex mixture, Norman-Johnson Solution 1 or 2 (Van Pelt, 1970).

ARTIFICIAL INSEMINATION

Successful AI in non-human primates was first reported by Mastroianni & Rosseau (1965) and Mastroianni *et al.* (1967), who first recovered fertilized ova after AI in rhesus monkeys with intrauterine devices *in situ*. Marston & Kelly (1968) collected fertilized ova after intrauterine insemination with epididymal spermatozoa, and Bennett (1967a) reported similar results for squirrel monkeys in which the animals were induced to ovulate with injections of human chorionic gonadotropin (HCG). The delivery of viable offspring following AI was first reported by Dede & Plentl (1966). A conception rate of 12·1% was achieved by AI compared to 13·7% by natural mating (Table III). In parallel experiments, these authors recorded a conception rate of 13·3% by combining AI with induction of ovulation with HCG. Since these initial experiments, a number of studies have been conducted which have emphasized the site of insemination in relation to the conception rate and, to a lesser extent, the portion of the ejaculate and number of spermatozoa used in AI.

Four sites of seminal deposition have been employed. Vaginal insemination is the most common and simulates natural mating most closely. Although the actual insemination procedure varies with the investigator, all or part of the ejaculate is inserted deep in the vagina with either forceps (Dede & Plentl, 1966; Marston & Kelly, 1968) or a syringe (Settlage *et al.*, 1973) or both (Mastroianni & Rosseau, 1965; Mastroianni *et al.*, 1967; Valerio, Leverage, Bensenhaver *et al.*, 1971). Intracervical

TABLE III

Comparison of site of insemination in Macaca mulatta

Site of insemination	Breeding schedule	No. of inseminations	No. of pregnancies	Conception rate[a]	Pregnant females (%)
Artificial					
Vaginal					
Dede & Plentl (1966)	Calendar	33	4	12·1	—
Valerio, Leverage, Bensenhaver *et al.* (1971)	Calendar	124	5	4·03	—
Settlage *et al.* (1973)	Calendar	47	10	21·0	50 (10/20)
Intrauterine					
Valerio, Leverage, Bensenhaver *et al.* (1971)	Calendar	34	7	20·6	70 (7/10)
Czaja *et al.* (1975)	Sex skin	218	87	40·0	63 (69/109)
Intraperitoneal					
Van Pelt (1970)	Calendar	8	3	38·0	38 (3/8)
Natural					
Dede & Plentl (1966)	Calendar	160	22	13·7	—
Valerio, Leverage, Bensenhaver *et al.* (1971)	Calendar	6156	799	12·97	—
Valerio, Leverage, Bensenhaver *et al.* (1971)	Calendar	31	7	22·6	77·8 (7/9)
Settlage *et al.* (1973)	Calendar	96	16	16·6	40 (16/40)

[a] (No. of conceptions/No. of matings) × 100.

insemination is accomplished by partially penetrating the cervix with a long, blunt-tipped needle attached to a syringe or with a pipette for injection of the semen. Intrauterine insemination is accomplished using a teflon catheter (Valerio, Leverage, Bensenhaver *et al.*, 1971) or a long, 18-gauge needle (Czaja *et al.*, 1975) which is attached to a syringe and introduced through the cervix into the uterus for deposition of the semen. Intraperitoneal insemination is accomplished by injection of the semen by means of a syringe and 22-gauge needle through the body wall into the lower abdominal cavity.

Vaginal Insemination

Valerio, Leverage, Bensenhaver *et al.* (1971) reported conception rates of 4·03% and 9·25% from AI compared to 12·97% and 18·78% from natural mating in *Macaca mulatta* and *M. fascicularis*, respectively. Their data suggest that the conception rate following natural mating is considerably higher than that achieved with AI. However, Dede & Plentl (1966) and Settlage *et al.* (1973) reported conception rates from AI in rhesus monkeys to be equal to or even higher than those from natural mating (Table III). The methodology of the three studies was essentially the same except that Valerio, Leverage, Bensenhaver *et al.* (1971) allowed the ejaculate to liquefy for 10 min at 37°C before vaginal deposition. Although experiments have not been done to confirm the optimal time interval from collection to insemination, it is possible that the 10-min time interval and/or the conditions of storage during liquefaction adversely affected the spermatozoa and resulted in a lower conception rate.

Intracervical Insemination

Leverage, Valerio & Schultz (1971) reported a conception rate of 21·2% in *M. mulatta* and *M. fascicularis* for intracervical AI compared to a conception rate of 13% for natural mating.

Intrauterine Insemination

Valerio, Leverage, Bensenhaver *et al.* (1971) and Czaja *et al.* (1975) achieved a high conception rate with intrauterine insemination in the rhesus monkey (Table III). Although there was a considerable difference in the overall conception rate between the two studies, the data indicate that this approach is quite successful. It should be emphasized that while other aspects of the two studies varied only slightly, the main difference observed in conception rates may be explained on the basis that optimal mating time was more accurately predicted on the basis of sex skin by Czaja *et al.* (1975) than the calendar method of mating employed by Valerio, Leverage, Bensenhaver *et al.* (1971).

Intraperitoneal Insemination

Van Pelt (1970) reported a conception rate of 38% from intraperitoneal AI in the rhesus monkey, suggesting that this approach is comparable to or exceeds the conception rate achieved in other studies utilizing more conventional methodology. This approach may have less practical application than the others, but it may be uniquely applicable to studies concerned with fertilization and early embryonic development.

Comparison of these data suggests that intrauterine and intraperitoneal insemination sites yield the highest conception rates. There does not appear to be a significant difference in the conception rate for vaginal and intracervical insemination. From the practical standpoint, vaginal and intracervical insemination are considerably simpler and less time consuming, while the intrauterine and intraperitoneal inseminations offer distinct advantages for experimental studies, especially those concerned with the temporal aspects of early embryonic development.

SEMEN PRESERVATION

Cooling

Short-term preservation of semen through cooling to 4–5°C has been attempted in numerous species, but the results have rarely equalled those obtained with bull semen, the most extensively studied (Leverage, Valerio, Schultz et al., 1972). The semen of the human and the rhesus monkey does not survive well when stored for periods up to 24 hours at 4°C (Leverage, Valerio, Schultz et al., 1972), and so this method of storage seems unsuitable for primate semen at present.

Freezing

Numerous reports have appeared in the literature describing freeze preservation of mammalian cells (Sherman, 1964, 1965; Mazur, 1966). Among the many factors which must be considered in utilizing freeze preservation of germinal cells are freeze rate, dilution rate, medium, type and quantity of cryoprotectant, storage temperature, thaw rate and cell sensitivity. In the period since the initial study in 1953, human semen has been successfully frozen with a subsequent high spermatozoal motility upon thawing (Bunge & Sherman, 1953; Freund & Wiederman, 1966), and the thawed semen has produced viable offspring either by homologous or heterologous AI (Behrman & Sawada, 1966; Friberg & Gemzell, 1973).

In the non-human primate, successful freezing of semen with recovery of near pre-freeze levels of motility has been recently described (Table IV). Although most freezing procedures used for non-human primate semen are similar to those used for other species, one notable

TABLE IV

Methods and results of freezing simian semen

Species	Semen collection method	Freezing method	Sperm concentration (ml⁻¹)	Number of trials	Motility (%) Pre-freeze	Motility (%) Post-freeze	Survival (%)
Baboon	Rectal electroejaculation	N₂ Vapor	10×10^6 Motile	11	69·5	44·1	64·5
	Spontaneous ejaculation	N₂ Vapor	10×10^6 Motile	3	76·7	48·3	63·1
Squirrel monkey	Rectal electroejaculation	Solid CO₂	75×10^6 Total	11	63·6	51·8	87·9
				13	68·1	53·8	83·4
Rhesus monkey	Penile	Equil. + N₂ vapor[a]	150–200	20	0	0	0
		No equil. + N₂ vapor		35	73	NG[b]	53
		No equil. + N₂ vapor	$\times 10^6$ Total	50	69	47	68
Rhesus monkey	Rectal	N₂ Vapor	1:10 Ratio	7	54	27	50
Stumptail monkey	Rectal	N₂ Vapor	1:10 Ratio	3	50	25	50
Patas monkey	Rectal	N₂ Vapor	1:10 Ratio	7	45	23	51
African green monkey	Rectal	N₂ Vapor	1:10 Ratio	8	53	28	53
Chimpanzee	Rectal	N₂ Vapor	1:10 Ratio	2	50	27	54

[a] Equilibrated at 4°C for up to 24 hours prior to freezing over liquid nitrogen.
[b] Not given.
Adapted from: Kraemer & Vera Cruz (1969); Denis *et al.* (1976); Roussel & Austin (1967b), Leverage, Valerio, Schultz *et al.* (1972).

difference is that primate semen (at least, human and rhesus monkey) cannot be equilibrated for 12–24 hours at 4–5°C with recovery of motility (Leverage, Valerio, Schultz *et al.*, 1972). These authors reported that primate spermatozoa frozen in this manner yielded no motility upon thawing, whereas at least 50% motility was noted with spermatozoa that were frozen without prior equilibration at 5°C.

There are two basic methods by which simian semen has been successfully frozen: one involves a slow freezing process (1–5°C min^{-1}) until −20°C is reached (Leverage, Valerio, Schultz *et al.*, 1972) and the other utilizes a rapid freeze technique with either liquid nitrogen vapor (Roussel & Austin, 1967b; Kraemer & Vera Cruz, 1969) or placement of semen droplets on solid carbon dioxide (Denis *et al.*, 1976). The solution used to extend the semen prior to freezing is similar to that successfully used for bovine spermatozoa and consists of 20–25% fresh egg yolk (avian) and glycerol (4–14%) which act as cryoprotectants. These agents help to maintain the proper osmotic balance as the medium passes from liquid to solid. Other components that have been successfully used as extender solutions are: a mixture of sodium citrate, sodium bicarbonate and potassium phosphate (Leverage, Valerio, Schultz *et al.*, 1972); sodium glutamate (Roussel & Austin, 1967b; Kraemer & Vera Cruz, 1969) and lactose (Denis *et al.*, 1976). The concentration of spermatozoa for freezing has varied quite extensively among investigators, ranging from approximately 10 million motile sperm per ml before freezing (Kraemer & Vera Cruz, 1969) to 150–200 total sperm per ml (Leverage, Valerio, Schultz *et al.*, 1972). In some instances, dilution has been made on a fixed ratio basis (i.e. 1 ml semen to 9 ml diluent) (Roussel & Austin, 1967b) as the result of prior estimates of spermatozoal concentration. In order to maximize the number of spermatozoa available for freezing, some investigators have subjected the ejaculate to enzymatic digestion in order to free spermatozoa trapped in the coagulum.

The major obstacle to freeze-preservation of simian semen appears to be the maintenance of its fertilizing capacity upon thawing. Even though recovery of near pre-treatment motility is obtainable, no term pregnancy utilizing this preparation in an AI program has been reported. In two reports, frozen simian semen has produced pregnancy, but these aborted within the first trimester. Leverage, Valerio, Schultz *et al.* (1972) reported that of 48 inseminations with frozen rhesus semen, only one resulted in a conception and that the pregnancy failed after about 40 days of gestation. In 1975, Cho, Honjo & Makita reported a conception rate of 15% (2/13 inseminations) after AI with freeze-preserved semen of the cynomolgus monkey. These pregnancies also terminated spontaneously within six to eight weeks. It is apparent, therefore, that recovery of motility does not necessarily reflect the fertilizing capacity of the spermatozoa or the potential viability of the zygote. These same two groups of investigators have also placed at least partial blame for the failure to achieve successful term pregnancy on the observed damage to the acrosomes of frozen simian spermatozoa. Similar damage has been observed in ram (Healey, 1969; Nath, 1972), boar, stallion and chinchilla spermatozoa, but not bull spermatozoa (Healey, 1969), again indicating the apparent hardiness of bull spermatozoa. Electron microscopy studies (Leverage, Valerio, Schultz *et al.*, 1972; Cho *et al.*, 1975) have indicated that moderate to severe damage occurs in the acrosome in the form of fenestrations or tears in the acrosomal membranes with loss of acrosomal matrix and perhaps damage to the nucleus itself. Since the enzymes contained in the acrosome are vital to spermatozoal penetration of the zona pellucida and, therefore, eventual fusion with the oocyte, the loss of this material may be the factor which precludes successful fertilization. If, however, the spermatozoon is only partially damaged and fuses with the oocyte, the damage associated with the spermatozoal nuclear material is apparently responsible for the subsequent death of the embryo. It appears, therefore, that in order to achieve successful term pregnancy in the non-human primate, the degree of damage to the acrosome and the nucleus must be reduced in order to achieve viable offspring.

POTENTIAL UTILITY OF ARTIFICIAL INSEMINATION

Austin (1972) has discussed ways in which reproduction in non-human primates can be controlled and procedures which have been successfully applied in this regard. Artificial insemination can be utilized for both domestic breeding and experimental research. The potential usefulness of AI in these two areas is summarized below.

Breeding

1. Reduction in number of males needed in breeding colony.
2. More effective breeding of socially incompatible animals.
3. Improvement of genetic stock.
4. Use of reluctant, incapacitated or oligospermic males.
5. Establishment of semen banks for future breeding.
6. Prevention of disease transmission.

Experimental

1. Control of exact timing of gestation.
2. Determination of time required for capacitation.
3. Determination of number of spermatozoa required for fertilization.
4. Comparison of male fertility.
5. Determination of gamete survival time in the female reproductive tract.
6. Control of genetic make-up of animals.

METHODS OF PREGNANCY DETECTION

Physical Indicators

Although alterations in the physical characteristics of non-human primates indicative of pregnancy become apparent during gestation, the timing and degree of change shows substantial species and individual variability. The most general observation, evident after 60–100 days, is an increase in lower abdominal circumference and a sustained increase in the intensity of the red/purple coloration of the skin in the perineal region. While gestational amenorrhea is a consistent feature after the first 60 days, early diagnosis of pregnancy by the absence of menses is complicated by the presence of vaginal bleeding associated with implantation (the placental sign first described in the rhesus monkey by Hartman (1932) and more rarely observed in other macaques and baboons). The placental sign, when it occurs, unfortunately appears at approximately the same time and may have the same duration as normal menstrual flow in an infertile cycle, which prevents certain diagnosis of pregnancy on the basis of this observation.

The most common method of early pregnancy diagnosis, palpation of the uterus through the rectal wall, has proven to be reliable if the investigator has had substantial experience and has prior knowledge of uterine size and tone in the specific animal under investigation in the absence of a conceptus. Utilizing this procedure, pregnancy can be determined after 22–30 days of gestation in macaques and baboons, although a second palpation 15–20 days later will be required to substantiate the initial observation if questionable. In view of the marked biological variability in the external markers of early pregnancy and the experience needed to identify these alterations, it is not surprising that more specific methods have been sought, particularly if detection before Days 25–30 is required.

Urine and Serum Indicators

Following the observation that monkeys secrete chorionic gonadotropin during the early stages of pregnancy (Hamlett, 1937; van Wagenen & Simpson, 1955), several groups (Tullner & Hertz, 1966; Arslan, Meyer & Wolf, 1967; Tullner, 1968; Hodgen, Dufau et al., 1972) demonstrated that significant quantities of this placental hormone could be identified in either blood or urine between 15 and 35 days after successful mating. The bioassays used, however, were too cumbersome and time consuming for simple pregnancy tests.

The goal of a rapid and specific test for non-human primate chorionic gonadotropin early in the course of gestation was achieved (Hodgen, Ross et al., 1974) following the production of an antiserum to ovine luteinizing hormone which reacted with the antigenic determinants common to the chorionic gonadotropin of humans, gorillas, orang-utans, chimpanzees, baboons and macaques. Using small quantities of unextracted urine (minimum of 0·2 ml), the hemagglutination inhibition test for urinary chorionic gonadotropin can be completed in two hours. A positive test was found in one animal on Day 14, but it is most reliable on Days 18–23 after fertilization. Our experience, and that of Cole & Dal Corobbo (1977), indicates that if the reagents are used as directed during the stable period and if two urine samples collected 24 hours apart are tested, no false positives are recorded and a very high percentage (89–94%) of true negatives and positives are identified. This kit is available through the National Institutes of Health.*

Steroid Levels

The circulating levels of progesterone in the luteal phase of fertile and non-fertile cycles are indistinguishable until Days 22–24 of the cycle (9–12 days following ovulation). At this point, the corpus luteum of the menstrual cycle begins to regress and progesterone levels decrease, whereas in the fertile cycle this decrease does not occur and, in fact, serum levels may increase to values two- to three-fold that seen in the infertile cycle (Neill, Johansson & Knobil, 1969; Hodgen, Dufau et al., 1972; Bosu, Johansson & Gemzell, 1973; Atkinson et al., 1975). Estrogens show a similar pattern with significantly greater concentrations present in the peripheral plasma of pregnant animals 12–15 days after ovulation (Atkinson et al., 1975).

Using a rapid radioimmunoassay for total estrogen and progesterone which provided serum concentrations within 4–6 hours of sample collection, we have demonstrated that elevated steroid concentrations on Days 14–17 can be used as an index for early pregnancy in the bonnet monkey (Fig. 1). In this retrospective study, it was determined that estrogen levels must exceed 200 pg ml^{-1} for three of the four days and progesterone levels must exceed 2·5 ng ml^{-1} for the same period before reliable detection of pregnancy can be assured at this point of gestation (Fig. 1A). In no case did the 23 non-pregnant animals demonstrate both estrogen and progesterone concentrations greater than the suggested criteria. Although six of the non-pregnant monkeys showed progesterone concentrations which exceeded the criteria, estrogen values were clearly much lower (Fig. 1B). Since two of the ten pregnant animals (included in Fig. 1A) and four non-pregnant animals exceeded the estrogen minimum criteria but did not exceed the progesterone criteria (Fig. 1C), it is not possible to rule out pregnancy in animals with elevated estrogens alone. Consequently, the rate of false positive pregnancy identification when estrogen alone is increased is 4/33 or 12% of the total animals studied. We also attempted to determine whether the analysis of estrogen and progesterone in a single serum sample obtained on Day 27 of the menstrual

---

* Enquiries should be addressed to the Hormone Distribution Officer, Office of the Director, NIAMDD, Building 31A, Room 9A47, NIH, Bethesda, Maryland 20014, USA.

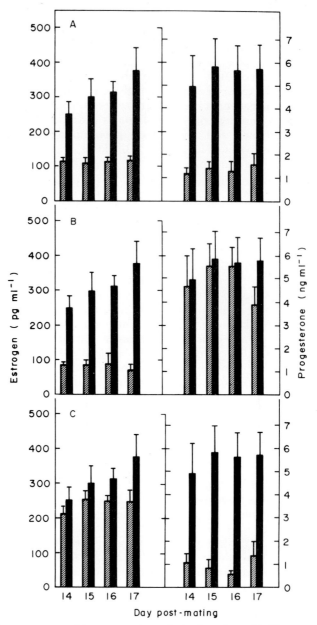

FIG. 1. Mean serum steroid concentrations obtained on Days 14–17 post-mating in 33 bonnet monkeys. The data from pregnant animals (solid bars, $n = 10$) are repeated in A, B and C. The low concentrations of both estrogen (left) and progesterone (right) in 13 non-pregnant animals (stippled bars) are shown in A. Similar values for non-pregnant animals demonstrating either elevated progesterone or estrogen are shown by stippled bars in B ($n = 6$) and C ($n = 4$) respectively (vertical lines indicate s.e.).

cycle (corresponding to 15–17 days of gestation) in the rhesus monkey would permit identification of pregnant animals. Although both steroids were markedly elevated (Table V), the large variability in progesterone values suggested that luteal function fluctuates more in the rhesus monkey than in the bonnet monkey. While it would appear that a single elevated serum estrogen and progesterone value is an adequate indicator of pregnancy in the rhesus monkey, multiple sample analysis in early gestation should prove a more reliable index in this species as well.

TABLE V

Serum steroid concentrations in pregnant and non-pregnant rhesus monkeys on Day 27 (Day 15–17 post-mating) of the menstrual cycle

	Estrogen (pg ml^{-1}) (mean ± s.e.)	Progesterone (ng ml^{-1}) (mean ± s.e.)
Pregnant ($n = 6$)	222 ± 15[a]	$4 \cdot 75 \pm 1 \cdot 27$[b]
Non-pregnant ($n = 5$)	107 ± 21[a]	$0 \cdot 85 \pm 0 \cdot 23$[b]

[a] $P < 0 \cdot 01$, $t = 4 \cdot 36$.
[b] $P < 0 \cdot 05$, $t = 3 \cdot 04$.

ACKNOWLEDGEMENTS

We would like to acknowledge Dr K. Matayoshi, R. Parker, G. Stoner and S. Virk for their technical contributions to this study. This study was supported by the National Institute of Health grant PR00169 and contract HO-1-2088.

REFERENCES

Adams, C. E. (1976). Bibliography on artificial insemination in the rabbit. *Bibliography Reprod. Res. Inf. Serv. Camb.* No. 93.

Arslan, M., Meyer, R. K. & Wolf, R. C. (1967). Chorionic gonadotropin in the blood and urine of pregnant rhesus monkeys (*Macaca mulatta*). *Proc. Soc. exp. Biol. Med.* **125**: 349–352.

Atkinson, L. E., Hotchkiss, J., Fritz, G. R., Surve, A. H., Neill, J. D. & Knobil, E. (1975). Circulating levels of steroids and chorionic gonadotropin during pregnancy in the rhesus monkey, with special attention to the rescue of the corpeus luteum in early pregnancy. *Biol. Reprod.* **12**: 335–345.

Austin, C. R. (1972). Experimental control of reproduction in primates. In *Breeding primates*: 184–197. Beveridge, W. I. B. (ed.). Basel: Karger.

Behrman, S. J. & Ackerman, D. R. (1969). Freeze preservation of human sperm. *Am. J. Obstet. Gynec.* **103**: 654–661.

Behrman, S. J. & Sawada, Y. (1966). Heterologous and homologous inseminations with human semen frozen and stored in a liquid-nitrogen refrigerator. *Fert. Steril.* **17**: 457–466.

Bennett, J. P. (1967a). Artificial insemination of the squirrel monkey. *J. Endocr.* **37**: 473–474.

Bennett, J. P. (1967b). Semen collection in the squirrel monkey. *J. Reprod. Fert.* **13**: 353–355.

Bosu, W. T. K., Johansson, E. D. B. & Gemzell, C. (1973). Peripheral plasma levels of oestrogens, progesterone and 17α-hydroxyprogesterone during gestation in the rhesus monkey. *Acta endocr., Copenh.* **74**: 348–360.

Brotherton, J. (1965). The counting and sizing of spermatozoa from ten animal species using a Coulter counter. *Andrologia* **7**: 169–185.

Bunge, R. G. & Sherman, J. K. (1953). Fertilizing capacity of frozen human spermatozoa. *Nature, Lond.* **172**: 767–768.

Cho, F., Honjo, S. & Makita, T. (1975). Fertility of frozen-preserved spermatozoa of cynomolgus monkeys. In *Contemporary primatology. Int. congr. Primatol.* **5**: 125–133. Kondo, S., Kawai, M. & Ehara, A. (eds). Basel: Karger.

Cole, M. P. & Dal Corobbo, M. D. (1977). Evaluation of the sub-human primate tube test for pregnancy in the rhesus monkey (*Macaca mulatta*). *Program Am. Soc. Primatol.* April 16–19.

Czaja, J. A., Eisele, S. G. & Goy, R. W. (1975). Cyclical changes in the sexual skin of female rhesus: Relationships to mating behavior and successful artificial insemination. *Fedn Proc. Fedn Am. Socs exp. Biol.* **34**: 1680–1684.

Dede, J. A. & Plentl, A. A. (1966). Induced ovulation and artificial insemination in a rhesus colony. *Fert. Steril.* **17**: 757–764.

Denis, L. T., Poindexter, A. N., Ritter, M. B., Seager, S. W. J. & Deter, R. L. (1976). Freeze preservation of squirrel monkey sperm for use in timed fertilization studies. *Fert. Steril.* **27**: 723–739.

Freund, M. & Wiederman, J. (1966). Factors affecting the dilution, freezing and storage of human semen. *J. Reprod. Fert.* **11**: 1–17.

Friberg, J. & Gemzell, C. (1973). Inseminations of human sperm after freezing in liquid nitrogen vapors with glycerol or glycerol-egg-yolk-citrate as protective media. *Am. J. Obstet. Gynec.* **116**: 330–334.

Fussell, E. N., Franklin, L. E. & Frantz, R. C. (1973). Collection of chimpanzee semen with an artificial vagina. *Lab. Anim. Sci.* **23**: 252–255.

Fussell, E. N., Roussel, J. D. & Austin, C. R. (1967). Use of the rectal probe method for electrical ejaculation of apes, monkeys and a prosimian. *Lab. Anim. Care* **17**: 528–530.

Greer, W. E., Roussel, J. D. & Austin, C. R. (1968). Prevention of coagulum in monkey semen by surgery. *J. Reprod. Fert.* **15**: 153–155.

Hamlett, G. W. D. (1937). Positive Friedman tests in the pregnant rhesus monkey. *Am. J. Obstet. Gynec.* **37**: 287–290.

Hartman, C. G. (1932). Studies in the reproduction of the monkey *Macacus* (*Pithecus*) *rhesus* with special reference to menstruation and pregnancy. *Contr. Embryol.* **23**: 1–161.

Healey, P. (1969). Effect of freezing on the ultrastructure of the spermatozoon of some domestic animals. *J. Reprod. Fert.* **18**: 21–27.

Healey, P. & Sadleir, R. M. F. S. (1966). The construction of rectal electrodes for electroejaculation. *J. Reprod. Fert.* **11**: 299–301.

Hodgen, G. D., Dufau, M. L., Catt, K. J. & Tullner, W. W. (1972). Estrogens, progesterone and chorionic gonadotropin in pregnant rhesus monkeys. *Endocrinology* **91**: 896–900.

Hodgen, G. D., Ross, G. T., Turner, C. K., Barker, D. L. & O'Connor, A. M. (1974). Pregnancy diagnosis by a hemagglutination inhibition test for urinary macaque chorionic gonadotropin (mCG). *J. clin. Endocr. Metab.* **38**: 927–930.

Hoskins, D. D. & Patterson, D. L. (1967). Prevention of coagulum formation with recovery of motile spermatozoa from rhesus monkey semen. *J. Reprod. Fert.* **13**: 337–340.

Jones, R. C. (1971). Uses of artificial insemination. *Nature, Lond.* **229**: 534–537.

Kraemer, D. C. & Vera Cruz, N. C. (1969). Collection, gross characteristics and freezing of baboon semen. *J. Reprod. Fert.* **20**: 345–348.

Leverage, W. E., Valerio, D. A. & Schultz, A. P. (1971). Paper presented at session of Am. Assoc. lab. Anim. Sci., cited by Settlage, D. S. F., Swan, S. & Hendrickx, A. G. (1973). Comparison of artificial insemination with natural mating technique in rhesus monkeys, *Macaca mulatta. J. Reprod. Fert.* **32**: 129–132.

Leverage, W. E., Valerio, D. A., Schultz, A. P., Kingsbury, E. & Dorey, C. (1972). Comparative study on the freeze preservation of spermatozoa. Primate, bovine and human. *Lab. Anim. Sci.* **22**: 882–889.

Marston, J. H. & Kelly, W. A. (1968). Time relationships of spermatozoon penetration into the egg of the rhesus monkey. *Nature, Lond.* **217**: 1073–1074.

Martin, D. E., Graham, C. E. & Gould, K. G. (1977). Successful pregnancy in the chimpanzee using artificial insemination. *Program. Am. Soc. Primatol.* April 16–19.

Mastroianni, L., Jr. & Manson, W. A., Jr. (1963). Collection of monkey semen by electroejaculation. *Proc. Soc. exp. Biol. Med.* **112**: 1025–1027.

Mastroianni, L., Jr. & Rosseau, C. U. (1965). Influence of the intrauterine coil on ovum transport and sperm distribution in the monkey. *Am. J. Obstet. Gynec.* **93**: 416–420.

Mastroianni, L., Jr., Suzuki, S., Manabe, Y. & Watson, F. (1967). Further observations on the influence of the intrauterine device on ovum and sperm distribution in the monkey. *Am. J. Obstet. Gynec.* **99**: 649–661.

Matheson, G. W., Carlborg, L. & Gemzell, C. (1969). Frozen human semen for artificial insemination. *Am. J. Obstet. Gynec.* **104**: 495–501.

Mazur, P. (1966). Theoretical and experimental effects of cooling and warming velocity on the survival of frozen and thawed cells. *Cryobiology* **2**: 181–192.

Nath, J. (1972). Correlative biochemical and ultrastructural studies on the mechanism of freezing damage to ram semen. *Cryobiology* **9**: 240–246.

Neill, J. D., Johansson, E. D. B. & Knobil, E. (1969). Patterns of circulating progesterone concentrations during the fertile menstrual cycle and the remainder of gestation in the rhesus monkey. *Endocrinology* **84**: 45–48.

Perloff, W. H., Steinberger, E. & Sherman, J. K. (1964). Conception with human spermatozoa frozen by nitrogen vapor technic. *Fert. Steril.* **15**: 501–504.

Perry, E. J. (ed.). (1960). *The artificial insemination of farm animals*, 3rd rev. edn. New Brunswick, New Jersey: Rutgers University Press.

Roussel, J. D. & Austin, C. R. (1967a). Enzymic liquification of primate semen. *Int. J. Fert.* **12**: 288–290.

Roussel, J. D. & Austin, C. R. (1967b). Preservation of primate spermatozoa by freezing. *J. Reprod. Fert.* **13**: 333–335.

Roussel, J. D. & Austin, C. R. (1968). Improved electroejaculation of primates. *J. Inst. Anim. Techns* **19**: 22–32.

Settlage, D. S. F. (1971). *Establishment of normal parameters in semen analysis of highly fertile rhesus monkeys* (Macaca mulatta). M.S. Thesis: University of California, Davis.

Settlage, D. S. F. & Hendrickx, A. G. (1974a). Electroejaculation technique in *Macaca mulatta* (rhesus monkeys). *Fert. Steril.* **25**: 157–159.

Settlage, D. S. F. & Hendrickx, A. G. (1974b). Observations on coagulum characteristics of the rhesus monkey electroejaculate. *Biol. Reprod.* **11**: 619–623.

Settlage, D. S. F., Swan, S. & Hendrickx, A. G. (1973). Comparison of artificial insemination with natural mating technique in rhesus monkeys, *Macaca mulatta*. *J. Reprod. Fert.* **32**: 129–132.

Sherman, J. K. (1964). Low temperature research on spermatozoa and eggs. *Cryobiology* **1**: 103–129.

Sherman, J. K. (1965). Practical applications and technical problems of preserving spermatozoa by freezing. *Fedn Proc. Fdn Am. Socs exp. Biol.* **24**: 5288–5296.

Swanson, E. W. & Bearden, H. J. (1951). An eosin-nigrosin stain for differentiating live and dead bovine spermatozoa. *J. Anim. Sci.* **10**: 981–987.

Tullner, W. W. (1968). Urinary chorionic gonadotropin excretion in the monkey (*Macaca mulatta*), early phase. *Endocrinology* **82**: 874–875.

Tullner, W. W. & Hertz, R. (1966). Chorionic gonadotropin levels in the rhesus monkey during early pregnancy. *Endocrinology* **78**: 204–207.

Valerio, D. A., Ellis, E. B., Clark, M. L. & Thompson, G. E. (1969). Collection of semen from macaques by electroejaculation. *Lab. Anim. Care* **19**: 250–252.

Valerio, D. A., Leverage, W. E., Bensenhaver, J. C. & Thornett, H. D. (1971). The analysis of male fertility, artificial insemination and natural matings in the laboratory breeding of macaques. In *Medical primatology 1970*: 515–526. Goldsmith, E. I. & Moor-Jankowski, J. (eds). Basel: Karger.

Valerio, D. A., Leverage, W. E. & Munster, J. H. (1970). Semen evaluation in macaques. *Lab. Anim. Care* **20**: 734–740.

Van Pelt, L. F. (1970). Intraperitoneal insemination of *Macaca mulatta*. *Fert. Steril.* **21**: 159–162.

Van Pelt, L. F. & Keyser, P. E. (1970). Observations on semen collection and quality in macaques. *Lab. Anim. Care* **20**: 726–733.

van Wagenen, G. & Simpson, M. E. (1955). Gonadotrophic hormone excretion of the pregnant monkey (*Macaca mulatta*). *Proc. Soc. exp. Biol. Med.* **90**: 346–348.

Weisbroth, S. & Young, F. A. (1965). The collection of primate semen by electroejaculation. *Fert. Steril.* **16**: 229–235.

Symp. zool. Soc. Lond. (1978) No. 43, 241–248

Artificial Insemination in Foxes

J. AAMDAL, J. FOUGNER and K. NYBERG

The Veterinary College of Norway, Oslo, Norway

SYNOPSIS

Experiments with artificial insemination in foxes were commenced in 1969. Collection of semen from male foxes by digital manipulation is described. Semen from foxes has been diluted with Illini Variable Temperature (IVT) diluent if intended for insemination of fresh semen, or with Tris-diluent for freezing. The number of spermatozoa per insemination dose was 150–200 million. A method for intrauterine insemination has been worked out.

The conception rate after insemination with fresh semen was 80%, and after frozen semen was 72%.

By using artificial insemination in fox breeding it is possible to extend the use of male foxes both by dividing the ejaculate into three or four insemination doses and by prolonging the breeding period of the male.

INTRODUCTION

The breeding of foxes for fur production is of considerable importance in some countries (Canada, Finland, Norway, Poland, USSR). Two areas which need to be developed are the increased use of particular males and the storage of genes of special mutants. Artificial insemination, particularly with frozen semen, could achieve both objectives. There are several published reports of collection and insemination of fresh semen (Starkov, 1933–34; Venge, 1959; Chronopulo, 1961; Creed, 1964). A report of the insemination of foxes with frozen semen was published in 1972 (Aamdal, Andersen & Fougner, 1972).

In this paper methods for collection, dilution, freezing and insemination are described.

MATERIALS AND METHODS

Collection and Processing of Semen

As the foxes kept for fur production are semi-wild animals, the collection of semen presents some problems.

Two different methods for collection of semen from male foxes are described in the literature. Chronopulo (1961) collected semen by electroejaculation, whilst Starkov (1933–34) used a type of digital manipulation. In our experiments both methods have been tried. We found it possible to produce ejaculation by electric impulses, but the procedure took a long time and appeared to be painful to the animals. Moreover, the number of spermatozoa obtained was generally low. Collection of semen by digital manipulation has also given some problems, but we found it more suitable than electroejaculation.

IVT extender (VanDemark & Sharma, 1957) has been found suitable for dilution of fox semen which is to be inseminated in the fresh state. No reduction in motility was noticed during the first 24 hours after collection. We have, however, no insemination experiment where stored fox semen has been used. For deep freezing of fox semen different extenders have been tried. We concluded that Tris-extender (Davis, Bratten & Foote, 1963) was suitable for fox semen. The following procedure was used for the freezing and thawing of fox semen: the semen was diluted 1 : 3–1 : 4 with Tris-extender, equilibrated for three hours, and frozen in 0·5 ml plastic straws in nitrogen vapour (Jondet, 1964). It was thawed at 70°C in 8 s (Aamdal & Andersen, 1968). More than 60% of spermatozoa were motile after thawing.

Insemination Technique

After some experimentation, an intrauterine insemination method was worked out (Fougner, Aamdal & Andersen, 1973). A plastic tube is inserted into the vagina of the vixen. Then the cervix is fixed from the outside of the abdomen and a catheter is threaded through the cervical canal and into the uterus (Fig. 1). Satisfactory use of the technique requires a knowledge of the anatomy of the reproductive organs in vixens. Technicians are trained on excised reproductive organs from vixens in oestrus.

RESULTS

Of 40 male blue foxes, 35 accepted digital manipulation (Fig. 2). Silver foxes and mutants of silver foxes are often more shy and nervous than blue foxes, and collection of semen of this species is sometimes more difficult.

The composition of fox semen is similar to that of dog semen, with a sperm-rich fraction and a larger fraction from the accessory organs, mainly from the prostate gland. The mean volume of the sperm-rich fraction in 35 ejaculates from five male blue foxes was 0.62 ml (Aamdal, 1972; Table I). The mean total number of spermatozoa obtained was 648 million.

Our experience showed that collections must be limited to two or three times per week. More frequent collection led to a reduction in the number of spermatozoa ejaculated and the animals became less willing to accept digital manipulation.

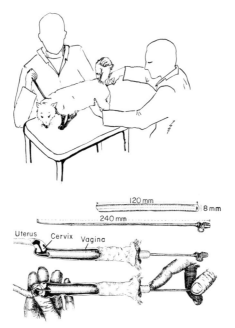

Fig. 1. Insemination technique and equipment

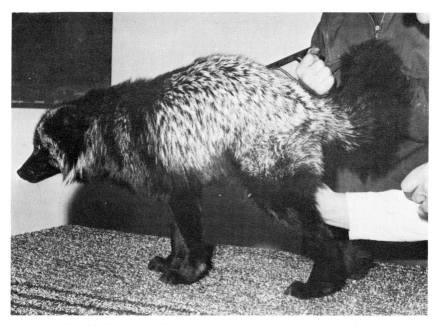

Fig. 2A. Collection of semen by digital manipulation. Preparation.

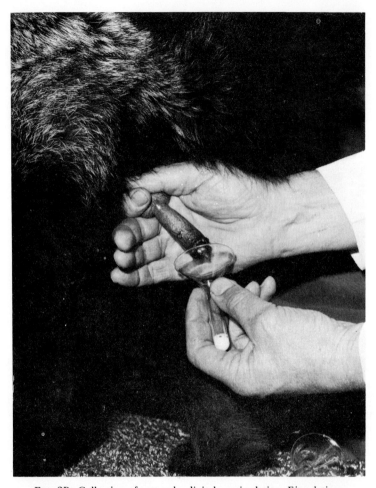

FIG. 2B. Collection of semen by digital manipulation. Ejaculation.

The breeding period for the male fox is relatively short. In an investigation we found that 6085 blue fox vixens were mated by 1298 male foxes, an average of 4·7 vixens per male (Aamdal & Fougner, 1973). Our experiments showed, however, that regular semen collection, two to three times a week, may be continued for six to 12 weeks in male blue foxes. For silver foxes and mutants of silver fox, our experience is more limited, but it appeared to be possible to prolong the male's breeding period considerably in this species as well.

By surgical exploration of the abdomen of 10 blue fox vixens in oestrus it was found that ovulation occurs between the third and fourth day of oestrus.

TABLE I

Volume, density and total number of spermatozoa in 35 ejaculates from five male blue foxes (Aamdal, 1972)

Number of ejaculates	Mean volume of sperm-rich fraction (ml)	Spermatozoal density ($\times10^6$ ml^{-1})	Mean number of spermatozoa ($\times10^6$)
35	0·62 range (0·25–1·3)	1310·76 range (401–2775)	647·9 range (100·1–1377·5)

Intravaginal insemination with fresh semen has resulted in pregnancies (Starkov, 1933–34). Our experiments with intravaginal insemination with diluted semen, however, gave confusing results. Out of four vixens inseminated intravaginally with 300 million fresh spermatozoa no fertilized ova could be recovered. In a further four vixens no spermatozoa could be detected in the uterus two hours after intravaginal insemination with 300 million spermatozoa. Of a total of 30 vixens inseminated intravaginally with 150 million frozen spermatozoa twice during the oestrous period, none became pregnant.

In natural mating the male fox deposits the ejaculate through the cervix of the vixen. By examination of the reproductive organs of two vixens two hours after mating, 50 million and 100 million spermatozoa, respectively, were found in the uterus, and a few in the oviducts. Experiments have been performed to find the minimum number of spermatozoa per insemination dose necessary to achieve maximal numbers of fertilized ova (Table II). The table shows that by using intrauterine insemination of 150 million frozen spermatozoa 90% of the ova were fertilized when two inseminations were performed. With only one insemination the percentage of fertilized ova was reduced.

Artificial insemination of vixens has been used under practical conditions to some extent in Norway. The insemination work is performed by technicians after a short training course. Inseminations are made on the third and fourth day of oestrus in blue fox vixens and on the second and third day in silver fox vixens.

Artificial insemination with fresh semen has been used on some blue fox farms with good results (Table III). Of 54 blue fox vixens almost 80% gave birth to litters and this result is equal to that of natural mating (Aamdal & Fougner, 1973).

Inseminations with frozen semen have given a slightly lower conception rate. Of 107 blue fox vixens 70% gave birth to cubs (Table IV). It is suggested that the reason for lower conception rates with frozen semen is that too few living spermatozoa were inseminated.

TABLE II

The results of intrauterine inseminations of vixens with frozen semen; the vixens were killed 144 hours later and fertilized ova were counted

Number of vixens	Day of insemination[a]	Approximate number of spermatozoa inseminated	Number of vixens with fertilized ova	Number of fertilized ova	Number of ova not fertilized	Percentage fertilized
5	3 and 4	150×10^6	5	60	5	92
5	4	150×10^6	5	35	8	81
4	3	75×10^6	4	27	13	68

[a] The onset of oestrus was taken as Day 1.

TABLE III

Survey of conception rate of blue fox vixens following intrauterine insemination with fresh semen; two inseminations per oestrous period (Days 3 and 4) with 100–150 million spermatozoa per dose

Technician	Number of vixens	Number pregnant	Percentage pregnant
1	14	11	78·5
2	40	32	80·0

TABLE IV

Survey of conception rate of blue fox vixens following intrauterine insemination with frozen semen; two inseminations per oestrous period (Days 3 and 4) with 100–150 million spermatozoa per dose

Technician	Number of vixens	Number pregnant	Percentage pregnant	Mean litter size
1	47	37	74	6·5
2	60	42	70	7·2

CONCLUSIONS

These experiments and field trials showed that it is possible to collect semen from most male foxes by digital manipulation. Electroejaculation was found to be less satisfactory especially with repeated collections.

At the present time, artificial insemination is used on three fox farms in Norway. By using artificial insemination in fox breeding, it is possible to use each ejaculate for three or four inseminations, and also to extend the breeding period of the male fox. Moreover, the use of frozen semen makes it possible to breed from an individual male fox over a wide geographical area, and to store mutant genes for many years.

However, the collection of semen by digital manipulation and the intrauterine insemination technique both require training on the part of the operators. The detection of the onset of oestrus in vixens, and thus the timing of inseminations, present some problems.

REFERENCES

Aamdal, J. (1972). Investigation in the reproduction of the male blue fox. In *Riproduzione animale e fecondazione artificale*: 1–6. Bologna: Edizione Agricole.

Aamdal, J. & Andersen, K. (1968). Fast thawing of semen in straws. *Zuchthygiene*
    3: 22–24.

Aamdal, J., Andersen, K. & Fougner, J. A. [1972]. Insemination with frozen
    semen in the blue fox. *VII Int. Congr. Anim. Reprod. Artif. Insem.* [Munich] **2**:
    1713–1716.

Aamdal, J. & Fougner, J. A. (1973). Fruktbarhet i blårevavlen. [Fertility in the blue
    fox breeding.] *Nord Vet.-Med.* **25**: 504–508.

Chronopulo, N. P. (1961). Spermagewinnung von Fuchsrüden durch Elek-
    troejakulation. *Krolikovod. Zverovod.* **1**: 16.

Creed, R. F. S. [1964]. Collection of semen from the red fox. *V Int. Congr. Anim.
    Reprod. Artif. Insem.* [Trento]: 557–561.

Davis, J. S., Bratten, R. W. & Foote, R. H. (1963). Livability of bovine spermatozoa
    at −5, −25 and −85°C in Tris-buffered and citrate-buffered yolk-glycerol
    extenders. *J. Dairy Sci.* **46**: 333–336.

Fougner, J. A., Aamdal, J. & Andersen, K. (1973). Intrauterine insemination with
    frozen semen in the blue fox. *Nord Vet.-Med.* **25**: 144–149.

Jondet, R. [1964]. Congelation rapide du sperme de taureau conditionné en
    paillettes. *V Int. Congr. Anim. Reprod. Artif. Insem.* [Trento] **4**: 463–468.

Starkov, I. D. (1933–34). [Artificial insemination in the Fox.] *Itogi. rab. vsesojuz.
    sovešč. iskusst. osem.* **1**: 122–130, [In Russian.] (Abstract in *Anim. Breed. Abstr.*
    (1935) **3**: 287.)

VanDemark, N. L. & Sharma, V. D. (1957). Preliminary fertility results from
    preservation of bovine semen at room temperatures. *J. Dairy Sci.* **40**: 438–
    439.

Venge, O. (1959). Reproduction in foxes. *Våra Pälsdjur* **30**: 29–49.

Symp. zool. Soc. Lond. (1978) No. 43, 249–260

Successful Artificial Insemination in the Chimpanzee

D. E. MARTIN,* C. E. GRAHAM and K. G. GOULD

Yerkes Regional Primate Research Center, Emory University, Atlanta, Georgia USA

SYNOPSIS

A successful pregnancy in a chimpanzee was achieved using artificial insemination of semen collected from a fertile, but non-copulating, male. The female was multiparous, and with regular menstrual cyclicity. Genital sexual swelling was monitored in order to estimate the optimal date for insemination. Semen was collected from the male by auto-masturbation, using an orange slice as a reward. The semen coagulated within the urethra permitting retrieval of the entire specimen after ejaculation, but partial liquefaction occurred within 30 min after collection, at which time the entire specimen was applied onto the external cervical os of the female. She recovered uneventfully from the light ketamine anesthesia. Daily urine collections were begun when it became evident that the menstrual period following the cycle of insemination would not occur. Three immunological pregnancy tests were employed to determine that pregnancy had been initiated. Gestation proceeded uneventfully, and 244 days following the single artificial insemination a normal, healthy female weighing 2 kg was delivered. The infant presently is in excellent health and developing at a normal rate.

INTRODUCTION

For several reasons there has been interest at the Yerkes Primate Research Center in the development of a technique for artificial insemination in the chimpanzee. First, because of the increasing scarcity of this species it is both desirable and necessary to increase the efficiency of breeding in the colony. Secondly, several males in our colony produce adequate quantities of apparently normal spermatozoa, but do not breed because of behavioral abnormalities. Such males should be utilized to increase the breeding gene pool. Finally, the technique of artificial insemination is valuable in reproduction research as a means of providing timed gestations producing offspring of known parentage. This latter consideration requires a single insemination rather than the multiple insemination procedure more generally followed with most species, and used for the only other reported successful artificial insemination in the chimpanzee (Hardin, Liebherr & Fairchild, 1975). We report here the

* Present address: College of Allied Health Sciences, Georgia State University, Atlanta, Georgia, USA

details of our first successful chimpanzee artificial insemination having an exactly known conception date, and using a fertile, but non-copulating, semen donor.

CASE HISTORY

The female chimpanzee in this study, Banana, had an estimated birth date of April 1946, and had been housed at the Yerkes Primate Research Center since 1961. She was multiparous, having delivered normal female infants on 13 August 1961, 1 July 1963 and 7 October 1970. A set of female twins were stillborn on 22 August 1968. The male chimpanzee, Hoboh, arrived at the Yerkes Primate Research Center in 1971, and had no accompanying records to indicate an approximate age. The mean serum testosterone level, from five blood samples taken over the seven-month period prior to the insemination, was $2 \cdot 77$ ng ml^{-1} and indicated that he was either an adolescent male or at the low end of the normal range for a mature male (Martin, Swenson & Collins, 1977). Although he was caged frequently with female partners, no mating behavior was observed, and with some females he was an incompatible partner. He did produce ejaculates by auto-masturbation which were within the normal range for mature chimpanzees (Warner, Martin & Keeling, 1974) in terms of volume ($1 \cdot 3$–$2 \cdot 5$ ml), spermatozoal morphology (Fig. 1) and concentration (450×10^6–970×10^6 ml^{-1}). These ranges were based upon analyses of nine ejaculates collected at weekly intervals. Hoboh was therefore considered likely to be a suitable donor for artificial insemination. Both animals were housed and fed at the Yerkes Primate Research Center in accordance with established guidelines of animal care (Keeling & Roberts, 1972).

The menstrual cycle of the female had been observed routinely for several years. Using procedures described by Graham (1970), the genital sexual swelling was monitored on a scale of zero to four, the score of four representing maximal tumescence. Studies by Graham, Keeling *et al.* (1973), and our unpublished data on luteinizing hormone (LH) levels, indicate that ovulation in the chimpanzee probably occurs most frequently during the 24-hour interval prior to the last day of maximum tumescence. Figure 2 depicts the cycle of insemination as well as the previous two cycles, all of which were typical cycles for this animal.

FIG. 1. A. Light micrograph of four normal chimpanzee spermatozoa, unstained by an eosin-nigrosin stain, and hence presumed alive at the time of staining. (Mean spermatozoal dimensions (±s.e.): head length $4 \cdot 7$ μm, head width $2 \cdot 9$ μm, midpiece length $6 \cdot 3$ μm, total length $57 \cdot 0$ μm), ×2100. B. Scanning electron micrograph of chimpanzee spermatozoon with normal morphology, showing the acrosomal membrane covering the anterior two-thirds of the head, and the mitochondrial gyres beneath the plasma membrane of the midpiece, ×11 000.

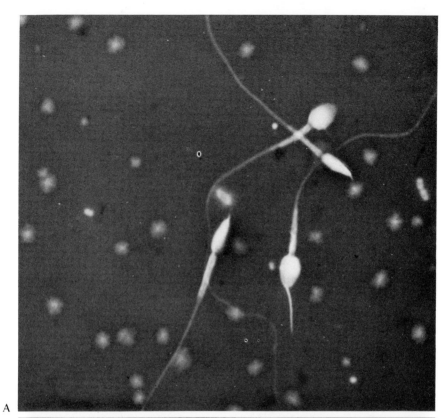

A

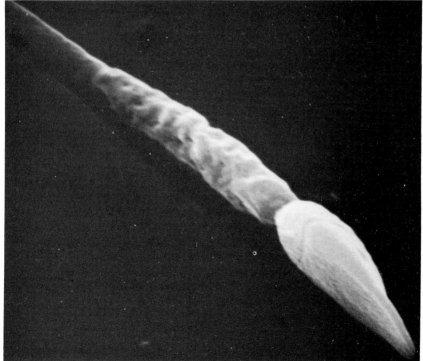

B

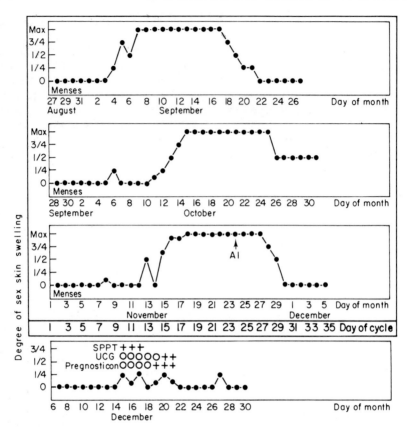

FIG. 2. Sexual swelling pattern of female chimpanzee, Banana, for the two menstrual cycles preceding artificial insemination, the cycle during which the insemination was performed, and the period during which pregnancy was established. Positive (+) and negative (○) results of three immunological pregnancy tests performed on daily urine collections are indicated.

Defining Day 0 as the last day of maximum tumescence, it can be seen from Fig. 2 that the actual insemination date was Day −3.

Semen was obtained from the male by auto-masturbation, using an orange slice as a reward. Since chimpanzee semen is ejaculated as a coagulated mass, retrieval of the entire specimen was possible. It was transferred to a small covered container in a 37°C incubator where partial liquefaction occurred within 30 min. An eosin-nigrosin stain performed on the fluid fraction (Martin & Davidson, 1976) indicated 75% live cells. The spermatozoal concentration was $450 \times 10^6 \, \mathrm{ml}^{-1}$ of fluid with 70% motility. The total fluid volume recovered was 0·9 ml and the residual coagulum volume was 1·2 ml. Large numbers of spermatozoa were present in the coagulated portion, but an accurate estimate was not possible.

Approximately 20 min elapsed after ejaculation while the female was lightly anesthetized with ketamine hydrochloride. Insemination was performed with the female in the prone position, with a posterior elevation of 20–30°. The marked perineal sexual swelling (Fig. 3) in the chimpanzee near the time of ovulation makes the effective vaginal length much greater than in the human. This necessitated the previous construction of a long vaginal speculum, designed on the pattern of the Graves speculum and fabricated from a 50 ml plastic graduated cylinder in order to gain ready access to the region of the cervical os. Illumination of the vaginal vault was obtained by means of a battery-operated headlamp.

The sperm-rich seminal fluid obtained from the liquefied portion of the ejaculate was applied to the external cervical os using a 5 ml syringe connected to plastic tubing containing the fluid mixture. The end of the tubing was positioned at the os using forceps. The remaining coagulum was then inserted with forceps deep into the vaginal vault. The animal was returned to its home cage, and it recovered uneventfully from the anesthesia.

Analysis of a blood sample obtained from the female on the day of the insemination (using the NIH human LH radioimmunoassay kit) showed a plasma LH concentration of $23 \cdot 6$ mIU ml^{-1}. Unpublished observations from our laboratory indicate that plasma LH levels greater than 20 mIU ml^{-1} represent a significant elevation. This suggests that insemination occurred at some time during either the rise or fall of the pre-ovulatory LH peak.

Daily urine collections were begun after it became evident that the menstrual period following the cycle of insemination had not occurred. Three immunological tests were employed to determine whether pregnancy had been established. One of these, the Subhuman Primate Pregnancy Test (SPP Test), was developed by Hodgen & Ross (1974) and validated for the chimpanzee in our laboratory (Woodard, Graham & McClure, 1976). The test was originally used for macaque chorionic gonadotropin, but fortuitously the antiserum raised against the β-subunit of ovine LH also cross-reacts with chimpanzee chorionic gonadotropin. It reacted positively from the beginning of urine collection, on the eighteenth day following sexual swelling. Another, the Pregnosticon Accusphere Test (Organon, Inc, West Orange, NJ) produced positive results four days after the SPP Test (Day 22) while a third test, the Urinary Chorionic Gonadotrophin Test (Wampole Laboratories, Stamford, CT) produced positive results five days after the SPP Test. All three of these tests are routinely utilized in our laboratory for pregnancy diagnosis.

Gestation proceeded uneventfully, and 244 days following artificial insemination, a normal healthy female was delivered weighing 2 kg. The infant presently is in excellent health and developing at a normal rate.

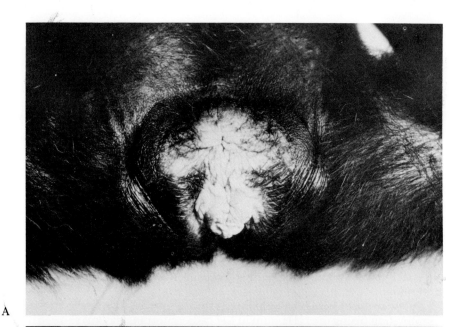

A

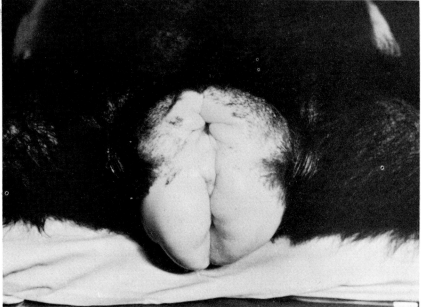

B

FIG. 3. A. Sexual swelling of intact chimpanzee scored as zero tumescence (completely detumesced). B. Maximal swelling, scored as "4" on a rating scale of "1" to "4", on same animal as A. Reproduced from Graham, Collins *et al.* (1972) with permission of the J. B. Lippincott Company.

DISCUSSION

The case described is the ninth precisely timed successful pregnancy on record in the Yerkes Primate Center colony, which dates back to 1930. We are not aware of any others at chimpanzee colonies elsewhere in the world. The first eight precisely timed records were obtained during the early days of the colony, between 1935 and 1938, and resulted from single-day natural matings, where no matings occurred during the menstrual cycles either preceding or following. This permitted calculation of exact gestation periods with a mean of 227·6 days (Table I). The approximate length of the pregnancy (239–244 days) described by Hardin *et al.* (1975) after a recent successful artificial insemination of a chimpanzee, and the gestation period of the present infant (244 days), are both slightly longer than that for natural matings, although both infant birth weights (1·8 kg for Hardin's infant, 2 kg for ours) were within the normal range for births in the Yerkes Primate Center colony (Keeling & Roberts, 1972).

Although we are able to report that pregnancy has occurred following a single insemination, not all the problems associated with the procedure have been solved. Appropriate timing of a single insemination to achieve a pregnancy presumes accurate knowledge of when ovulation occurs, how long the ovum remains fertilizable and how long inseminated spermatozoa retain their fertilizing ability. Such information is not available with certainty in the chimpanzee. Multiple inseminations during a given cycle would undoubtedly give a greater chance of conception, but this would not allow precise determination of gestation length. Moreover, no information would be obtained as to the optimum time of insemination.

The data from the eight pregnancies resulting from single matings at the Yerkes Primate Center together with the present data (Table I) suggest that conception can occur over a rather wide time span in relation to the last day of maximal genital swelling. This wide range is probably not due to a large variability in the relationship of ovulation and detumescence of the genital sexual swelling. Laparoscopic data from the chimpanzee (Graham, Keeling *et al.*, 1973) and from the pigtail macaque, *Macaca nemestrina* (Blakley, Blaine & Morton, 1977), indicate that ovulation and the beginning of detumescence both occur close together in time, perhaps within 24 hours.

This, in turn, suggests that inseminated spermatozoa deposited in advance of ovulation may retain their fertilizing ability for a period of three to five days, which would increase the chances of a single insemination initiating a pregnancy. No direct evidence is available for spermatozoal survival in the female tract of the chimpanzee, but progressively motile spermatozoa have been found in the oviductal ampulla up to 85 hours following intercourse in the human female (Hafez, 1976).

The single plasma LH value for the chimpanzee female, Banana, obtained at the time of insemination, is difficult to interpret. Its magnitude clearly indicates that insemination occurred during either the rising

TABLE I

Single chimpanzee matings resulting in a normal pregnancy

Infant	Birth date	Father	Mother	Insemination date	Relationship to last day of max. genital swelling[a]	Gestation period (days)
Hal	07 Oct 1935	Jack	Josie	16 Feb 1935	0	233
Ami	26 Jun 1936	Jack	Mona	31 Oct 1935	−5	230
Dina	27 Mar 1937	Bokar	Pati	02 Aug 1936	−1	237
Cap	13 Apr 1937	Bokar	Dita	16 Aug 1936	−4	240
Jule	08 May 1937	Bokar	Fifi	17 Oct 1936	−2	203
Fin	10 Nov 1937	Bokar	May	27 Mar 1937	−4	228
Kola	15 Dec 1937	Jack	Cuba	06 May 1937	−1	223
Art	20 Jul 1939	Bokar	Fifi	05 Dec 1938	0	227
Artifee	24 Jul 1976	Hoboh	Banana	24 Nov 1975	−3	244

[a] Day 0, last day of maximum tumescence; Day −1, penultimate day of maximum tumescence.

TABLE II

Artificial insemination of primates resulting in successful pregnancy

Species	Mode of semen collection	Route of insemination	Fresh or frozen semen	Reference
Rhesus monkey	Electroejaculation	Intravaginal	Fresh	Valerio et al. (1971)
Rhesus monkey	Electroejaculation	Intravaginal	Fresh	Dede & Plentl (1966)
Rhesus monkey	Electroejaculation	Intravaginal	Fresh	Settlage, Swan & Hendrickx (1973)
Rhesus monkey	Electroejaculation	Intraperitoneal	Fresh	Van Pelt (1970)
Rhesus monkey	Electroejaculation	Intrauterine	Fresh	Czaja, Eisele & Goy (1975)
Rhesus monkey	Electroejaculation	Intrauterine	Frozen	Anonymous (1972)
Squirrel monkey	Electroejaculation	Intravaginal	Fresh	Bennett (1967)
Chimpanzee	Masturbation	Intravaginal	Fresh	Hardin et al. (1975)
Human	Masturbation	Intrauterine	Frozen	Barwin (1974)
Human	Masturbation	Intracervical	Frozen	Barkay, Zuckerman & Heiman (1974)
Human	Masturbation	Cervical cap	Fresh	Strickler, Keller & Warren (1975)
Human	Masturbation	Cervical cap	Fresh	Dixon, Buttram & Schum (1976)
Human	Masturbation	Intravaginal	Frozen	Perloff, Steinberger & Sherman (1964)
Human	Masturbation	Intravaginal	Frozen	Friberg & Gemzell (1973)
Human	Masturbation	Cervical cap	Frozen	Jondet et al. (1975)

or falling phase of the pre-ovulatory LH surge. However, unpublished data by our group, based upon 300 LH determinations from plasma obtained randomly throughout the menstrual cycle in several normal cycling chimpanzees, indicated that the LH peak most often occurs on Day 0 (the last day of maximal genital tumescence), with the shoulders of the peak encompassing a time span from Day -2 to Day $+1$. Assuming the menstrual cycle of this female was normal, there being no contrary indications, it is more likely that the day of insemination was on the rising phase of the LH surge. The precise interval following the pituitary discharge of LH before oocytes are released from mature follicles has yet to be determined for the chimpanzee, but it is likely to be similar to the human, about 36 hours after the LH peak (Edwards, 1973). Ovulation would therefore most often occur shortly after the last day of maximal tumescence, although in this particular instance the early rise of the LH surge suggests that ovulation may have occurred somewhat earlier.

A final aspect to consider is the various sites available for deposition of spermatozoa at insemination. Intrauterine deposition of whole semen as a routine procedure was deemed inadvisable owing to the possibility of spermatozoal antibody production with consequent immunization of animals to seminal proteins (Shulman, 1976). If this occurred, it could compromise future productivity of the breeding colony. Accordingly, intravaginal insemination is routinely employed. This, in turn, increases the influence of the cervix and its secretion on the success of the insemination, a problem not encountered in the insemination of some domestic species where intrauterine insemination is employed.

Artificial insemination is a technique with great potential application to great apes with regard to both animal husbandry and scientific research. Although techniques of semen collection, storage and insemination have been developed largely in domesticated animals, modifications of these techniques are being applied successfully to primates (see Table II and Hendrickx et al., this volume, p. 219). We have demonstrated the feasibility of this technique in the chimpanzee. Further development for routine use in chimpanzee breeding should help alleviate the shortage of these animals in research facilities. Additionally, it will allow us to obtain new information about chimpanzee reproduction of value to our natural cage breeding program and to the development of the species for research.

ACKNOWLEDGEMENTS

This research was supported in part by NIH grants RR00165 and RR00992 to the Yerkes Regional Primate Research Center of Emory University, Atlanta, Georgia 30322, USA and Ford Foundation Grant 690–0645A. We thank Dr J. D. Neill of the Department of Physiology, Emory University, for the LH radioimmunoassays.

REFERENCES

Anonymous (1972). Artificial insemination techniques aid breeding of non-human primates. *Primate Rec.* **3**: 7–9.

Barkay, J., Zuckerman, H. & Heiman, M. (1974). A new, practical method of freezing and storing human sperm and a preliminary report on its use. *Fert. Steril.* **25**: 399–406.

Barwin, B. N. (1974). Intrauterine insemination of husband's semen. *J. Reprod. Fert.* **36**: 101–106.

Bennett, J. P. (1967). Artificial insemination of the squirrel monkey. *J. Endocr.* **37**: 473–474.

Blakley, G. A., Blaine, C. R. & Morton, W. R. (1977). Correlation of perineal detumescence and ovulation in the Pigtail macaque (*Macaca nemestrina*). *Lab. Anim. Sci.* **27**: 352–355.

Czaja, J. A., Eisele, S. G. & Goy, R. W. (1975). Cyclical changes in the sexual skin of female rhesus: relationships to mating behavior and successful artificial insemination. *Fedn Proc. Fedn Am. Socs exp. Biol.* **34**: 1680–1684.

Dede, J. A. & Plentl, A. A. (1966). Induced ovulation and artificial insemination in a rhesus colony. *Fert. Steril.* **17**: 757–764.

Dixon, R. E., Buttram, V. C. & Schum, C. W. (1976). Artificial insemination using homologous semen: a review of 158 cases. *Fert. Steril.* **27**: 647–654.

Edwards, R. G. (1973). Studies on human conception. *Am. J. Obstet. Gynec.* **117**: 587–601.

Friberg, J. & Gemzell, C. (1973). Inseminations of human sperm after freezing in liquid nitrogen vapors with glycerol or glycerol-egg-yolk-citrate as protective media. *Am. J. Obstet. Gynec.* **116**: 330–334.

Graham, C. E. (1970). Reproductive physiology of the chimpanzee. In *The chimpanzee* **3**: 183–220. Bourne, G. H. (ed.). Basel: Karger.

Graham, C. E., Collins, D. C., Robinson, H. & Preedy, J. R. K. (1972). Urinary levels of estrogens and pregnanediol and plasma levels of progesterone during the menstrual cycle of the chimpanzee: relationship to the sexual swelling. *Endocrinology* **91**: 13–24.

Graham, C. E., Keeling, M., Chapman, C., Cummins, L. B. & Haynie, J. (1973). Method of endoscopy in the chimpanzee: relations of ovarian anatomy, endometrial histology, and sexual swelling. *Am. J. Phys. Anthrop.* **38**: 211–215.

Hafez, E. S. E. (1976). Transport and survival of spermatozoa in the female reproductive tract. In *Human semen and fertility regulation in man*: 107–129. Hafez, E. S. E. (ed.). St. Louis: Mosby.

Hardin, C. J., Liebherr, G. & Fairchild, O. (1975). Artificial insemination in chimpanzees. *Int. Zoo Yb.* **15**: 132–134.

Hodgen, G. D. & Ross, G. T. (1974). Pregnancy diagnosis by a hemagglutination inhibition test for urinary macaque chorionic gonadotropin (MCG). *J. clin. Endocr. Metab.* **38**: 927–930.

Jondet, M., Millet, D., Cornuau, J., Drapier, E., Picaud, C. & Netter, A. (1975). Utilisation du sperme congelé pour l'insémination humaine hétérologue. *Gynécologie* **26**: 285–288.

Keeling, M. E. & Roberts, J. R. (1972). Breeding and reproduction of chimpanzees. In *The chimpanzee* **5**: 127–152. Bourne, G. H. (ed.). Basel: Karger.

Martin, D. E. & Davidson, M. W. [1976]. Differential live–dead stains for bovine and primate spermatozoa. *VIII Int. Congr. Anim. Reprod. Artif. Insem.* [Krakow] **4**: 919–922.

Martin, D. E., Swenson, R. B. & Collins, D. C. (1977). Correlation of serum testosterone levels with age in male chimpanzees. *Steroids* **29**: 471–481.

Perloff, W. H., Steinberger, E. & Sherman, J. K. (1964). Conception with human spermatozoa frozen by nitrogen vapor technic. *Fert. Steril.* **15**: 501–504.

Settlage, D. S. F., Swan, S. & Hendrickx, A. G. (1973). Comparison of artificial insemination with natural mating techniques in rhesus monkeys, *Macaca mulatta. J. Reprod. Fert.* **32**: 129–132.

Shulman, S. (1976). Sperm antibodies in serum of men and women and in cervical mucus. In *Biological and clinical aspects of reproduction*: 185–193. Ebling, F. J. G. & Henderson, I. W. (eds). Amsterdam: Excerpta Medica.

Strickler, R. C., Keller, D. W. & Warren, J. C. (1975). Artificial insemination with fresh donor semen. *New Engl. J. Med.* **293**: 848–853.

Valerio, D. A., Leverage, W. E., Bensenhaver, J. C. & Thornett, H. D. (1971). The analysis of male fertility, artificial insemination and natural matings in the laboratory breeding of macaques. *Med. primatol.* **1970**: 515–525.

Van Pelt, L. F. (1970). Intraperitoneal insemination of *Macaca mulatta. Fert. Steril.* **21**: 159–162.

Warner, H., Martin, D. E. & Keeling, M. E. (1974). Electroejaculation of the great apes. *Ann. Biomed. Eng.* **2**: 419–432.

Woodard, D. K., Graham, C. E. & McClure, H. M. (1976). Comparison of hemagglutination inhibition pregnancy tests in chimpanzee and orangutan. *Lab. Anim. Sci.* **26**: 922–927.

Symp. zool. Soc. Lond. (1978) No. 43, 261–269

Studies on Handling Spermatozoa from the African Elephant, *Loxodonta africana*

R. C. JONES

University of Newcastle, New South Wales, Australia

SYNOPSIS

The motility of elephant spermatozoa was reduced by dilution into Krebs–Henseleit–Ringer (KHR) at 10 and 5°C rather than 30°C. However, elephant spermatozoa were not as susceptible to cold shock as spermatozoa from many scrotal mammals.

Spermatozoa freshly collected from the epididymis were immotile or only a small proportion showed weak motility. A 50-fold dilution of freshly collected samples with epididymal plasma slightly increased the proportion of motile cells, but a 50-fold dilution with phosphate-buffered saline or KHR induced motility in a high proportion of cells. However, only a small proportion of cells were motile in epididymal semen stored for one hour before dilution. Dilution rates higher than 50-fold with KHR reduced the survival of spermatozoa during incubation at 37°C. Spermatozoa survived best in diluents with a high sodium to potassium ratio, but varying the ratio did not seem to affect the induction of motility by diluting semen.

In a factorial experiment the effects on the survival of spermatozoa at 37°C of diluent pH (values of 5·5, 7·0 and 8·5) and osmotic pressure (150, 225, 330, 375 and 450 mosmol kg^{-1}) were tested and it was found that these factors did not have independent effects. In general spermatozoa survived best in media at a pH of 8·5. However, at pH 5·5 the best survival occurred in diluents with the lowest osmotic pressure, at pH 7·0 the optimal osmotic pressure was about 250 mosmol kg^{-1} and at pH 8·5 the optimal osmotic pressure was about 275 mosmol kg^{-1}.

INTRODUCTION

As elephant bulls may at times be aggressive towards their handlers they are not usually kept in zoos. Consequently, there is a need to breed the cows artificially, and so towards this objective the Zoological Society of London commenced work in 1972 to investigate the possibility of collecting and preserving elephant semen (Jones, 1973; Jones, Rowlands & Skinner, 1974; Jones, Bailey & Skinner, 1975).

A major concern on initiating the elephant studies was that nothing was known of the physiology of elephant semen which would provide a basis for determining how it might best be handled. This concern was particularly pertinent since, unlike other animals which had been used for artificial breeding, the elephant has a primitive reproductive tract with testes and their excurrent ducts located in the abdominal cavity. Further, the only relevant report available at the time indicated that spermatozoa collected from the excurrent ducts of the testes were immotile and that it was not possible to induce much motility (Short, Mann & Hay, 1967). Consequently, as it was most convenient to collect spermatozoa from dead animals the object of the initial studies was to determine how to obtain spermatozoa from the male reproductive tract and induce motility. It was found that large numbers of spermatozoa could be collected from the distal end of the epididymis where they are stored (Jones, Rowlands & Skinner, 1974; Jones, Skinner & Rowlands, 1974) and it was confirmed that either the spermatozoa were immotile or only a small proportion of cells showed weak flagellation. Preliminary studies showed that motility could be induced by diluting the epididymal semen in buffered saline (Jones, 1973), but the studies were not sufficiently extensive to determine what factor(s) induced the motility. Consequently, this report describes further work on the induction of sperm motility as well as some basic studies on handling of semen and the composition of diluents.

MATERIALS AND METHODS

The studies were carried out in Kruger National Park, South Africa, on animals which were being culled to prevent overstocking. Bulls from breeding and bachelor herds were immobilized with succinylcholine (Scoline, Glass Laboratories Ltd), shot, bled from the jugular vein and eviscerated. The testes and excurrent ducts were removed within about 15 min of death and the vasa deferentes and distal part of the epididymides (external diameter more than 1·5 mm) were dissected free of the mesentery which supported them. The spermatozoa were removed by dissecting lengths of 100–200 mm of duct free of the connective tissue stroma and stripping them by pulling the duct between the thumb and forefinger of one hand. Care was taken to avoid exposing samples to direct sunlight or changes in temperature. The semen was mixed with a Pasteur pipette which was also used to add it to the diluents in 15 ml polyethylene tubes and to mix it again. The tubes were stored in a vacuum flask at 37°C and transported to a field laboratory for further incubation and microscopical examination. All of the diluents (except those in the experiment studying the effects of rapid cooling) were warmed to 37°C about one hour before use and kept at that temperature for the incubation period.

Solutions were prepared from A.R. grade chemicals and twice distilled water at three or five times the final concentration and sub-samples were tested for pH and osmolality. They were stored at 5°C and diluted within five hours of use. The epididymal plasma used in the third experiment was prepared by concentrating the spermatozoa from freshly collected semen by centrifuging at 1000 g for 15 min, removing the supernatant and storing it at −20°C until it was warmed for use. The Krebs–Henseleit–Ringer contained 110 mM sodium chloride, 5 mM potassium chloride, 1 mM potassium dihydrogen phosphate, 1 mM magnesium sulphate,

2 mM sodium bicarbonate, 10 mM disodium hydrogen phosphate, 10 mM sodium dihydrogen phosphate and 10 mM fructose. The composition of the other diluents used in the studies is described with the design of the experiment.

All of the tubes storing semen in an experiment were coded and randomized so that on microscopic examination the observer was unaware of which treatment was being examined. At hourly intervals during incubation (except for the first experiment), sub-samples of diluted semen were examined as a thin film between a slide and coverslip on a microscope warm stage at 37°C and scored for rate of progressive motility (Emmens, 1947) and percentage of motile spermatozoa. As both these scores ranked treatments in much the same manner, for the sake of brevity only the latter are reported. The statistical significance of treatment effects were assessed by analyses of variance using orthogonal polynomial coefficients (scores of percentage motile were transformed to angles for these analyses); the error mean square used as the denominator for calculation of the F ratios was composed of first order and, where appropriate, higher order interactions. The means and their standard errors, which are shown in the tables and figures, were calculated from untransformed data (the standard errors were calculated in the same way as for the statistical tests).

RESULTS

The first experiment tested the effect of diluting (20-fold) semen into Krebs–Henseleit–Ringer at 20°C, 10°C and 0°C. The experiment was replicated with semen from three different elephants and the diluted samples were examined microscopically after 1·5 hours. It was found (Table I) that lowering the temperature of the diluent caused a decrease in scores of percentage motile spermatozoa ($P < 0.05$).

TABLE I

Susceptibility of epididymal spermatozoa to rapid cooling

Diluted 20-fold in Krebs–Henseleit–Ringer at	Mean scores of percentage motile after 1·5 hours
30°C	60·0
10°C	46·7
0°C	30·0
s.e. of means	±5·1
P	<0·05

Replicated with semen from three elephants.

The second experiment examined the effect of storing epididymal semen undiluted, storing it undiluted and diluting it 50-fold just prior to examination, and diluting it 2-, 10-, 50-, 250- and 1250-fold for storage (Table II). The undiluted semen was immotile or showed weak motility of a small proportion of cells after storage for one, two or three hours. The 50-fold dilution of this semen just prior to examination increased the mean scores of percentage motile ($P < 0.05$), but not to the value of the mean score for semen which was diluted 50-fold before storage. Increasing the dilution rate from 2-fold to 50-fold increased motility, but higher rates of dilution decreased the response ($P < 0.001$).

TABLE II

Effects of diluting epididymal semen in Krebs–Henseleit–Ringer

Semen stored	Mean scores of percentage motile after two hours at 37°C
Undiluted	10·2
Undiluted, then diluted 50-fold before examination	23·0
Diluted 2-fold	6·2
Diluted 10–fold	20·0
Diluted 50-fold	42·0
Diluted 250-fold	10·0
Diluted 1250-fold	13·0
s.e. of means	±4·0
P (1 *versus* 2)	<0·05
P (dilution rate)	<0·001

Replicated with semen from five elephants.

The increase in sperm motility caused by dilution in the previous experiment may be interpreted as the dilution of an inhibitor in epididymal plasma which acts to prevent contraction of the microtubules of the sperm axoneme. Consequently, this possibility was tested in the third experiment in which the effect on sperm motility of diluting spermatozoa with epididymal plasma was tested. Further, the induction of sperm motility by dilution could also be initiated by a change in electrical potential across the cell membrane due to a change in the concentrations of sodium and potassium on either side of the membrane. Consequently, the effects on motility were tested by varying the concentrations of these ions in the semen diluent. Semen was diluted 50-fold in 5 mM phosphate buffer (pH 7·0) containing varying proportions of 154 mM sodium chloride and 154 mM potassium chloride, or in a solution of the organic buffer HEPES (Jones & Foote, 1972).

The results of this study (Table III) showed that a smaller proportion of spermatozoa were motile when diluted in epididymal plasma than in any of the other diluents ($P < 0.01$). The mean percentage motile scores were also lower for semen samples stored in HEPES compared to the inorganic electrolytes ($P < 0.05$). However, the mean percentage motile scores were improved by replacing potassium by sodium in the buffered solutions of inorganic electrolytes ($P < 0.001$).

TABLE III

Effects of diluting epididymal semen 50-fold in media containing various proportions of 154 mM NaCl and KCl, an organic buffer and epididymal plasma

Diluent	Mean scores of percentage motile after two hours at 37°C
0% NaCl, 100% KCl	42·5
10% NaCl, 90% KCl	47·5
50% NaCl, 50% KCl	65·0
90% NaCl, 10% KCl	77·5
100% NaCl, 0% KCl	82·5
235 mM HEPES	41·3
Epididymal plasma	21·3
s.e. of means	±5·9
P (1–5, linear)	
P (mean 1–5 *versus* mean 6)	
P (mean 1–6 *versus* mean 7)	

Replicated with semen from six elephants.

As an average of 21·3% of spermatozoa were motile after dilution in epididymal plasma and incubation for two hours it is questionable whether this degree of activity may have occurred spontaneously in undiluted semen. Since three of the replicates for Experiments 2 and 3 were run concurrently this possibility was tentatively assessed by comparing the mean responses for semen stored undiluted or diluted in epididymal plasma for these three replicates. After two hours' storage these means were respectively 10·0 and 25·0% which suggests that dilution with epididymal plasma did induce some motility.

The fourth experiment used a 50-fold dilution rate to investigate the interaction of diluent pH and osmolality using a 5×3 factorial design and replication with semen from six different elephants. The osmolalities of the final diluents were 150, 225, 300, 375 and 450 mosmol kg^{-1} (osmotic pressures relative to 160 mM sodium chloride were respectively 0·5, 0·75, 1·0, 1·25 and 1·5) and the pH values were 5·5, 7·0 and 8·5. All diluents contained 20 mM phosphate buffer, 4 mM potassium chloride, 6 mM fructose and sufficient sodium chloride to achieve the appropriate osmolality. The pH of the diluents was varied by altering the proportions of acid and dibasic sodium phosphate making up the 20 mM phosphate buffer. The analysis of variance of responses made after two hours incubation showed that, overall, spermatozoa survived best in solutions of pH 8·5 (mean scores for solution of pH 5·5, 7·0 and 8·5 were respectively 9·2, 19·5 and 32·0%; $P < 0·001$) and an osmolality of 300 mosmol kg^{-1} (mean scores for solutions from the lowest to the highest osmolality were respectively 15·1, 16·7, 30·6, 24·8 and 13·5%; $P < 0·001$). However, there was an interaction between diluent pH and osmolality (Fig. 1, $P < 0·001$) which showed that at pH 5·5, spermatozoa survived better at lower than at higher osmolalities; at pH 7·0 diluents with osmolalities between 225 and 300 mosmol kg^{-1} maintained optimal survival; however, at pH 8·5 optimal survival occurred in solutions with an osmotic pressure of about 375 mosmol kg^{-1}.

DISCUSSION

The finding in these studies that rapid cooling was detrimental to elephant spermatozoa confirms the earlier work on the elephant (Jones, Bailey & Skinner, 1975) and is in agreement with reports on spermatozoa from other mammals (Wales & White, 1959; White & Wales, 1960; Dott, 1968). However, considering that the present study used a higher dilution rate (20-fold *versus* 1·5-fold) and a longer incubation period at 5°C (90 min *versus* 5 min) than Wales & White (1959), who compared semen from a number of mammals (ram, bull, dog, rabbit and man) and the fowl, it is tempting to suggest that elephant spermatozoa are, like fowl spermatozoa, less sensitive to rapid cooling than spermatozoa from scrotal mammals. However, further studies are required, particularly since Wales & White (1959) studied ejaculated semen and the elephant studies were carried out on epididymal semen which is known to be the more resistant to cold shock (Wales & White, 1959; White & Wales, 1961). Furthermore, lipid analyses have indicated that elephant spermatozoa, in their phospholipid-bound fatty acid composition, most resemble spermatozoa of those mammalian species which are most sensitive to cold shock (Darin-Bennett *et al.*, 1976).

The detrimental effect of high dilution rates on elephant spermatozoa is consistent with reports of other species (Emmens & Swyer, 1949; Blackshaw, 1953; White & Wales, 1961). However, it is not possible directly to assess the relative susceptibility to dilution of spermatozoa

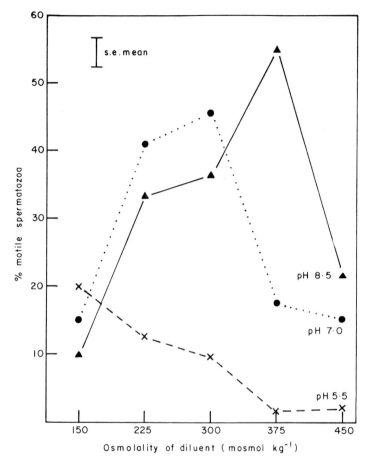

FIG. 1. The interaction of pH and diluent osmolality for scores of percentage motile spermatozoa after incubation for 2 hours at 37°C. ×--×, pH 5·5; ●···●, pH 7·0; ▲-▲, pH 8·5.

from the elephant and other mammals because of differences in experimental conditions. Nevertheless, it may be concluded that elephant spermatozoa are no more sensitive to high dilution than ram spermatozoa.

In these studies, a greater proportion of samples of undiluted epididymal semen contained motile spermatozoa (one out of five samples contained no motile spermatozoa) than in earlier studies (eight out of nine samples contained no motile spermatozoa; Jones, Bailey & Skinner, 1975). One explanation for this difference may be that the spermatozoa in these studies may have been immotile immediately upon collection (when it was not possible to examine them) and have gained some motility after storage for one hour. Alternatively, the difference may be due to differences in method of collecting the samples. The earlier studies used semen flushed from the ducts with air whilst in the present studies it was squeezed from the ducts and could be slightly contaminated with blood and interstitial fluid.

The increase in the proportion of motile spermatozoa induced by diluting epididymal semen with epididymal plasma indicates that it is unlikely that the plasma contains a substance which completely inhibits motility *in vivo*. However, it remains possible that an inhibitory substance could be loosely bound to the spermatozoa so that its concentration in plasma is low and the effect of dilution with the other media used was to wash the substance from the cells. An alternative explanation for the induction of motility by dilution is that it would increase the concentration of oxygen, or decrease the concentration of carbon dioxide, in the medium supporting the spermatozoa. Such a mechanism may explain why Short, Mann & Hay (1967) could only induce motility in 1–15% of epididymal spermatozoa using vesicular secretion collected from animals which had (presumably) been dead for some time, and why in these studies epididymal plasma (which had little exposure to air) did not induce motility in a large proportion of cells.

The effects of osmotic pressure on elephant spermatozoa are in general agreement with the results of Blackshaw & Emmens (1951) if it is taken into account that those workers calculated the "relative tonicities" of their diluents considering that 154 mM sodium chloride (actually 285 mosmol kg^{-1}) and 100 mM sodium phosphate buffer (actually about 210 mosmol kg^{-1}) were isotonic with spermatozoa (i.e. what they considered was isotonic with spermatozoa was hyposmotic to 300 mosmol kg^{-1}). However, as for other species studied the optimum osmotic pressure of the semen diluent depends upon its pH. At pH 7·0 the optimum osmotic pressure of diluents for elephant spermatozoa (approximately 250 mosmol kg^{-1}) is the same as for bull spermatozoa (Jones & Foote, 1972). In general, however, the tendency for elephant spermatozoa to survive best in diluents of high pH and osmotic pressure is more similar to the characteristics of human spermatozoa than either bull or ram spermatozoa (Blackshaw & Emmens, 1951).

These studies indicate that elephant spermatozoa are as resistant to handling as bull spermatozoa and so are worthy of further study to assess their use for artificially breeding animals in captivity.

ACKNOWLEDGEMENTS

I am indebted to Professor J. D. Skinner, Mammal Research Institute, University of Pretoria, South Africa and Dr U. de V. Pienaar and his staff at Kruger National Park, South Africa, for their generous help and support.

REFERENCES

Blackshaw, A. W. (1953). The motility of ram and bull spermatozoa in dilute suspension. *J. gen. Physiol.* **36**: 449–462.

Blackshaw, A. W. & Emmens, C. W. (1951). The interaction of pH, osmotic pressure and electrolyte concentration on the motility of ram, bull and human spermatozoa. *J. Physiol.* **114**: 16–26.

Darin-Bennett, A., Morris, S., Jones, R. C. & White, I. G. (1976). The glyceryl-phosphorylcholine and phospholipid pattern of the genital duct and spermatozoa of the African elephant, *Loxodonta africana. J. Reprod. Fert.* **46**: 506–507.

Dott, H. [1968]. Effect of rapid cooling on the proportion of eosinophilic bull, ram and rabbit ejaculated spermatozoa in diluted and undiluted semen. *VI Int. Congr. Anim. Reprod. Artif. Insem.* [Paris] **2**: 1235–1237.

Emmens, C. W. (1947). The motility and viability of rabbit spermatozoa at different hydrogen ion concentrations. *J. Physiol.* **106**: 471–481.

Emmens, C. W. & Swyer, G. I. M. (1949). Observations on the motility of rabbit spermatozoa in dilute suspension. *J. gen. Physiol.* **32**: 121–138.

Jones, R. C. (1973). Collection, motility and storage of spermatozoa from the African elephant, *Loxodonta africana. Nature, Lond.* **243**: 38–39.

Jones, R. C., Bailey, D. W. & Skinner, J. D. (1975). Studies on the collection and storage of semen from the African elephant, *Loxodonta africana. Koedoe* **18**: 147–164.

Jones, R. C. & Foote, R. H. (1972). Effect of osmolality and phosphate, 'TRIS', 'TES', 'MES', and 'HEPES' hydrogen ion buffers on the motility of bull spermatozoa stored at 37 or 5°C. *Aust. J. biol. Sci.* **25**: 1047–1055.

Jones, R. C., Rowlands, I. W. & Skinner, J. D. (1974). Spermatozoa in the genital ducts of the African elephant, *Loxodonta africana. J. Reprod. Fert.* **41**: 189–192.

Jones, R. C., Skinner, J. D. & Rowlands, I. W. (1974). The role of the urogenital ducts of the African elephant, *Loxodonta africana. J. Reprod. Fert.* **36**: 441–442.

Short, R. V., Mann, T. & Hay, M. F. (1967). Male reproductive organs of the African elephant, *Loxodonta africana. J. Reprod. Fert.* **13**: 517–536.

Wales, R. G. & White, I. G. (1959). The susceptibility of spermatozoa to temperature shock. *J. Endocr.* **19**: 211–220.

White, I. G. & Wales, R. G. (1960). The susceptibility of spermatozoa to cold shock. *Int. J. Fert.* **5**: 195–201.

White, I. G. & Wales, R. G. (1961). Comparison of epididymal and ejaculated semen of the ram. *J. Reprod. Fert.* **2**: 225–237.

Symp. zool. Soc. Lond. (1978) No. 43, 271–287

Observations on the Artificial Breeding of Red Deer

A. KRZYWIŃSKI and Z. JACZEWSKI

Institute of Genetics and Animal Breeding, Polish Academy of Sciences, Popielno, Wejsuny, Poland

SYNOPSIS

Six methods of semen collection from red deer stags using an artificial vagina are described. These methods enable semen collection from both tame and wild individuals, using either a very tame or a stuffed hind (dummy) as a "teaser" female. In some of the methods a specially modified artificial vagina is inserted into artificial hind-quarters which are strapped to a hind in oestrus or to a hind treated with chlorpromazine and pheromones.

Eighty ejaculates were collected from nine red deer males of different ages and different degrees of tameness. The semen was collected in different seasons making possible a preliminary assessment of changes in the semen quality occurring during the year. During the increased rutting activity (September–December) in a four-year-old stag the ejaculate volume varied from 1·0 to 4·5 ml (usually about 2·0 ml) and the spermatozoal concentration ranged from $1·85 \times 10^9$ to $3·95 \times 10^9$ ml^{-1} (usually about $2·5 \times 10^9$ ml^{-1}). In the spring at the time of antler shedding the percentage of secondary abnormalities increased rapidly. During the period of intensive antler growth (May–June) ejaculates were devoid of spermatozoa.

The detection of oestrus was performed with a vasectomized 12-year-old sexually experienced stag which was introduced daily into the enclosures with the hinds. The mean length of the oestrous cycle was $18·76$ days $\pm 1·75$ (s.d.).

Two ejaculates were diluted in a glycerolated fructose-yolk-citrate extender and afterwards frozen as pellets on solid carbon dioxide and stored in liquid nitrogen.

In the breeding season 1976–77, 12 hinds were inseminated with either one or two pellets of frozen semen using a speculum and insemination equipment normally employed in the artificial insemination of cattle. An attempt was always made to deposit the semen into the cervix, but sometimes without success, probably on account of the inappropriate size of the insemination equipment. Three hinds became pregnant and calved, producing healthy fawns (one female and two males) which are developing normally. The recorded durations of gestation after artificial insemination were: 232, 231 and 235 days.

The methods described might also be applied in future to other species of non-domesticated ruminants. These methods not only would assist scientific research into the comparative physiology of reproduction but also offer a practical means of breeding some species of wild ruminants.

INTRODUCTION

The species, *Cervus elaphus* (*sensu lato*), includes many subspecies and is distributed in both northern and southern hemispheres (Flerov, 1952; Whitehead, 1972). There is a large body of literature on aspects of reproduction in *Cervus elaphus* under natural conditions, but more precise observations relating to reproductive physiology (Jaczewski & Gałka 1970a,b; Guinness, Lincoln & Short, 1971; Fletcher, 1974;

Jaczewski & Krzywińska, 1975) have been rendered possible by the development of deer farming in enclosures as practised for many years in Russia (Druri & Mitjušev, 1963) and, more recently, in Great Britain (Blaxter, 1974) and New Zealand (Drew & McDonald, 1976). The advantages of controlled breeding in red deer lie not only in the ability to select for fine antlers as hunters' trophies, but also in the production of meat and, in Russia and China, medicaments (Razmachnin & Ryvkin, 1976).

On the red deer farm at Popielno semen was initially collected by electroejaculation (Jaczewski & Morstin, 1973; Jaczewski & Jasiorowski, 1974; Jaczewski et al., 1976). This method, however, requires the stag to be immobilized and sometimes also tranquillized, which reduces the semen quality or even renders semen collection impossible. Spermatozoa collected by electroejaculation are always diluted by a large volume of seminal plasma.

On account of the disadvantages of electroejaculation, an alternative method, the artificial vagina, was investigated (Krzywiński, 1976). Previously, the use of an artificial vagina for semen collection in Cervidae had been restricted to reindeer alone (Dott & Utsi, 1971, 1973). Krzywiński (1976) collected 13 ejaculates from three red deer stags and showed that semen collected with the artificial vagina was of a much higher quality than that collected by electroejaculation.

The present investigations were designed to modify the method of semen collection so that it could be applied even to the more aggressive and the more timid individuals. Simultaneously, more accurate observations were made of the recurrence of oestrus and its detection in the hind. Finally, some ejaculates were frozen and hinds in oestrus were inseminated to ascertain whether the technique of artificial insemination could be applied to red deer.

MATERIAL AND METHODS

The deer farm at Popielno is located on the border of the Piska Forest. About 40 red deer are kept on the farm in enclosures 50×20 m or 25×20 m. From mid-May to mid-October the animals are each fed about 10 kg of green fodder, 1 kg of oats, 0·5 kg of concentrated food for ruminants and branches of trees. During the remainder of the year hay (1·5 kg) and beetroot (4 kg) are substituted for the green fodder. Naturally, stags and pregnant or lactating females receive more food than yearlings and barren females.

Experimental Animals

All males present on the farm during the years 1975–77 were used for semen collection with the artificial vagina.

The stag "Amor", born 1964, had been used for many experimental purposes and as a result was not aggressive towards humans. In July 1976 this stag was vasectomized and was subsequently used for oestrus detection in hinds.

Three stags, "Maciek", "Minister" and "Mikado", were born in 1972. "Maciek" and "Minister" were found in the forest, raised separately among people, and were therefore very tame before puberty. "Mikado", born on the farm, was separated from its mother when 14 days old and raised on cow milk together with a group of other fawns. It was less tame than "Maciek" and "Minister". "Maciek" sired offspring when two years old, and was used for semen collection by electroejaculation in 1975 (Jaczewski *et al.*, 1976). In October 1975 its antlers were sawn off and after two days of trials semen was collected with the artificial vagina held by an attendant (Krzywiński, 1976). This stag was very aggressive towards people and towards other deer but its libido was very high.

"Minister" sired two hinds when 17·5 and 20·5 months old, respectively. This stag was extremely aggressive towards people and towards other deer; when two and a half years old it killed a castrated male and a fawn. Its aggressiveness was greater than its sexual drive and it frequently attacks even the hinds in oestrus.

"Mikado" sired a hind when two years old. Semen was collected from "Mikado" in October 1975 by electroejaculation (Jaczewski *et al.*, 1976) and, in November 1975, by means of an artificial vagina fixed to a hind (Krzywiński, 1976).

Three stags, "Odys", "Ozyrys" and "Olimp", were born in 1974 on the farm at Popielno. They were separated from their mothers when three days old and raised on cow milk. Their tameness was limited because they were accustomed to one person only. "Odys" and "Ozyrys" were already aggressive as yearlings.

Three males, "Pokorny", "Parys" and "Pant", were born on the farm in 1975. They were also separated from their mothers and raised on cow milk, but during this upbringing their contacts with people were more frequent and they were considerably more tame than stags born in 1974. "Pant" was subordinate to "Pokorny" and "Parys" and as a yearling it manifested practically no sexual drive.

Two hinds, "Wala", born 1958, and "Maska", born 1968, were used as "teasers" for semen collection. Both hinds were raised among people and were very tame. They were trained to walk on a halter and to accept a special harness attaching the artificial hind-quarters (Fig. 1). "Wala" was smaller than "Maska" and therefore more suitable for semen collection because the artificial vagina fixed on a hind is located a little higher than the hind's vagina.

Artificial insemination was performed on 12 hinds: six raised on cow milk and quite tame and six raised by their mothers and more wild.

Methods of Semen Collection

The basic method of semen collection with an artificial vagina installed in the deer-skin artificial hind-quarters fixed to a hind was described by

Fig. 1. Semen collection from stag "Pokorny" with "teaser" "Maska" in January 1977, Method 2b (see text). The artificial hind-quarters are fixed to "Maska" by harness (photograph by B. Gałka).

Krzywiński (1976). Details were also given of the "membrane collector" which is inserted into the artificial vagina. The membrane collector is a glass receptacle, partially sealed by a rubber membrane to retain the semen after ejaculation. Further experiments proved that this method can be used with certain modifications depending on the age and aggressiveness of the stag. All the aggressive stags had their antlers sawn off, and only "Mikado", "Parys" and "Pant" retained their antlers. In addition a new method was introduced in which a stuffed hind was used for semen collection. Altogether, six variations of the method of semen collection with an artificial vagina were used.

1. Into the artificial vagina held by an attendant during the jump of a stag on a hind restrained on a halter by a man.
    a. Hind after application of chlorpromazine (Tranquiline*) and pheromones (urine from an oestrous hind) (Krzywiński, 1976). (The dose of chlorpromazine differed depending on both the body weight and the degree of excitement of the hind. It varied from about $1 \cdot 6$ mg kg^{-1} i.v. and $1 \cdot 6$ mg kg^{-1} i.m. to about $2 \cdot 1$ mg kg^{-1} i.v. and $3 \cdot 1$ mg kg^{-1} i.m.) This combination of intramuscular and intravenous administration was found to be very satisfactory for experiments requiring a more prolonged sedation.)
    b. Hind in oestrus.
2. Into the artificial vagina inserted in artificial hind-quarters fixed to a hind restrained on a halter by a man (Fig. 1).
    a. Hind after application of chlorpromazine and pheromones (Krzywiński, 1976).
    b. Hind in oestrus.
3. Into the artificial vagina held by an attendant during the jump of a stag on a stuffed hind (dummy) sprayed with pheromones.
4. Into the artificial vagina inserted in natural position in a stuffed hind (Fig. 2).
5. Into the artificial vagina inserted into artificial hind-quarters fixed to a tethered hind after application of chlorpromazine and pheromones.
6. Into the artificial vagina inserted into the artificial hind-quarters fixed to a very tame hind in oestrus and let free into the enclosure of a stag.

Semen collection is often undertaken on cold days (October–November) and it is very important to maintain a temperature of 40–42°C in the artificial vagina for long periods. It is therefore necessary to ensure that the artificial vagina has a large thermal capacity and is thermally insulated, especially for Methods 4, 5 and 6.

---

* The Polish drug Tranquiline contains per ml: chlorpromazine (25 mg), ascorbic acid (2 mg), sodium metabisulphite (1 mg) and sodium chloride (6 mg).

Fig. 2. Semen collection from stag "Mikado" using the dummy "teaser" in October 1976, Method 4 (see text) (photograph by B. Gałka).

The quality of the semen was measured according to the standard methods used for domestic bulls (Bielański, 1972). The spermatozoal counts were made with the aid of a haemocytometer. The percentage of primary and secondary abnormalities were evaluated according to Blom (1949), using the method of Hoppe & Jaśkowski (1969). The percentage of progressively motile spermatozoa was estimated according to the method of Bielański (1972). To obtain semen of good quality it is necessary to protect the membrane collector from cold, otherwise thermal shock may occur. As the chief aim of the present work was the elaboration of methods of semen collection, the membrane collector was often not protected from cold and in such cases the motility of the semen and the percentage of live cells were not estimated. In some cases the artificial vagina was too long for younger stags and the semen was not ejaculated into the membrane collector but was smeared on the interior of the vagina. In such cases the spermatozoal concentration and the semen volume were estimated only approximately.

Freezing of Semen

To collect an ejaculate suitable for freezing, the procedure was as follows: before collection, 5 ml of the glycerolated fructose-yolk-citrate extender* was placed into the membrane collector, which was protected from cold by a water jacket at 37°C. The ejaculate was collected directly into this solution and in these cases the colour and the consistency of semen could not be estimated. This procedure was necessary because the ejaculate of a stag at the height of the rutting period has the consistency of thick honey and sticks to the collector walls, thus making subsequent dilution very difficult.

The collected semen was cooled in a refrigerator to about 5°C over two hours and then transported to an Artificial Insemination Centre. After a total equilibration time of four hours the semen quality was evaluated and the semen was then diluted to obtain 15–20 million living spermatozoa in a single pellet (0·1 ml). The semen was frozen in wells on solid carbon dioxide and transferred to liquid nitrogen.

Detection of Oestrus in Hinds

The stag "Amor" was vasectomized on 25 September 1976. On the 16 and 20 October ejaculates were collected by Method 4 and contained only a few spermatozoal heads without tails. The stag was kept in a separate enclosure and from 20 October was given access every day (with the exception of a few Sundays) to the enclosures with the hinds. Some efforts

* The extender was prepared as follows: 2·9% sodium citrate—72 ml., fructose—1·25 g, egg yolk—20 ml, glycerol—8 ml, streptomycin—0·1 g.

were made to prevent the stag from mating with the hinds but nevertheless copulation took place in a few cases. In January and February 1977 "Amor's" libido gradually decreased and the stag became less suitable for the detection of oestrus.

In previous years oestrus had been detected simply by observing the behaviour of the hinds. The hinds in oestrus become more active (Ozoga & Verme, 1975), often approaching the fence separating them from a stag. Such signs as rubbing the neck and body against the male and licking the male are often observed. In very tame hinds, a standing response could be evoked by applying pressure to the back of the female. The stag in a neighbouring enclosure also approaches the fence and smells and licks the hind. However, this method based on observations is not very accurate and if the oestrous display is absent or weak it may pass unnoticed.

Artificial Insemination of Hinds

The hinds in oestrus were introduced into a separate enclosure and immobilized with succinylcholine chloride. Insemination was always performed on a recumbent animal using a speculum and insemination pipettes (diameter 6 mm) normally employed in the artificial insemination of cattle.

Either one or two semen pellets were thawed in an ampoule containing 1 or 2 ml physiological salt solution (0·9% NaCl) which had previously been warmed in the hand. The second pellet (when required) was added about 1 min after the first. The time between removing the pellet from liquid nitrogen and insemination varied from 10 to 60 min (in two cases it was only about 1 min). An attempt was always made to deposit the semen into the cervix but sometimes without success, probably owing to the inappropriate size of the insemination pipettes. In such cases the semen was deposited intravaginally.

RESULTS

Semen Collection

A total of 80 ejaculates was collected from nine red deer stags. The number of ejaculates collected from each of the stags and the methods used to obtain them are given in Table I. Two ejaculates were collected from "Amor" after the vasectomy (see p. 277). "Maciek" proved to be the best stag for semen collection; during the peak of the rutting season it was possible on one occasion to collect two ejaculates one day and a further ejaculate in the morning of the following day. The shortest interval between the two collections was 15 min. "Minister" was extremely aggressive and all methods but Method 6 were useless. Of the other stags, all but

TABLE I

Semen collection from red deer stags

Stag	Year	No. of ejaculates	Method of collection
Amor	1976	2[a]	4
Maciek	1975	9	1(a)
	1976	18	1(a), 3, 4
	1977	10	1(a), 1(b), 3
Minister	1976	2	6
Mikado	1975	1	2(a)
	1976	2	4
	1977	5	5, 6
Odys	1975	3	2(a)
	1976	1	2(a)
	1977	5	2(b), 5
Ozyrys	1976	1	2(b)
	1977	5	1(b), 2(a), 2(b), 5
Olimp	1976	1	2(a)
	1977	4	1(b), 2(b)
Pokorny	1976	3	2(a), 2(b)
	1977	3	2(b)
Parys	1976	1	6
	1977	6	2(a), 2(b)

[a] Ejaculates collected after vasectomy.

"Pant" were induced to serve an artificial vagina by one or other of the variations described. "Pant" was subordinated to "Pokorny" and "Parys" and being kept together with them in one enclosure this stag manifested practically no sexual drive.

These investigations showed that Method 6 was suitable for all stags, even those that are more wild. The disadvantages of this method are that a very tame hind in oestrus must be used, and the artificial vagina must be very well insulated. Method 5 was almost as universally applicable as Method 6 but it was unserviceable for very aggressive stags, which may attack the hind.

Method 4 was appropriate only for older, sexually experienced stags; it was unsuitable for younger males. This observation is consistent with the observations on stallions made by Wierzbowski (1959), that only the older males showed normal sexual reflexes with a dummy. It should be noted that a very aggressive stag may destroy the dummy.

Method 2 was very convenient for semen collection from younger and tame males. In this method it is very easy to control the temperature of the artificial vagina and to remove the membrane collector containing the semen quickly.

Methods 1 and 3 were suitable only for a very tame stag with a quiet disposition and a strong sexual drive. These methods are naturally the most convenient but their applications are limited probably because they can be used only for rather exceptional stags and with a very skilled staff. It should be noted that any change of personnel during semen collection is highly undesirable.

The use of pheromones during semen collection is very important, especially for younger males. In the investigations described above urine and mucus from a hind in oestrus were usually employed. It was also observed that the odour left by a stag stimulated the next male.

Chlorpromazine is not the best tranquillizer for hinds and should not be administered too frequently. A more suitable drug should be found in the future.

Characteristics of the Collected Semen

The characteristics of the semen obtained from the stag "Maciek" during a two-year period are given in Table II. Some of the data presented are only approximate, because not all ejaculates were properly collected (see p. 277). The data from other stags were rather scanty (Table I) but they indicated that the seasonal changes in semen quality were similar for other stags.

The semen of young males retains the colour and consistency of semen collected during the rutting season for a much longer period. For example, the semen collected on 5 March 1977 from "Parys" was yellow in colour and its consistency was that of thick honey, while the semen collected on the same date from "Maciek" was creamy-white and milky in consistency. This observation accords with the well-known fact that antler shedding in young stags occurs much later than in older ones. The volume of the ejaculate obtained from the older stags ("Maciek", "Mikado") was about 2·0 ml and from younger ones about 0·5–1·0 ml.

As can be seen from Table II, after the rutting season the consistency and colour of semen changes gradually and the percentage of secondary abnormalities increases. These changes are particularly marked after the shedding of antlers. Sexual excitation of a stag and semen collection are possible even during the period of intense antler growth (May–June), but the ejaculates at this stage are devoid of spermatozoa.

Recurrence of Oestrus in Hinds

The occurrence of oestrus in the hinds is shown in Table III. In Table III both natural mating and artificial insemination with frozen semen are indicated.

The length of the oestrous cycle, based on the data given in Table III, was 18·76 days ± 1·75 (mean ± s.d.). In two cases ("Nemezja" and "Nimfa") a very long interval between two oestrous periods seems to indicate that one oestrus passed unnoticed; these cases have been excluded from the calculations. The result is almost identical with the figure of 18·3 days ± 1·7 (mean ± s.d.) for the oestrous cycle of red deer given by Guinness *et al.* (1971).

On the farm at Popielno, experiments on the induction of antler growth in hinds were conducted simultaneously with the investigations of oestrus (Jaczewski, 1976, 1977). During these experiments some hinds were treated with testosterone during the winter. In January 1977 "Lula" was given 0·5 g and "Melodia" 1·0 g of testosterone, which caused the appearance of a rather long oestrous period. These observations seem to be consistent with the results reported by Fletcher (1975) who induced oestrus in hinds with testosterone after pretreatment with progesterone. In spite of these experiments on the induction of antler growth, "Lula" delivered a normal fawn on 2 July 1977.

Artificial Insemination

For artificial insemination with frozen semen two ejaculates from "Maciek", obtained on 1 October 1976 and on 2 December 1976, were used. The first ejaculate contained 20–30% progressively motile cells after thawing, while the second one contained a little over 50% progressively motile cells. The first ejaculate was used for all inseminations up to, and including, 16 December, and the second ejaculate was used for all subsequent inseminations.

As can be seen from Table III, three hinds ("Lula", "Mimoza" and "Zosia") gave birth to normal, healthy fawns following insemination with frozen semen. "Śniezynka", female, born 2 July after 232 days of gestation, weighed 11·9 kg; "Szron", male, born 17 July after 231 days of gestation, weighed 12·4 kg and "Sopel", male, born 27 July after 235 days of gestation, weighed 12·2 kg.

DISCUSSION

The experiments described above showed that it is possible to collect semen from red deer stags with the artificial vagina and to inseminate hinds with frozen semen. Much individual variation was observed between stags during semen collection with the artificial vagina, but the methods described above enabled the collection of semen not only from tame stags but also from those less tame.

The red deer is a very interesting animal for investigations of the comparative physiology of reproduction. This ruminant shows very distinct seasonal physiological changes during the year. The seasonal fluctuations in the semen characteristics accord with the evidence from histological investigations of the testes (Lincoln, 1971). The collection of semen is possible even in the period of antler growth in spite of a weak libido, but during the period of intense antler growth (May–June) the ejaculate is completely devoid of spermatozoa.

TABLE II
Semen characteristics of the stag "Maciek" during the year

Date	Method	Volume (ml)	Consistency	Colour	Spermatozoal concentration ($\times 10^6$ ml^{-1})	pH	Percentage progressively motile spermatozoa	Percentage abnormal cells		Antler cycle
								primary	secondary	
23.10.75	1a	~3	Thick honey	Amber-yellow	n.e.	n.e.	Cold shock	n.e.	n.e.	
24.10.75	1a	~5	Thick honey	Amber-yellow	n.e.	n.e.	Cold shock	n.e.	n.e.	
6.11.75	1a	~4	Thick honey	Amber-yellow	n.e.	n.e.	Cold shock	n.e.	n.e.	
16.11.75	1a	2	Thick honey	Amber-yellow	n.e.	n.e.	Cold shock	n.e.	n.e.	
28.11.75	1a	1·6	Thick honey	Amber-yellow	3810	6·5	60	1·0	2·8	
28.11.75	1a	3·5	Thick honey	Amber-yellow	720	7·0	Cold shock	1·2	4·8	
29.11.75	1a	1·5	Honey	Yellow	700	6·8	Cold shock	1·4	4·6	
4.12.75	1a	3·0	Honey	Amber-yellow	2325	6·8	40	1·4	3·8	
5.12.75	1a	1·5	Honey	Yellow	1290	6·8	Cold shock	2·2	5·2	
19.01.76	1a	2	Cream	Yellow	1500	7·0	Cold shock	1·8	9·8	
22.01.76	1a	1·5	Cream	Creamy	n.e.	6·9	40	n.e.	n.e.	
12.02.76	1a	1	Cream	Creamy-white	2080	6·8	Cold shock	2·0	19·0	
16.02.76	1a	2	n.e.	n.e.	2000	n.e.	30	n.e.	n.e.	
5.03.76	1a	1·5	Milk	Creamy-white	1550	7·0	Cold shock	n.e.	n.e.	
15.03.76	1a	1·5	Watery with granulation	Watery-white	1500	6·9	Cold shock	n.e.	n.e.	

23.03.76	1a	1·2	n.e.	n.e.	1500	n.e.	25	3·28	48·6
26.08.76	1a	1·4	Thick honey	Amber-yellow	2650	6·7	Cold shock	1·8	3·2
6.09.76	1a	2·0	Thick honey	Amber-yellow	3300	6·8	60	1·8	9·2
6.09.76	1a	1·8	Thick honey	Amber-yellow	550	6·5	50	0·4	5·2
21.09.76	1a	1·0	Thickhoney	Amber-yellow	2750	6·7	Cold shock	1·2	3·8
21.09.76	1a	1·0	Thick honey	Amber-yellow	1700	6·7	Cold shock	0·4	19·4
1.10.76	1a	~1·5	n.e.	n.e.	2600	n.e.	n.e.	n.e.	n.e.
20.10.76	4	~1·8	Thick honey	Amber-yellow	3950	6·7	Cold shock	0·8	10·8
26.10.76	3	~2·0	Thick honey	Amber-yellow	2975	6·8	Cold shock	1·8	9·2
2.11.76	3	4·5	Honey	Amber-yellow	1850	6·6	Cold shock	1·2	8·4
2.12.76	3	~2·1	n.e.	n.e.	2800	n.e.	n.e.	n.e.	n.e.
14.12.76	3	2·2	Honey	Yellow	2300	6·7	Cold shock	2·4	10·0
4.01.77	3	2·5	Honey	Milky-yellow	2050	6·8	Cold shock	1·2	8·4
4.01.77	3	2·0	Honey	Milky-yellow	1050	6·7	Cold shock	0·8	11·4
15.02.77	3	2·0	n.e.	Milky-yellow	1900	n.e.	50	n.e.	n.e.
15.02.77	3	1·6	n.e.	Milky-yellow	920	n.e.	65	n.e.	n.e.
22.02.77	3	1·6	Milk	White	810	6·7	Cold shock	1·9	14·3
5.03.77	1b	~1·0	Watery-Milk	Watery-white	2280	6·7	Cold shock	1·5	15·7
4.04.77	3	0·8	Watery with granulation	Watery-white	795	6·9	Cold shock	—	—
19.04.77	1a	n.e.	Watery-Milk	Watery-white	5	7·2	Cold shock	0·4	90·6
11.05.77	1a	~1·5	Watery	Watery-white	Aspermic	6·9	—	—	—
15.06.77	1a	0·8	Watery and 1/5 jelly	Watery-white	Aspermic	n.e.	—	—	—

, Hard antlers; , antlers shedding; , growing antlers; n.e., not estimated.

TABLE III

Occurrence of oestrus detected by vasectomized stag and results of artificial insemination

Hind	Number and date of oestrus								Birth date after artificial insemination
	I	II	III	IV	V	VI	VII	VIII	
Zosia	28.10.76 29.10 ⓘ	15.11.76 ⓘ	4.12.76 ⓘ						♂ 27.07.1977
Nemezja	3.11.76 ⓘ	18.11.76 ⓘ	5.12.76 ⓘ	23.12.76 ⓘ	?	28.01.77	15.02.77	5.03.77	
Muzyka	3.11.76 ⓘ	20.11.76 M							
Lula	12.11.76 ⓘ			19–22.01.77 T					♀ 2.07.1977
Maska	25.11.76 ⓘ	14.12.76 M	3.01.77	24.01.77	13.02.77	5.03.77			
Wala	27.11.76 ⓘ 28.11 ⓘ	17.12.76	6.01.77	25.01.77 26.01.77	16.02.77 17.02.77				
Mimoza	28.11.76 ⓘ								☿ 17.07.1977
Bella	16.12.76 ②ⓘ	3.01.77	24.01.77						
Melodia	16.12.76 ②ⓘ	3.01.77		18–21.01.77 T			23.02.77		
Słonka	17.12.76	6.01.77	25.01.77 ⓘ ?						
Nimfa	20.12.76 ⓘ	3.01.77 ⓘ		16.02.77					
Wanda	4.12.76 ⓘ	23.12.76 ⓘ	11.01.77 ⓘ						

ⓘ, artificial insemination; ②ⓘ, two inseminations during one oestrus; M, normal mating; T, oestrus induced by an injection of testosterone.

Three out of 12 hinds inseminated with frozen semen became pregnant and calved (25%). This percentage is not low if one takes into consideration that artificial insemination of wild ruminants is still in its infancy. By comparison, after intracervical inseminations in goats only 21·6% of the animals became pregnant and this was increased to 71·2% after intrauterine insemination (Fougner, 1976). In ewes, after deep intracervical insemination, 19·7% of the animals lambed and 50·9% lambed after intrauterine insemination (Fukui & Roberts, 1976).

In the experiments presented here not all the hinds were inseminated intracervically, the others being inseminated intravaginally, which probably diminished the chances of fertilization. Many details need to be investigated and improved in future work on the artificial insemination of red deer (e.g. the most favourable time of insemination, the size of the speculum and insemination equipment appropriate for red deer, etc.).

These techniques of artificial insemination will be of practical value both in zoological gardens and on red deer farms. It may even be possible to use artificial insemination in the management of wild red deer. In breeding deer for fine heads specific introductions and crossing are occasionally used. Such procedures, however, have, amongst many other disadvantages, problems on account of the social structure of red deer herds (Reuss, 1968) and the difficulty of transporting adult animals (Raesfeld, 1971; Szederjei & Szederjei, 1971). On the other hand, it is possible to trap the females of the local population which can then easily be inseminated with semen of genetically superior stags, and released.

Using artificial insemination stags with record antlers could be employed for breeding long after their death. It may also be possible to investigate interspecific crossing of deer. Moreover, the methods of semen collection described here might be applicable to other species of wild ruminants, and thereby assist further research on the comparative reproduction of mammals.

ACKNOWLEDGEMENTS

The part of this paper concerned with the semen collection with the artificial vagina and the artificial insemination of red deer is part of a dissertation submitted by A. Krzywiński in partial fulfilment of the requirements for the degree of Doctor of Animal Sciences in the Institute of Genetics and Animal Breeding of the Polish Academy of Sciences. We are greatly indebted to Dr J. Dziliński, ZHW, Olsztyn, for his helpful co-operation.

REFERENCES

Bielański, W. (1972). *Rozród Zwierzat*. Warszawa: PWRiL.
Blaxter, K. L. (1974). Deer farming. *Mammal Rev.* **4**: 119–122.

Blom, E. (1949). Über Sperma-Untersuchungs-Methoden. *Wiener Tierärztl. Mschr.* **36**: 49–55, 111–112, 161–168.

Dott, H. M. & Utsi, M. N. P. (1971). The collection and examination of semen of the Reindeer, *Rangifer tarandus. J. Zool., Lond.* **164**: 419–424.

Dott, H. M. & Utsi, M. N. P. (1973). Artificial insemination of Reindeer, *Rangifer tarandus. J. Zool., Lond.* **170**: 505–508.

Drew, K. R. & McDonald, M. F. (Eds) (1976). *Deer farming in New Zealand. Progress and Prospects.* Wellington: Editorial Services Limited.

Druri, I. V. & Mitjušev, P. V. (1963). *Olenevodstvo.* Moskva-Leningrad: Izdatelstvo Selsckochozjajstvennoj Literatury Žurnalov i Plakatov.

Flerov, K. K. (1952). Kabargi i Oleni. *Fauna SSSR*, N.S. No. 55: 1-256.

Fletcher, T. J. (1974). The timing of reproduction in Red deer, *Cervus elaphus* in relation to latitude. *J. Zool., Lond.* **172**: 363–367.

Fletcher, T. J. (1975). *The environmental and hormonal control of reproduction in male and female Red deer,* Cervus elaphus L., Ph.D. Thesis: Cambridge.

Fougner, J. A. [1976]. Uterine insemination with frozen semen in goats. *VIII Int. Congr. Anim. Reprod. Artif. Insem.* [Krakow] **4**: 987–990.

Fukui, Y. & Roberts, E. M. [1976]. Studies of non-surgical intra-uterine insemination of frozen pelleted semen in the ewe. *VIII Int. Congr. Anim. Reprod. Artif. Insem.* [Krakow] **4**: 991–993.

Guinness, F., Lincoln, G. A. & Short, R. V. (1971). The reproductive cycle of the female red deer, *Cervus elaphus* L. *J. Reprod. Fert.* **27**: 427–438.

Hoppe, R. & Jaśkowski, L. (1969). Instrukcja w sprawie sposobu badania i oceny przydatności rozpłodowej buhajów. Warszawa: PWRiL.

Jaczewski, Z. (1976). The induction of antler growth in female Red deer. *Bull. Acad. pol. Sci.* **24**: 61–65.

Jaczewski, Z. (1977). The artificial induction of antler cycles in female red deer. *Deer* **4**: 83–86.

Jaczewski, Z. & Gałka, B. (1970a). Effects of administration of testosteronum propionicum on the antler cycle in red deer. (VIII Int. Congr. Game Biol., Helsinki 1967). *Riistat. Julk.* **30**: 303–308.

Jaczewski, Z. & Gałka, B. (1970b). Effect of human chorionic gonadotrophin on the antler cycle in red deer. *Trans. Int. Congr. Game Biol. Moskwa,* **9**: 217–218.

Jaczewski, Z. & Jasiorowski, T. (1974). Observations on the electroejaculation in Red deer. *Acta Theriol.* **19**: 151–157.

Jaczewski, Z. & Krzywińska, K. (1975). The effect of testosterone on the behaviour of castrated females of red deer, *Cervus elaphus* L. *Pr. Mater. Zoot.* **8**: 37–45.

Jaczewski, Z. & Morstin, J. (1973). Collecction of the semen of the red deer by electroejaculation. *Pr. Mater. Zoot.* **3**: 83–86.

Jaczewski, Z., Morstin, J., Kossakowski, J. & Krzywiński, A. [1976]. Freezing the semen of red deer stags. *VIII Int. Congr. Anim. Reprod. Artif. Insem.* [Krakow] **4**: 994–997.

Krzywiński, A. [1976]. Collection of red deer semen with the artificial vagina. *VIII Int. Congr. Anim. Reprod. Artif. Insem.* [Krakow] **4**: 1002–1005.

Lincoln, G. A. (1971). The seasonal reproductive changes in the Red deer stag, *Cervus elaphus. J. Zool., Lond.* **163**: 105–123.

Ozoga, J. J. & Verme, L. J. (1975). Activity patterns of white-tailed deer during estrus. *J. Wildl. Mgmt* **39**: 679–683.

Raesfeld, F. (1971). *Das Rotwild.* Hamburg und Berlin, Paul Parey.

Razmachnin, V. E. & Ryvkin, L. M. (1976). Roga dikich kopytnych—cennoe lekarstvennoe syr'e. In *Ochotowedenie*: 197–266. Gavrin, V. F. (ed.). Moskva: Lesnaja Promyšlennost'.

Reuss, H. (1968). Wege zur Verbesserung der Rothirsch-qualität. *Wiss. Anblick* No. 11.

Szederjei, A. & Szederjei, M. (1971). *Geheimnis des Weltrekordes der Hirsch.* Budapest: Terra.

Wierzbowski, S. (1959). Odruchy płciowe ogierów. *Rocz. Nauk roln., Kraków*, **73**-B-4: 753–788.

Whitehead, G. K. (1972). *Deer of the world.* London: Constable.

Symp. zool. Soc. Lond. (1978) No. 43, 289–301

The Collection, Handling and Some Properties of Marsupial Semen

J. C. RODGER and I. G. WHITE

University of Sydney, Australia

SYNOPSIS

Eight species of Australian marsupials were electroejaculated using a multipolar rectal probe of dimensions suitable for each species. The electrical stimulus was a half-time square wave direct current (DC) pulse which could be delivered at a rate of 5–100 pulses s^{-1} over an amplitude of 0–10 V. A stimulus of 30–40 pulses s^{-1} and 6–10 V proved most suitable. The results obtained varied considerably between species but can be grouped in the following categories.

1. Erection but no significant release of fluid or spermatozoa (bandicoot, Tasmanian devil and native cat).
2. Copious seminal plasma ejaculated but no or few spermatozoa (tammar wallaby, red and grey kangaroo).
3. Semen with high spermatozoal numbers (brush-tailed possum).

Attempts to electroejaculate conscious or sedated animals were relatively unsuccessful in comparison with the response from animals anaesthetized prior to stimulation. Of the anaesthetics tested (ether, halothane and pentobarbitone sodium) ether was selected for routine collection of possum semen.

Ejaculates from 40 successful collections out of 51 attempts from six possums during 1976 showed the following characteristics. There was a mean ejaculate volume of 2·7 ml with a mean spermatozoal concentration of $4·8 \times 10^7$ ml^{-1} (range 0·7 to 17×10^7) giving a mean total number of spermatozoa per ejaculate of $12·9 \times 10^7$ (range 1.6 to $39·6 \times 10^7$).

Semen collected from possums is amenable to handling techniques commonly applied to other mammalian species. It can be diluted in calcium-free Krebs Ringer phosphate and washed by centrifuging at 750 g without any apparent loss in motility (dilution rate ×5, washed twice). Possum spermatozoa are also tolerant of cold shock which simplifies handling techniques. However, the spermatozoa are rapidly immobilized, usually after less than one hour of aerobic incubation at 37°C, by resuspension after washing at a concentration above that normally found in ejaculates. Under similar incubation conditions but at lower concentration, possum spermatozoa retain their motility with or without exogenous substrate for up to six hours. Marsupial ejaculates always contain large numbers of spherical bodies of prostatic origin. It may be these which produce the deleterious effects of spermatozoal concentration as immotile spermatozoa are often associated in clumps around these bodies.

Ejaculated marsupial semen has been examined biochemically and physiologically. Biochemical studies have shown that *N*-acetylglucosamine rather than fructose is the characteristic seminal sugar of many Australian marsupials. Possum spermatozoa utilize oxygen at around 100 μl per 10^8 spermatozoa per 3 hours at 37°C, which is considerably greater than most domestic or laboratory animals. This high rate of consumption is not dependent upon the supply of an exogenous substrate, either *N*-acetylglucosamine or glucose. However when these substrates are supplied they were utilized at a high rate both being consumed at 5–7 μmol per 10^8 spermatozoa per 3 hours.

INTRODUCTION—MARSUPIAL REPRODUCTION

The Male

The male marsupial has only recently attracted the attention of reproductive physiologists as a major field of study. At present, work is continuing in two main areas: spermatozoal maturation during epididymal transit (e.g. Harding, Carrick & Shorey, 1975; Cummins, 1976) and semen biochemistry and physiology (Rodger and White; see Rodger, 1976). Prior to these developments marsupial reproductive physiology was dominated by studies of the female, especially of embryonic diapause in macropods and its regulation (see review by Tyndale-Biscoe, Hearn & Renfree, 1974). Although many marsupial species have been bred in captivity, some for many years, the artificial breeding of any marsupial species has yet to be achieved.

Anatomy of the male tract (Fig. 1)

In almost all breeding marsupials the testes and epididymides are scrotal. The marsupial epididymis is a relatively large organ only loosely applied to the surface of the testis. The cauda region is conspicuous, so that its shape is often clearly visible in the scrotum. The prostate gland, the major accessory sex gland of all marsupials, is often quite large and divided into two or three main regions or segments. The prostate gland of marsupials is disseminate; the glandular elements do not pass beyond the urethral muscle but lie within it, although the glandular region may be quite large. The only other accessory sex glands found in marsupials are bulbo-urethral (Cowper's) glands, and there may be up to three pairs which all empty into the urethra near the base of the penis (Rodger & Hughes, 1973).

Seasonality of breeding

Most marsupial species show peaks and troughs of breeding activity in the wild, but these are often not reflected in changes in the male reproductive tract. In only a few species (e.g. ringtail possum, *Pseudocheirus peregrinus*; greater glider, *Schoinobates volans*, and Tasmanian devil, *Sarcophilus harrisii*) is there an obvious seasonal variation in spermatogenic activity (Tyndale-Biscoe, 1973). Although no such testicular changes occur in the brush-tailed possum, *Trichosurus vulpecula*, there is a marked seasonal change in the size and secretory activity of the prostate gland (Gilmore, 1969).

The Female

Studies of male reproductive physiology must be complemented by appropriate knowledge of the female, and the manipulation of female reproductive activity, if artificial breeding is to be achieved. As mentioned previously, many marsupials are seasonal breeders and this is especially

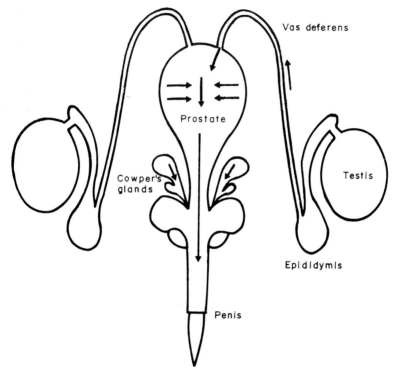

FIG. 1. The generalized male reproductive tract of marsupials. Arrows indicate direction of flow of fluids and spermatozoa.

true of the female, which may be anoestrous for long periods of the year. Thus the detection of oestrous activity and its induction in non-breeding animals is of critical importance to their artificial breeding.

The oestrous cycle and prediction of oestrus

Most marsupial species are polyoestrous and when isolated from males continue to cycle regularly over the breeding season for that particular species. The length of the oestrous cycle varies considerably between species, from 22–46 days, and between individuals of a species; however, a cycle length of around 28 days is common. In only one species, the swamp wallaby, *Wallabia bicolor*, does gestation extend beyond the length of a normal oestrous cycle, although in some other macropods (e.g. the grey kangaroo, *Macropus giganteus*) the luteal phase of the cycle appears to have been extended to accommodate a lengthened gestation period (Tyndale-Biscoe *et al.*, 1974). The oestrous cycle of many marsupials has been followed by taking daily "vaginal" smears from the anterior portion of the urogenital sinus in the region of entry of the lateral vaginal canals.

Marsupial vaginal smears do not give as clear-cut an indication of stage of the cycle as can be seen in smears from laboratory rodents. However, cyclical fluctuations in numbers of cornified epithelial cells, non-cornified epithelial cells and polymorphonuclear leucocytes allow the oestrous cycle to be plotted, oestrus observed and its occurrence in the next cycle predicted (e.g. Sharman, 1955; Hughes, 1962; Pilton & Sharman, 1962). Unfortunately there is no distinctive pro-oestrous vaginal smear in any of the marsupial species so far examined. Vaginal smear techniques are thus of limited use as indicators of impending oestrus in marsupials. Oestrus can be detected in the marsupial "mice", *Sminthopsis larapinta* and *S. crassicaudata*, by the appearance of large numbers of epithelial cells in the urine (Godfrey, 1969; Smith & Godfrey, 1970).

Oestrus may be induced naturally in many marsupials simply by removal of a suckling pouch young. The removal of pouch young initiates follicular development and oestrus occurs approximately six to ten days later, depending on species. In many macropod species there is a post-partum oestrus (e.g. tammar wallaby, *Macropus eugenii*) and the occurrence of this oestrus can be predicted simply by observation of the pouch of pregnant females. This technique has recently been used to obtain natural mating for a spermatozoal transport study in the tammar wallaby (Tyndale-Biscoe & Rodger, 1978).

Induction of ovulation and oestrus

Ovulation can be induced in marsupials by gonadotrophin treatment; however, in only the American opossum, *Didelphis virginiana*, has such treatment resulted in fertile matings which produced offspring (Nelson & White, 1941; Nelson & Maxwell, 1941). Follicle growth was induced in anoestrous quokkas, *Setonix brachyurus*, by a combination of pregnant mare serum gonadotrophin (PMSG) and horse anterior pituitary extract, but ovulation did not occur. In contrast, similar treatment of pro-oestrous animals resulted in superovulation (Tyndale-Biscoe, 1961). In the dasyurid "mouse", *Sminthopsis crassicaudata*, PMSG and human chorionic gonadotrophin (HCG) treatment induced follicle development, ovulation and mating, but the ova shed were not fertilized. This failure of fertilization may not have been due to abnormalities in the ova but rather asynchrony of mating and ovulation which could possibly be overcome by artificial insemination (Smith & Godfrey, 1970). Ovulation has been induced by PMSG alone in the brush-tailed possum and superovulation did occur. However, mating studies have not been carried out in this species (Harding, 1969).

The artificial stimulation of ovulation and mating in non-breeding marsupials or of species that fail to breed in captivity is thus feasible. However, much more work is required before it could be used as a routine technique. Superovulation also appears to be feasible in marsupials, even in monovular species. In all marsupials litter size is strictly controlled by the number of suitable teats available at parturition. Nevertheless, excess young produced as a result of superovulation could possibly be transferred to foster mothers if they were collected very shortly after birth.

Collection of Semen from Marsupials; Early Work

Prior to the work described in the following section of this paper, marsupial semen had been collected by electroejaculation from three marsupial species. However, the semen collected was not the subject of systematic biochemical or physiological examination. The aim of these studies was the detection of spermatozoa in semen or urine as an indicator of male sexual activity. Howarth (1950) used the hypodermic needle-rectal probe method of Gunn (1936) and a stimulus of 30 V alternating current (AC) at 50 cycles s^{-1} to collect semen from the brush-tailed possum. This technique resulted in ejaculates of similar quality to those described in the present paper. A bipolar rectal electrode and a DC stimulus of 12 V, 75 pulses s^{-1} has also been used to collect from brush-tailed possums but with less satisfactory results (Gilmore, 1969). Both Howarth and Gilmore anaesthetized their animals; Howarth used ether but Gilmore did not specify the anaesthetic employed. Semen was collected from two macropod species by Sadleir (1965). He used a bipolar rectal probe and a DC stimulus (10·5 V) and obtained ejaculates containing spermatozoa from animals within five minutes of death by shooting. Electroejaculation has also been attempted on long-nosed bandicoots, *Parameles nasuta*, but without success (Bolliger, 1946; R. L. Hughes, pers. comm.). The first study to deal in detail with the electroejaculation of marsupial species and to examine the biochemistry of the ejaculated semen obtained was that of Rodger & White (1975) and this work, together with later unpublished material, is discussed at length in the next section.

SEMEN COLLECTION FROM MARSUPIALS: ELECTROEJACULATION

Animal Restraint

Wild animals in general are not easily approached or handled and are not readily trained. As a result the artificial vagina and masturbation, common methods of semen collection, are not often used with wild species. Electroejaculation, however, is suitable for the collection of semen from such animals provided an appropriate physical or chemical restraint is used.

The technique of electroejaculation has been applied to a number of marsupial species and a variety of restraints tested (Table I). Three species, brush-tailed possum, tammar wallaby and red kangaroo,

TABLE I

Anaesthetics and tranquillizers tested as restraints for electroejaculation of marsupials

Species	Drug	Dose
Brush-tailed possum	Ether[a]	By mask
(*Trichosurus vulpecula*)	Halothane[a] (Fluothane, ICI)	2–4% vapour in oxygen
	Ketamine (Ketalar, Parke-Davis)	25–30 mg kg^{-1}
Tammar wallaby	Ether[a]	By mask
(*Macropus eugenii*)	Halothane[a]	2–4% vapour in oxygen or nitrous oxide
	Phencyclidine (Sernylan, Parke-Davis)	3 mg kg^{-1}
	Xylazine[a] (Rompun, Bayer)	4·5 mg kg^{-1}
	Acepromazine maleate (Acetyl-promazine, Boots)	1·4 mg kg^{-1}
Long-nosed bandicoot	Halothane[a]	2–3% vapour in oxygen
(*Perameles nasuta*)	Xylazine	7 mg kg^{-1}
	Acepromazine maleate	1·5 mg kg^{-1}
Short-nosed bandicoot	Ether[a]	By mask
(*Isoodon macrourus*)		
Tasmanian devil	Ether[a]	By mask
(*Sarcophilus harrisii*)	Pentobarbitone[a] (Nembutal, Abbott)	20–40 mg kg^{-1}
Native cat	Pentobarbitone[a]	20 mg kg^{-1}
(*Dasyurus viverrinus*)		

[a] Those which induced a sufficient degree of restraint.

Megaleia rufa, have been electrically stimulated while conscious, but physically restrained in bags. The first two species provided ejaculates when treated in this manner but the animals appeared to be extremely agitated.

A number of anaesthetics and tranquillizers have been tested as restraints for electroejaculation (Table I). Only one tranquillizer (xylazine) provided a sufficient degree of restraint, and this was only in one species, the tammar wallaby. In the other species tested, the long-nosed bandicoot, even very high doses of xylazine were unsuccessful. Electroejaculation thus requires a much deeper state of sedation than can be achieved with the tranquillizers tested (xylazine, acepromazine maleate, ketamine and phencyclidine). This is an interesting observation because xylazine and ketamine have been used as surgical anaesthetics for various marsupial species (Denny, 1974). All general anaesthetics tested (ether, halothane and pentobarbitone) provided satisfactory restraint. Ejaculates have also been collected from animals shot in the field and then electrically stimulated shortly before or some minutes after the heart ceased beating. This technique has been applied successfully to tammar wallabies, red kangaroos and grey kangaroos. Ether has been selected for routine collection from brush-tailed possums, because the technique is simple and the animals appear to suffer no ill effects.

Rectal Probes and Characteristics of the Electrical Stimuli

Most of the work described here utilized one of the two multipolar rectal probes illustrated in Fig. 2. A probe of slightly larger diameter than the smaller illustrated probe has been tested (Rodger & White, 1975) but present practice is to use the one illustrated here. Electroejaculation has also been achieved using a flank electrode (hypodermic needle) as one pole and the rectal probe as the other. The flank electrode was placed in either skin or muscle in the spinal area near the pelvis or the abdomen near the scrotum. The ejaculates collected in this way were essentially similar to those obtained with the multipolar rectal probe.

The electrical stimulus was a half time-square wave DC pulse which could be delivered at a rate of 5–100 pulses s^{-1} over an amplitude of 0–10 V. A stimulus of 30–40 pulses s^{-1} and 6–10 V has proved most suitable. The stimulator used has been described by Martin & Rees (1962). The pattern of application of the stimulus and positioning of the probe has been varied according to species and individual responsiveness. In general, stimulation is first begun with the rectal probe deep in the rectum, the electrodes lying at the anterior extremity of the male tract and the stimulus is a series of four or five periods of 3 s duration at 3 s intervals (30 pulses s^{-1}, 8 V). Contraction of the rectum following stimulation tends to force the probe posteriorly and this movement must be resisted. The initial stimulus is followed by a longer period of stimulation (10–15 s) and during this stimulus the probe is moved back and forwards between the earlier deep position and the posterior position where full erection begins to occur. Full erection requires stimulation over the last 3 or 4 cm of the rectum. Full erection in no way improves the sample collected and tends to make collection more difficult. After this second stimulation period there is a non-stimulation period of 10 s and the penis is massaged to increase flow of fluid along the urethra. The long stimulus is then repeated once or twice and ejaculation usually occurs. If the animal is still not responsive a variety of short and long stimuli are applied and the voltage increased to 10 V.

The total period of manipulation of probe and stimulation would not exceed 5–10 min and in highly responsive animals the period would be considerably shorter.

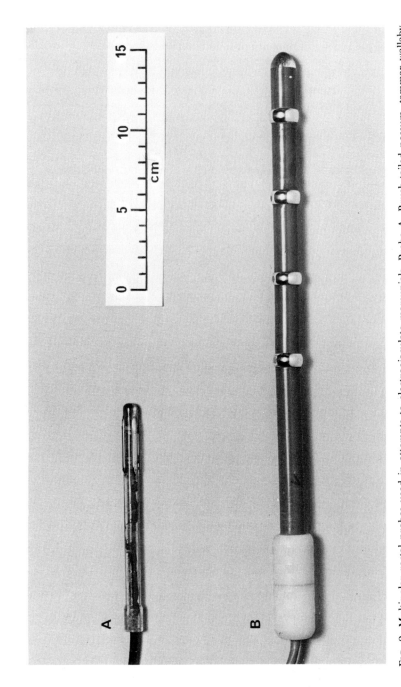

FIG. 2. Multipolar rectal probes used in attempts to electroejaculate marsupials. Probe A, Brush-tailed possum, tammar wallaby, bandicoots, Tasmanian devil and native cat. Probe B, Kangaroos.

Results of Electrical Stimulation

The response to the treatment just described varied considerably between species but can be grouped in the following three categories.

1. No ejaculate. Erection, but no significant release of fluid (0–0·3 ml). The fluid lacks spermatozoa and prostatic bodies (bandicoots, Tasmanian devil and native cat).
2. Seminal plasma. Copious seminal fluid ejaculated (5–10 ml) but no or few spermatozoa. This fluid must be almost entirely prostatic secretion (wallaby and kangaroos).
3. Semen. Ejaculates which are of consistently high volume and spermatozoal count (brush-tailed possum).

The characteristics of semen collected from six brush-tailed possums during 1976 (40 collections from 51 attempts) were as follows.

Volume: mean 2·7 ml, range 0·4 to 10·3 ml.

Spermatozoal concentration: mean $4·8 \times 10^7$ ml^{-1}, range 0·7 to $17·1 \times 10^7$ ml^{-1}.

Total number of spermatozoa per ejaculate: mean $12·9 \times 10^7$, range 1·6 to $39·6 \times 10^7$.

THE HANDLING OF MARSUPIAL SEMEN

Coagulation and Gel Clots

Possum semen, unlike that of kangaroos (Rodger & White, 1975) does not coagulate shortly after collection. Occasionally possum ejaculates do contain gel material which may partially clot the ejaculate but this is a relatively unusual occurrence and results in the loss of less than 5% of ejaculates. Presumably it is this gel clot which was described by Howarth (1950), as "a variable tendency to rapid clotting". When such a high gel-containing ejaculate is mixed with fluid ejaculates the combined material usually clots or a gel plug, containing the spermatozoa, results on centrifugation. It is thus advisable to discard such ejaculates.

Cold Shock

Marsupial spermatozoa are relatively tolerant of cold shock. Exposure of ejaculates to air temperatures down to 10°C or the placing of a test tube containing a thin film of possum semen in iced water does not appear to reduce spermatozoal motility when the semen sample is returned to 37°C. Although possum spermatozoa are tolerant of cold shock, as a general precaution against as yet undetected deleterious effects of such treatments, semen is routinely kept close to 37°C in a vacuum flask following collection and then only diluted with warmed solutions. No attempt has yet been made to freeze marsupial semen, but the resistance to cold shock of possum spermatozoa encourages such investigation.

Dilution, Washing and Incubation of Possum Spermatozoa

Dilution of possum semen five-fold with calcium-free Krebs-Ringer-phosphate (KRP) (Umbreit, Burris & Stauffer, 1972) is well tolerated as is centrifugation at 200 or 750 g (20°C). Thus possum spermatozoa can be washed free of seminal plasma for metabolic studies in a chemically defined medium. Although the fluid component of the semen can be washed from the spermatozoa the prostatic spheres or bodies, present in quite large numbers ($0 \cdot 5 - 1 \cdot 8 \times 10^7$ ml^{-1}), cannot. They remain with the spermatozoa on centrifugation.

The incubation requirements of possum spermatozoa are as yet poorly understood. Washed spermatozoa can retain motility in KRP for up to six hours when incubated in air at 37°C. However, there are considerable differences between semen samples, even from the same animal, with respect to this parameter. Most ejaculates are highly motile at ejaculation and the percentage of progressively motile cells is also high (greater than 90%). Despite this initially high motility some ejaculates are immotile after incubation for two hours at 37°C, irrespective of treatment. Survival of ejaculates is generally better if the spermatozoa are washed rather than left in the original diluted seminal plasma or resuspended in it following centrifugation. However, semen samples with high survival time are little affected by such treatment. This variability of ejaculates may be related to the concentration of prostatic bodies and to the close proximity of these bodies and the spermatozoa in centrifugation plugs. Resuspension of possum spermatozoa and prostatic spheres at concentrations above those found in ejaculates results in reduced sperm survival upon incubation at 37°C. This, together with the observation that immotile spermatozoa are often associated in clumps around prostatic bodies following incubation, suggests that they may be deleterious to spermatozoa when incubated at 37°C.

THE BIOCHEMISTRY OF MARSUPIAL SEMEN

Biochemistry of the Seminal Plasma

The majority of eutherian species, including most domestic mammals, have fructose as the major seminal sugar. This, however, is not the case in any marsupial species as yet examined (Rodger, 1976). The semen of brush-tailed possums, tammar wallabies and kangaroos is characterized by the presence of relatively high levels of free N-acetylglucosamine (Fig. 3; 150–500 mg 100 ml^{-1}). The source of this sugar is the prostate gland.

N – Acetylglucosamine

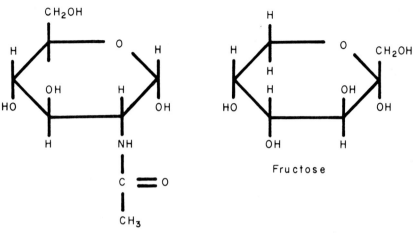

FIG. 3. The major mammalian seminal sugars: *N*-acetylglucosamine, marsupials; fructose, eutherians.

High levels of *N*-acetylglucosamine are also found in the prostate glands of bandicoots and ring-tail possums.

The Metabolism of Spermatozoa from the Brush-tailed Possum

Possum spermatozoa washed in the manner described above utilize oxygen at quite a high rate, approximately $100\ \mu l$ per 10^8 spermatozoa per 3 hours, when suspended in KRP and incubated in micro-Warburg flasks (Braun) at 37°C (Table II). Such a rate is considerably higher than obtained in this laboratory with eutherian spermatozoa, with similar handling and incubation. The rate for possum spermatozoa is up to twice that obtained with ram spermatozoa, four to six times the rate with rabbit and rat spermatozoa and 20 times that of human spermatozoa (Brown-Woodman & White, 1975, 1976; Holland & White, in preparation). Although the addition of exogenous substrates (glucose and *N*-acetylglucosamine, Table II) did not alter the rate of oxygen consumption by possum spermatozoa, both sugars were utilized during incubation ($4.7\ \mu mol$ glucose and $7.3\ \mu mol$ *N*-acetylglucosamine per 10^8 spermatozoa per 3 hours). The level of substrate utilization by possum spermatozoa is also considerably higher than that seen with eutherian spermatozoa. Presumably these sugars had a sparing effect on the unknown endogenous substrate. In many species oxygen consumption by spermatozoa is stimulated by addition of monosaccharides. Added glucose increases the oxygen consumption of ram spermatozoa two to three times and of rabbit spermatozoa four to 10 times (Brown-Woodman

TABLE II

Oxygen consumption of spermatozoa from the brush-tailed possum

Substrate	μl per 10^8 spermatozoa per 3 hours mean ± s.e.
Endogenous	$97\cdot8 \pm 19\cdot7$ (6)[a]
Glucose (2 mM)	$110\cdot6 \pm 14\cdot3$ (5)
N-Acetylglucosamine (2 mM)	$80\cdot4 \pm 20\cdot5$ (4)

[a] Number of times the experiment was repeated.

& White, 1976). However, oxygen consumption of human spermatozoa like that of possum spermatozoa is unaffected by added substrate (Murdoch & White, 1968).

ACKNOWLEDGEMENTS

We are indebted to Professor C. W. Emmens and Dr I. C. A. Martin for their criticism of the manuscript. The later part of the work reported here was supported financially by the Nuffield Foundation.

REFERENCES

Bolliger, A. (1946). Some aspects of marsupial reproduction. *Proc. R. Soc. N.S.W.* **80**: 2–13.

Brown-Woodman, P. D. C. & White, I. G. (1975). Effect of α-chlorohydrin on cauda epididymis and spermatozoa of the rat and general physiological status. *Contraception* **11**: 69–78.

Brown-Woodman, P. D. C. & White, I. G. (1976). A comparison of the inhibition of the metabolism of ram and rabbit spermatozoa by α-chlorohydrin *in vitro*. *Theriogenology* **6**: 29–37.

Cummins, J. M. (1976). Epididymal maturation of spermatozoa in the marsupial *Trichosurus vulpecula*: Changes in motility and gross morphology. *Aust. J. Zool.* **24**: 499–511.

Denny, M. J. S. (1974). Anaesthesia in kangaroos. *Aust. Mammal.* **1**: 294–298.

Gilmore, D. P. (1969). Seasonal reproductive periodicity in the male Australian brush-tailed possum (*Trichosurus vulpecula*). *J. Zool., Lond.* **157**: 75–98.

Godfrey, G. K. (1969). Reproduction in a laboratory colony of the marsupial Mouse *Sminthopsis larapinta* (Marsupialia: Dasyuridae). *Aust. J. Zool.* **17**: 637–654.

Gunn, R. M. C. (1936). Fertility in sheep. *Bull. Coun. Sci. Ind. Res., Aust.* No. 94.

Harding, H. R. (1969). *Studies on the periovulatory changes in the ovary, uterus and vagina of the marsupial* Trichosurus vulpecula *Kerr, and the effect of pregnant mare serum (PMS) and human chorionic gonadotrophin (HCG) in inducing ovulation.* B.Sc. Honours Thesis: University of New South Wales.

Harding, H. R., Carrick, F. N. & Shorey, C. (1975). Ultrastructural changes in spermatozoa of the brush-tailed possum, *Trichosurus vulpecula* (Marsupialia), during epididymal transit. I. The flagellum. *Cell Tissue Res.* **164**: 133–144.

Howarth, V. S. (1950). A method for the collection of the secretions of the individual accessory sex glands in a marsupial (*Trichosurus vulpecula*). *Med. J. Aust.* **1950(i)**: 566–567.

Hughes, R. L. (1962). Reproduction in the macropod marsupial *Potorous tridactylus* (Kerr). *Aust. J. Zool.* **10**: 193–224.

Martin, I. C. A. & Rees, D. (1962). The use of direct current pulses for the electroejaculation of the bull. *Aust. vet. J.* **38**: 92–98.

Murdoch, R. N. & White, I. G. (1968). Studies of the metabolism of human spermatozoa. *J. Reprod. Fert.* **16**: 351–361.

Nelson, O. E. & Maxwell, N. (1941). Induced oestrus and mating in the opossum, *Didelphys virginiana. Anat. Rec.* **81**: Abstract, p. 105.

Nelson, O. E. & White, E. L. (1941). A method for inducing ovulation in the anoestrous opossum (*Didelphys virginiana*). *Anat. Rec.* **81**: 529–535.

Pilton, P. & Sharman, G. B. (1962). Reproduction in the marsupial *Trichosurus vulpecula. J. Endocr.* **25**: 119–136.

Rodger, J. C. (1976). Comparative aspects of the accessory sex glands and seminal biochemistry of mammals. *Comp. Biochem. Physiol.* **55B**: 1–8.

Rodger, J. C. & Hughes, R. L. (1973). Studies of the accessory glands of male marsupials. *Aust. J. Zool.* **21**: 303–320.

Rodger, J. C. & White, I. G. (1975). Electroejaculation of Australian marsupials and analyses of the sugars in the seminal plasma from three macropod species. *J. Reprod. Fert.* **43**: 233–239.

Sadleir, R. M. F. S. (1965). Reproduction in two species of kangaroo (*Macropus robustus* and *Megaleia rufa*) in the arid Pilbara region of Western Australia. *Proc. zool. Soc. Lond.* **145**: 239–261.

Sharman, G. B. (1955). Studies on marsupial reproduction II. The oestrous cycle of *Setonix brachyurus. Aust. J. Zool.* **3**: 44–55.

Smith, M. J. & Godfrey, G. K. (1970). Ovulation induced by gonadotrophins in the marsupial, *Sminthopsis crassicaudata* (Gould). *J. Reprod. Fert.* **22**: 41–47.

Tyndale-Biscoe, C. H. (1961). *Studies in the reproductive physiology of* Setonix brachyurus. Ph.D. Thesis: University of Western Australia.

Tyndale-Biscoe, C. H. (1973). *Life of marsupials.* London: Arnold.

Tyndale-Biscoe, C. H., Hearn, J. P. & Renfree, M. (1974). Control of reproduction in macropodid marsupials. *J. Endocr.* **63**: 589–614.

Tyndale-Biscoe, C. H. & Rodger, J. C. (1978). Differential transport of spermatozoa into the two sides of the genital tract of a monovular marsupial, the tammar wallaby (*Macropus eugenii*) *J. Reprod. Fert.* **52**: 37–43.

Umbreit, W. W., Burris, R. H. & Stauffer, J. F. (1972). *Manometric and biochemical techniques.* 5th edn. Minneapolis: Burgess.

Symp. zool. Soc. Lond. (1978) No. 43, 303–316

Embryo Transfer and Embryo Preservation

C. POLGE

ARC Institute of Animal Physiology, Animal Research Station, Cambridge, Cambridgeshire, England

SYNOPSIS

Embryo transplantation is now well established as an experimental technique in laboratory and farm animals. It is also applied in practice, especially in cattle breeding, for more rapid multiplication of rare breeds. Although very little has yet been done in non-domestic animals, the induction of ovulation and transfer of embryos from wild to domestic rabbits has been used as an aid to breeding wild rabbits in captivity, and a few experiments have been carried out in sub-human primates and in marsupials.

Superovulation can be induced in prepuberal, cyclic or anoestrous animals and eggs might also be obtained directly from ovarian follicles. Collection and transfer of embryos is generally achieved by surgical methods, but can also be done non-surgically in some species. Embryos at pre-implantation stages of development survive normally when transferred to the reproductive tracts of synchronized recipients and pregnancy has also been established and maintained in prepuberal or ovariectomized animals following egg transfer and appropriate treatment with hormones. Although no young have been born after inter-specific egg transfer, the transfer of embryos between species which can hybridize might be a possibility.

Embryos of a number of species have been successfully preserved *in vitro* for several days either by culture at body temperatures or by cooling. Temporary transfer of embryos to the oviduct of a rabbit also provides an alternative method for short-term preservation. However, the only practical method for really long-term preservation of embryos is by deep freezing and storage in liquid nitrogen and this is now possible since live young have been produced following transfer of frozen and thawed embryos in the mouse, rat, rabbit, sheep, cow and goat. Deep freezing of embryos might therefore be applied in non-domestic animals for genetic banking and the preservation of endangered species.

INTRODUCTION

It is appropriate that a considerable part of this Symposium has been devoted to aspects of semen collection, preservation and insemination in various species. Certainly artificial insemination will be a most valuable technique in the artificial breeding of non-domestic animals and it is one that can probably be applied with reasonable chances of success at the present time. It is equally appropriate, however, that our horizons should not be limited to artificial insemination (AI) alone, but should be broadened to include all methods by which we might be able to control reproductive functions in animals by artificial means. In this respect, techniques concerned with the control of reproductive cycles, the

induction of ovulation and developments in methods for egg transplantation deserve special consideration.

Studies of oestrus synchronization and egg transplantation have
increased tremendously during recent years. For example, in the field of
egg transplantation alone, a good idea of the developing momentum can
be gained simply by the number of papers on this subject which has now
accumulated. In three recent surveys of the literature, over 850
references were found to papers concerning egg transfer published
between 1887 and 1977 (Adams & Abbott, 1971; Adams, 1975a, 1977)
and the great majority of these were published after 1950. So there is no
doubt that egg transplantation has now become firmly entrenched as an
experimental technique and it has proved a valuable aid to studies in the
fields of physiology of reproduction and developmental biology. More
recently, as the technique has been developed and applied in farm
animals, it has also become established as a practical method for animal
breeding on a commercial scale and there are now several egg transfer
companies operating in various parts of the world.

Detailed reviews on egg transplantation in laboratory animals include
those of Austin (1961), Chang & Pickworth (1969) and Dickmann (1971),
and in farm animals those of Dziuk (1969) and David et al. (1976). In
addition, there have been several comprehensive reports of symposia on
egg transplantation in cattle and other farm animals (Swensson & Hewett,
1975; Rowson, 1976; Betteridge, 1977) which present up-to-date
information on many aspects of the technique. On the other hand, the
number of species in which egg transfer has been performed is really
quite small (Table I) and even in the domestic species it is surprising that
no experiments appear to have been carried out in cats and dogs.
Relatively few attempts have been made to apply these techniques in truly
non-domestic animals. Notable exceptions are those of Adams (1975b) in
which egg transfer was used for breeding wild rabbits in captivity and
those of Tyndale-Biscoe (1963, 1970) who used embryo transfer to
examine factors affecting the resumption of development of quiescent
blastocysts in two Australian marsupials, the tammar wallaby and the

TABLE I

Animals in which egg transfer has been performed

Rabbit	Cow	Human	Tammar wallaby
Hare	Sheep	Baboon	Quokka
Mouse	Goat	Rhesus monkey	
Rat	Pig		
Hamster	Horse		
Guinea-pig	Donkey		
Ferret			
Mink			

quokka. In sub-human primates there is also one report of a successful egg transfer in the baboon (Kraemer, Moore & Kramen, 1976) and another of egg transfer in rhesus monkeys (Marston, Penn & Sivelle, 1977) in which several pregnancies were established. Hare eggs have also been used in some transfer experiments (Chang, 1965). However, the principles which have been developed in laboratory and farm animals provide an excellent background for further applications in other species and many of the objectives will be the same. For example, two of the main objectives for applying these techniques in cattle breeding are the quicker multiplication of rare breeds through superovulation and egg transfer and the increase of reproductive capacity through induced twinning. Perhaps the most important application in the future, however, will come from the possibility of long-term preservation of embryos by deep freezing; this will greatly extend the scope for "livestock" importation and minimize the risk of spreading disease since the embryos can be kept in quarantine before sending them to another country for transfer. In non-domestic animals it might be envisaged that egg transfer could be applied in order to build up stocks of rare species if suitable closely related and more plentiful ones could be found in which the transferred embryos would develop. Perhaps also it might be used as an aid to breeding some animals in captivity and certainly serious consideration should be given to the possibilities of "genetic banking" and the preservation of endangered species.

To be a practical proposition, egg transfer requires: the ability to obtain a reasonable number of embryos at a determined stage of development from the donor animals; simple and effective methods for their collection and transfer to recipients which should also be at an appropriate stage of the reproductive cycle; and effective techniques for storing the embryos *in vitro* with minimal loss of viability.

THE SUPPLY OF EMBRYOS

In practice, the supply of embryos from donor animals can be increased by using techniques for superovulation. In most species superovulation can be induced by the administration of gonadotrophic hormones, either in the form of anterior pituitary extracts or, more commonly, pregnant mare's serum gonadotrophin (PMSG). The endocrinology of superovulation is certainly not yet well understood and there are generally quite large variations in response between individual animals, between different strains within a species and between animals of different ages. Nevertheless, these variations do not detract from the advantage gained by using exogenous gonadotrophins, and the eggs produced following superovulation have generally been found to be normal and have the same chance of fertilization and survival as those ovulated in an unstimulated cycle. The gonadotrophins should be given at an appropriate stage of the cycle, close to the start of the follicular phase. Accurate knowledge of the stage of the cycle may be difficult to obtain in some non-domestic animals. However, in species in which luteolysis can be induced by prostaglandins, the gonadotrophins can be given at almost any time during the luteal phase so long as luteolysis is induced shortly afterwards. For superovulation in cattle, this technique is now used routinely and the optimal response is obtained when PMSG is administered during the mid-luteal phase and the prostaglandin given two days later (Newcomb & Rowson, 1976). Paradoxically, and perhaps not surprisingly, the horse is one species in which it has not been possible to induce superovulation with PMSG.

Superovulation can also be induced in prepuberal animals, and in calves and lambs it has been found that the response to gonadotrophins is increased if they are first "primed" with progesterone (Seidel, Larson & Foote, 1971; Trounson, Willadsen & Moor, 1977). Similarly, it is also possible to induce superovulation during anoestrus, and this fact may be of particular importance in non-domestic animals which may become anoestrous when kept in captivity. Indeed, the best demonstration so far of the use of egg transfer in order to stimulate reproduction of wild animals kept in captivity took advantage of techniques for induced ovulation (Adams, 1975b). Female wild rabbits did not breed under laboratory conditions because the does failed to mature sexually. However, they were induced to ovulate with gonadotrophins and were artificially inseminated with epididymal spermatozoa collected from wild bucks. The fertilized eggs so obtained were transferred to synchronous domestic rabbits and nearly half of the transferred eggs developed into normal young which retained all the behavioural characteristics of the wild species.

Circumstances might warrant the consideration of other methods of obtaining a suitable supply of eggs and perhaps the most promising would be to collect oocytes directly from ovarian follicles because the ovaries contain a vast number of eggs which would not normally be ovulated. This approach appeared feasible when it was found that a large proportion of immature oocytes resumed meiotic division when liberated from follicles and cultured *in vitro* (Edwards, 1965). However, despite the fact that nuclear maturation occurred under these conditions, only a very small proportion of the oocytes seemed to be fully competent to undergo normal fertilization and development. Very few young have been produced from oocytes matured *in vitro* that have been fertilized either *in vitro* or after transfer to the oviduct of an inseminated recipient (Cross & Brinster, 1970). Cytoplasmic factors therefore seemed to be lacking. By contrast, much more promising results have recently been obtained in experiments involving the culture of intact follicles. If sheep follicles are cultured in a medium suitably enriched with follicle stimulating hormone, luteinizing hormone and oestrogen, not only does nuclear maturation occur, but also the patterns of protein synthesis within the cytoplasm of the oocytes undergo changes similar to those observed in oocytes matured *in vivo* (Warnes, Moor & Johnson, 1977). When these oocytes from the cultured follicles are transferred to the oviducts of inseminated ewes, a high proportion are fertilized and develop into normal blastocysts. These blastocysts are then capable of full development to normal lambs when transferred to other recipients (Moor & Trounson, 1977). Thus, even if only the ovary of an animal is available, it may yet be possible to obtain eggs which can be fertilized and produce young.

COLLECTION AND TRANSFER OF EMBRYOS

Normally one would hope to be in a position to collect eggs and embryos at a determined stage of development from the reproductive tracts of the donor animals. This may be done by relatively simple surgical techniques involving laparotomy in the anaesthetized animal and flushing of the eggs from the oviducts or uterus. In some circumstances, laparoscopy might be used, as has been done in the human, for collecting follicular eggs (Steptoe & Edwards, 1976). Laparoscopic techniques are now being applied in a wide variety of species (Dukelow & Ariga, 1976). In some species it may also be possible to collect eggs from the uterus by non-surgical techniques similar to those which are being developed for use in cattle. By means of a suitable catheter introduced via the cervix, good recovery rates of uterine embryos can be obtained (Drost, Brand & Aarts, 1976). Similar non-surgical methods have been applied in the horse (Oguri & Tsutsumi, 1972; Allen & Rowson, 1975), but in many species the cervical canal may be too narrow to allow the introduction of a catheter. An ingenious way of collecting fertilized eggs non-surgically from rabbits was developed by Tsutsumi *et al.* (1976). The rabbits were treated with oestrogen following mating and this accelerated the transport of the fertilized eggs through the reproductive tract and they were expelled into the vagina from where they could be collected by flushing. Viable young were obtained from eggs collected by this means 2–4 days *post coitum* and transferred to recipients.

It might be of interest in relation to non-domestic species to note that embryos may also be collected from animals after slaughter, and used successfully. Indeed, in many of the laboratory animals, the embryos are normally collected in this way and in cattle, high survival rates following transfer were achieved with embryos collected from the reproductive tracts within 30 min of slaughter (Tervit & Rowson, 1972). In all probability, longer periods of time could elapse between death of the donor and collection of embryos without causing loss of embryonic viability.

Transfer of embryos to the reproductive tracts of recipients is generally achieved by surgical means, and under optimal conditions the survival rate of embryos is equivalent to that occurring during normal pregnancy. In species in which implantation occurs relatively early, pregnancy has been established following the transfer of embryos at any pre-implantation stage of development. In species in which implantation does not occur for some time following ovulation, it is necessary to carry out the transfer before the time when the presence of an embryo within the uterus is required in order to maintain luteal function. Embryos at very early stages of development, however, may be more sensitive to handling and storage *in vitro* than those obtained at somewhat more advanced stages such as the blastocyst (Trounson *et al.*, 1976).

Non-surgical techniques have also been used for embryo transfer in the cow and horse (Rowson & Moor, 1966; Allen & Rowson, 1975; Oguri & Tsutsumi, 1974) and have even been performed successfully in animals as small as the rat and mouse (Marsk & Larsson, 1974; Vickery, Erickson & Bennett, 1969). In early experiments in cattle in which embryos were transferred to the uterus via the cervix, the pregnancy rate was very low (Rowson & Moor, 1966). At that time the transfers were on the third or fourth day following oestrus and failure to maintain pregnancy in the recipients appeared to be associated with the expulsion of the embryos from the uterus (Rowson, Bennett & Harper, 1964). By contrast, better results are now being achieved by non-surgical techniques if the transfers are carried out later and a pregnancy rate of 59% has been obtained in cows to which transfers have been made on Day 9 of the cycle (A. O. Trounson & L. E. A. Rowson, pers. comm.).

One of the most important factors affecting success in egg transfer is the degree of synchrony between recipient and donor, and the survival of embryos is generally reduced if the stage of the reproductive cycle of the recipient is more than about one day out of phase with that of the donor. Deviation on the side of transferring "older" eggs to "younger" uteri is often better tolerated than the reverse situation. The effect on the embryo of asynchronous transfer has been well demonstrated in the rabbit in experiments which showed clearly that exposure of the early embryo for less than 24 hours to a more advanced progestational uterus was incompatible with its survival (Adams, 1971). The site of transfer, either to the Fallopian tube or uterus, may also be an important factor even in synchronous transfers. For example in cattle, embryonic survival is greatly reduced if tubal eggs are transferred to the uterus more than one day earlier than the time that they would normally enter the uterus (Newcomb & Rowson, 1975).

All these factors indicate the necessity of being able to predict or control quite accurately the reproductive state of donor and recipient animals. Methods for the synchronization of oestrus in a number of species have now been developed using either progestational agents or prostaglandins (see *Control of sexual cycles in domestic animals*, 1975). But in non-domestic animals kept in captivity there may be problems of ano-estrus or failure to attain sexual maturity which will cause difficulties in recipients as well as donors. As has been mentioned, ovulation may be induced by means of gonadotrophins in the prepuberal animal, but pregnancy may not be maintained following insemination or egg transfer if pituitary support is insufficient to maintain adequate steroid secretion. This can sometimes be overcome by injecting or implanting steroid hormones as demonstrated in the pig (Ellicott, Dziuk & Polge, 1973). It has even been shown that pregnancy can be induced in prepuberal animals before the time that an ovary is capable of ovulating. In rabbits, morulae were transferred to the uteri of infant does which had been primed with oestrogen before receiving subcutaneous implants containing progesterone, and 31% of the embryos developed to term (Abbott, 1973). Pregnancy has also been established following egg transfer to mature ovariectomized females when appropriate progesterone and oestrogen treatment is given (see Chang & Pickworth, 1969).

Greater problems will obviously exist if the object of egg transfer is to obtain quicker multiplication of a rare species, but there is no closely related species to which egg transfers might be made, as was the case of the wild and domestic rabbit already cited (Adams, 1975b). In experiments on interspecific egg transfer, some development of embryos within a different species has been observed, but no live young have been produced (see Chang & Pickworth, 1969). On the other hand, it might be reasonable to predict that interspecific egg transfer might be successful between species which are known to be capable of producing hybrid offspring. In nine reciprocal embryo transfers between mares and donkeys no foals were produced, although pregnancy was established in one mare which had received a donkey embryo. However, the transfer of mule or hinny zygotes to donkeys or mares did result in the birth of foals (Allen & Rowson, 1972). Hybrids of European bison and domestic cattle have been reported (Krasinska, 1971) and reciprocal egg transfer between these species would be of great interest, but apparently has not yet been attempted.

PRESERVATION OF EMBRYOS

The transfer of embryos involves their exposure to conditions *in vitro* for a certain length of time and in many circumstances it would be advantageous to be able to store them for extended periods after collection from the donor. Several approaches have been adopted in order to maintain normal development during storage or to prolong viability by arresting development. Much of the work on the preservation of mammalian embryos has recently been reviewed by Maurer (1976). The embryos of many species have continued to develop *in vitro* for up to five or six days when cultured at body temperature in suitable media and under optimal atmospheric conditions. Success has varied to some extent between species and according to the stage of embryonic development at which culture has been attempted. In the cow, sheep and pig, for example, eight-celled eggs or morulae appear to survive better in culture than earlier stages. In some strains of mice, it has been difficult to get the unicellular fertilized egg to divide in culture and attempts to culture hamster eggs beyond the two-cell stage have met with little success (Whittingham, 1975a). In all species, however, the viability of embryos, as measured by their ability to develop *in vivo* after transfer to a suitable recipient, falls off more rapidly than their ability to continue to cleave *in vitro*. Although some young have been born after transferring embryos cultured for as long as six days (Tervit & Rowson, 1974), viability is generally much reduced after only two to three days.

In some instances, viability can be extended by cooling the embryos below body temperature. Of rabbit morulae stored at 10°C for four days, 46% developed into young (Chang, 1948), and cooled rabbit embryos have been transported successfully between America and England (Marden & Chang, 1952). In general, storage temperatures of between 5 and 10°C have been found to be more satisfactory than 0°C. However, the sensitivity of embryos to cooling depends very much upon the species and also upon the stage of embryonic development within the species. Mouse embryos at any pre-implantation stage and even oocytes have been cooled successfully to 0°C (Whittingham, 1977b), but no pig embryos at any stage of development examined so far have survived cooling to temperatures below 10–15°C (Polge, Wilmut & Rowson, 1974). In the cow, few eight-celled embryos have survived cooling to 0°C, but later stages such as the late morula or blastocyst have survived extremely well (Wilmut, Polge & Rowson, 1975; Trounson *et al.*, 1976).

Another approach to the storage of embryos outside the original donor has been to transfer them to the oviduct of a recipient, which may even be of another species, and this then acts as a temporary host. The rabbit has been found to be very convenient as a temporary host and has been used in many experiments. Cow, sheep, pig, horse, mouse and ferret eggs have all been shown to be capable of development within the rabbit oviduct (Lawson, Rowson & Adams, 1972; Lawson, Adams & Rowson, 1972; Polge, Adams & Baker, 1972; Allen *et al.*, 1976; Brinster & Tenbroeck, 1969; Chang, 1966). The eggs will not normally continue development beyond the time at which they would hatch from the zona pellucida or implant. Nevertheless, the rabbit oviduct still provides a suitable environment for the storage of embryos for perhaps five to six days if the eggs are introduced into the oviduct at an early stage of development. Indeed, the rabbit oviduct may provide a better environment than has been developed so far for the culture *in vitro* of early stage embryos of some species. Cow embryos stored for three to four days in rabbit oviducts had a survival rate of 73% when transferred back to recipients and the rabbit genital tract has been used for long distance transport of both sheep and horse embryos (Adams *et al.*, 1961; Allen *et al.*, 1976).

In species in which implantation is normally delayed or can be induced, it may be possible to achieve extended viability of embryos within the reproductive tract. As delayed implantation is not common in the domestic species it would be of particular interest to examine some of the non-domestic animals in this respect. Implantation does not occur in mice which are ovariectomized on Day 4 of pregnancy, but the unimplanted blastocysts survive for many days within the uterus. Viability of the blastocysts is maintained for at least 15 days and may be extended to much longer periods if progesterone is given to the ovariectomized animals (Weitlauf, 1971). However, the ability to withstand such delay is a characteristic of the embryo and it has not been possible to maintain viability in the same way with embryos of species which do not normally display delayed implantation.

The possibility of being able to preserve embryos *in vitro* for very long periods of time will obviously depend upon the development of satisfactory techniques for maintaining their viability during freezing and thawing so that they may be stored in liquid nitrogen in the same way as spermatozoa. Early attempts to freeze mammalian embryos met with only very limited success (see Polge, 1977). On the other hand, the problem was obviously capable of solution because it had been clearly demonstrated that a proportion of immature oocytes in ovarian tissue could be preserved during freezing and thawing. Frozen and thawed mouse ovarian tissue was successfully grafted into the ovarian capsule of mice previously sterilized by irradiation and live young were later produced from the surviving oocytes (Parrott, 1960). It was not until 1972, however, that high survival rates were first achieved with ovulated eggs and early embryos that had been frozen in liquid nitrogen (Wilmut, 1972; Whittingham, Leibo & Mazur, 1972). The key to this success lay in the use of very much slower freezing and thawing rates than had been adopted previously. Early mouse embryos or blastocysts were treated with dimethyl-sulphoxide (DMSO) and frozen at a rate of less than 1°C per minute to temperatures of −60°C or below before storage in liquid nitrogen. A relatively high proportion of the embryos then survived thawing provided that the rate of rewarming was also quite slow, of the order of 4–10°C per minute. Live young were produced from frozen and thawed embryos following removal of DMSO and transfer to recipients. In numerous experiments with mouse embryos it has been shown that between 20 and 30 live offspring can be expected from every 100 embryos frozen (Whittingham, 1977b), and there is no evidence of a decrease in survival rate during storage at −196°C for up to four years.

Similar techniques have been used for the preservation of embryos of several other laboratory animals. Rabbit embryos frozen at the eight-cell stage appeared to survive well when examined in culture after thawing, but viability after transfer to recipients was very low (Bank & Maurer, 1974; Whittingham & Adams, 1976). More recent experiments in which morulae were frozen have now provided evidence that a survival rate equivalent to that of unfrozen embryos can be achieved (Tsunoda & Sugie, 1977a) and survival of younger embryos can be improved by the addition of serum to the medium in which the embryos are cooled

(Tsunoda & Sugie, 1977b). A few rat embryos have also survived to term after freezing and thawing (Whittingham, 1975b). Unfertilized hamster oocytes have been frozen and thawed, and subsequently fertilized *in vitro* (Tsunoda, Parkening & Chang, 1976), and in similar experiments with mouse oocytes, live young have been produced (Whittingham, 1977a).

Work on the preservation of embryos of farm animals is also now well advanced and the first calf resulting from the transfer of a frozen and thawed cow blastocyst was born in 1973 (Wilmut & Rowson, 1973). Cattle embryos are interesting in that the early stages of embryonic development, up to the morula stage, are quite sensitive to cooling and do not survive well after exposure to temperatures around 0°C. By contrast, embryos at the expanding blastocyst stage will tolerate cooling and are therefore suitable for freezing (Wilmut *et al.*, 1975; Trounson *et al.*, 1976). The techniques for freezing and thawing have been refined and, in a recent experiment, eight pregnancies were obtained following the transfer of 12 blastocysts which had been frozen and thawed (Willadsen, Polge & Rowson, 1978). Survival rates of a similar order have also been achieved with sheep and goat embryos (Willadsen *et al.*, 1976; Moore & Bilton, 1977). Pig embryos, on the other hand, still present a problem because no embryos at any stage of development examined so far have survived cooling to temperatures lower than +10°C and therefore it has not been possible to freeze them (Polge, Wilmut *et al.*, 1974).

Thus, with the exception of the pig, survival after freezing and thawing has now been obtained with the embryos of all species examined and it would be reasonable to predict that it should be possible to preserve the embryos of non-domestic animals. As experiments have continued, the conditions during freezing and thawing which affect embryonic survival are becoming better understood. It is no longer necessary, for example, to adhere to the stringent requirements of slow thawing in order to obtain maximum survival. By freezing embryos rapidly to −196°C from temperatures of around −30°C, the survival rate is high after rapid thawing (Willadsen, 1977). Procedures for the low temperature preservation of embryos can therefore be simplified and practical applications made easier.

"Embryo banking" is already considered a practical proposition for maintaining valuable mutant strains of mice in the laboratory, and it will undoubtedly be applied increasingly in the breeding of farm animals. The potential use of these techniques in non-domestic animals awaits appraisal, but some fascinating possibilities are opened up.

REFERENCES

Abbott, M. (1973). Pregnancy in the infant rabbit. *J. Reprod. Fert.* **35**: 620–621.
Adams, C. E. (1971). The fate of fertilized eggs transferred to the uterus or

oviduct during advancing pseudopregnancy in the rabbit. *J. Reprod. Fert.* **26**: 99–111.

Adams, C. E. (1975a). Bibliography on recovery and transfer of mammalian eggs. *Biblphy Reprod. Res. Inf. Serv., Cambr.* No. 89.

Adams, C. E. (1975b). Stimulation of reproduction in captivity of the wild rabbit, *Oryctolagus cuniculus. J. Reprod. Fert.* **43**: 97–102.

Adams, C. E. (1977). Bibliography on recovery and transfer of mammalian eggs. *Biblphy Reprod. Res. Inf. Serv., Cambr.* No. 101.

Adams, C. E. & Abbott, M. (1971). Bibliography on recovery and transfer of mammalian eggs, and ovarian transplantation. *Biblphy Reprod. Res. Inf. Serv., Cambr.* No. 45.

Adams, C. E., Rowson, L. E. A., Hunter, G. L. & Bishop, G. P. [1961]. Long distance transport of sheep ova. *IV Int. Congr. Anim. Reprod. Artif. Insem.* [The Hague] **2**: 381–382.

Allen, W. R. & Rowson, L. E. A. [1972]. Transfer of ova between horses and donkeys. *VII Int. Congr. Anim. Reprod. Artif. Insem.* [Munich] **1**: 483–488.

Allen, W. R. & Rowson, L. E. A. (1975). Surgical and non-surgical egg transfer in horses. *J. Reprod. Fert.* **23** Suppl.: 525–530.

Allen, W. R., Stewart, F., Trounson, A. O., Tischner, M. & Bielanski, W. (1976). Viability of horse embryos after storage and long-distance transport in the rabbit. *J. Reprod. Fert.* **47**: 387–390.

Austin, C. R. (1961). *The mammalian egg.* Oxford: Blackwell.

Bank, H. & Maurer, R. R. (1974). Survival of frozen rabbit embryos. *Expl Cell Res.* **89**: 188–196.

Betteridge, K. J. (ed.) (1977). Embryo transfer in farm animals, a review of techniques and applications. *Monogr. Can. Dept Agric.* No. 16.

Brinster, R. L. & Tenbroeck, T. J. (1969). Blastocyst development of mouse pre-implantation embryos in the rabbit fallopian tube. *J. Reprod. Fert.* **19**: 417–421.

Chang, M. C. (1948). Effects of low temperature on fertilized rabbit ova *in vitro* and the normal development of ova kept at low temperature for several days. *J. gen. Physiol.* **31**: 385–410.

Chang, M. C. (1965). Artificial insemination of snowshoe hares (*Lepus americanas*) and the transfer of their fertilized eggs to the rabbit (*Oryctolagus cuniculus*). *J. Reprod. Fert.* **10**: 447–449.

Chang, M. C. (1966). Reciprocal transplantation of eggs between rabbit and ferret. *J. exp. Zool.* **161**: 297–305.

Chang, M. C. & Pickworth, S. (1969). Egg transfer in the laboratory animal. In *The mammalian oviduct*: 389–405. Hafez, E. S. E. & Blandau, R. J. (eds). Chicago and London: University of Chicago Press.

Control of sexual cycles in domestic animals. Colloque—*Anns Biol. anim. Biochem. Biophys.* **15**(2) 1975. Institut Nationale de la Recherche Agronomique.

Cross, P. C. & Brinster, R. L. (1970). *In vitro* development of mouse oocytes. *Biol. Reprod.* **3**: 298–307.

David, J. S. E., Jones, W. A., Newcomb, R., Smith, G. F. & Wishart, D. F. (1976). *Embryo transfer with particular reference to cattle.* London: Society for the Study of Animal Breeding, British Veterinary Association.

Dickmann, Z. (1971). Egg transfer. In *Methods in mammalian embryology*: 133–145. Daniel, J. C. (ed.). San Francisco: W. H. Freeman & Co.

Drost, M., Brand, A. & Aarts, M. H. (1976). Device for non-surgical recovery of bovine embryos. *Theriogenology* **6**: 503–507.

Dukelow, W. R. & Ariga, S. (1976). Laparoscopic techniques for biomedical research. *J. med. Primatol.* **5**: 82–99.

Dziuk, P. J. (1969). Egg transfer in cattle, sheep and pigs. In *The mammalian oviduct*: 407–417. Hafez, E. S. E. & Blandau, R. J. (eds). Chicago and London: University of Chicago Press.

Edwards, R. G. (1965). Maturation *in vitro* of mouse, sheep, cow, pig, rhesus monkey and human ovarian oocytes. *Nature, Lond.* **208**: 349–351.

Ellicott, A. R., Dziuk, P. J. & Polge, C. (1973). Maintenance of pregnancy in prepuberal pigs. *J. Anim. Sci.* **37**: 971–973.

Kraemer, D. C., Moore, G. T. & Kramen, M. A. (1976). Baboon infant produced by embryo transfer. *Science, Wash.* **192**: 1246–1247.

Krasinska, M. (1971). Hybridization of European bison with domestic cattle. *Acta Theriol.* **16**: 413–422.

Lawson, R. A. S., Adams, C. E. & Rowson, L. E. A. (1972). The development of sheep eggs in the rabbit oviduct and their viability after re-transfer to ewes. *J. Reprod. Fert.* **29**: 105–116.

Lawson, R. A. S., Rowson, L. E. A. & Adams, C. E. (1972). The development of cow eggs in the rabbit oviduct and their viability after re-transfer to heifers. *J. Reprod. Fert.* **28**: 313–315.

Marden, W. G. R. & Chang, M. C. (1952). The aerial transport of mammalian ova for transplantation. *Science, N.Y.* **115**: 705–706.

Marsk, L. & Larsson, K. S. (1974). Simple method for non-surgical blastocyst transfer in mice. *J. Reprod. Fert.* **37**: 393–398.

Marston, J. H., Penn, R. & Sivelle, P. C. (1977). Successful autotransfer of tubal eggs in the rhesus monkey (*Macaca mulatta*). *J. Reprod. Fert.* **49**: 175–176.

Maurer, R. R. (1976). Storage of mammalian oocytes and embryos: a review. *Can. J. Anim. Sci.* **56**: 131–145.

Moor, R. M. & Trounson, A. O. (1977). Hormonal and follicular factors affecting maturation of sheep oocytes *in vitro* and their subsequent developmental capacity. *J. Reprod. Fert.* **49**: 101–109.

Moore, N. W. & Bilton, R. J. (1977). Frozen storage of embryos of farm animals: progress and implications. In *The freezing of mammalian embryos*: 203–211. (Ciba Found. Symp). Amsterdam: Elsevier.

Newcomb, R. & Rowson, L. E. A. (1975). Conception rate after uterine transfer of cow eggs in relation to synchronization of oestrus and age of eggs. *J. Reprod. Fert.* **43**: 539–541.

Newcomb, R. & Rowson, L. E. A. (1976). Multiple ovulation, egg transplantation: towards twinning. In *Principles of cattle production*: 59–83. Swan, H. & Broster, W. H. (eds). London: Butterworths.

Oguri, N. & Tsutsumi, Y. (1972). Non-surgical recovery of equine eggs and an attempt at non-surgical egg transfer in horses. *J. Reprod. Fert.* **31**: 187–195.

Oguri, N. & Tsutsumi, Y. (1974). Non-surgical egg transfer in mares. *J. Reprod. Fert.* **41**: 313–320.

Parrott, D. M. V. (1960). The fertility of mice with orthotopic ovarian grafts derived from frozen tissue. *J. Reprod. Fert.* **1**: 230–241.

Polge, C. (1977). The freezing of mammalian embryos: perspectives and possibilities. In *The freezing of mammalian embryos*: 3–13. (Ciba Found. Symp.). Amsterdam: Elsevier.

Polge, C., Adams, C. E. & Baker, R. D. [1972]. Development and survival of pig embryos in the rabbit oviduct. *VII Int. Congr. Anim. Reprod. Artif. Insem.* (Munich] **1**: 513–517.

Polge, C., Wilmut, I. & Rowson, L. E. A. (1974). The low temperature preservation of cow, sheep and pig embryos. *Cryobiology* **11**: 560.

Rowson, L. E. A. (ed.) (1976). *Egg transfer in cattle.* Agric. Res. Seminar, EUR 5491. European Communities Directorate General, scientific & technical information & information management, Luxembourg.

Rowson, L. E. A., Bennett, J. P. & Harper, M. J. K. (1964). The problem of non-surgical egg transfer to the cow uterus. *Vet. Rec.* **76**: 21–23.

Rowson, L. E. A. & Moor, R. M. (1966). Non-surgical transfer of cow eggs. *J. Reprod. Fert.* **11**: 311–312.

Seidel, G. E., Larson, L. L. & Foote, R. H. (1971). Effects of age and gonadotrophin treatment on superovulation in the calf. *J. Anim. Sci.* **33**: 617–622.

Steptoe, P. C. & Edwards, R. G. (1976). Re-implantation of a human embryo with subsequent tubal pregnancy. *Lancet* **1976 (i)**: 880–882.

Swensson, T. & Hewett, C. (eds) (1975). *Symposium on egg transplantation.* Publ. No. 75. Ascn. for Swedish Livestock Breeding & Production (SHS), Hallsta, Sweden.

Tervit, H. R. & Rowson, L. E. A. (1972). Viability of ova from slaughtered cattle. *VII Int. Congr. Anim. Reprod. Artif. Insem.* [Munich] **1**: 489–492.

Tervit, H. R. & Rowson, L. E. A. [1972]. Viability of ova from slaughtered cattle, *in vitro* for up to 6 days. *J. Reprod. Fert.* **38**: 177–179.

Trounson, A. O., Willadsen, S. M. & Moor, R. M. (1977). Reproductive function in prepuberal lambs: ovulation, embryo development and ovarian steroidogenesis. *J. Reprod. Fert.* **49**: 69–75.

Trounson, A. O., Willadsen, S. M., Rowson, L. E. A. & Newcomb, R. (1976). The storage of cow eggs at room temperature and at low temperatures. *J. Reprod. Fert.* **46**: 173–178.

Tsunoda, Y., Parkening, T. A. & Chang, M. C. (1976). *In vitro* fertilization of mouse and hamster eggs after freezing and thawing. *Experientia* **32**: 223–224.

Tsunoda, Y. & Sugie, T. (1977a). Survival of rabbit eggs preserved in plastic straws in liquid nitrogen. *J. Reprod. Fert.* **49**: 173–174.

Tsunoda, Y. & Sugie, T. (1977b). Effect of the freezing medium on the survival of rabbit eggs after deep freezing. *J. Reprod. Fert.* **50**: 123–124.

Tsutsumi, Y., Takeda, T., Yamamobo, K. & Tanabe, Y. (1976). Non-surgical recovery of fertilized eggs from the vagina of oestrogen-treated rabbits. *J. Reprod. Fert.* **48**: 393–395.

Tyndale-Biscoe, C. H. (1963). Blastocyst transfer in the marsupial, *Setonix brachyurus. J. Reprod. Fert.* **6**: 41–48.

Tyndale-Biscoe, C. H. (1970). Resumption of development by quiescent blastocysts transferred to primed, ovariectomised recipients in the marsupial, *Macropus eugenii. J. Reprod. Fert.* **23**: 25–32.

Vickery, B. H., Erickson, G. I. & Bennett, J. P. (1969). Non-surgical transfer of eggs through the cervix in rats. *Endocrinology* **85**: 1202–1203.

Warnes, G. M., Moor, R. M. & Johnson, M. H. (1977). Changes in protein synthesis during maturation of sheep oocytes *in vivo* and *in vitro. J. Reprod. Fert.* **49**: 331–335.

Weitlauf, H. M. (1971). Effect of progesterone on survival of blastocysts in uteri of ovariectomised mice. *J. Endocr.* **51**: 375–380.

Whittingham, D. G. (1975a). Fertilization, early development and storage of mammalian ova *in vitro.* In *Symposium on early mammalian development*: 1–24. Balls, M. & Wild, A. E. (eds). London: Cambridge University Press.

Whittingham, D. G. (1975b). Survival of rat embryos after freezing and thawing. *J. Reprod. Fert.* **43**: 575–578.

Whittingham, D. G. (1977a). Fertilization *in vitro* and development to term of unfertilized mouse oocytes previously stored at −196°C. *J. Reprod. Fert.* **49**: 89–94.

Whittingham, D. G. (1977b). Some factors affecting embryo storage in laboratory animals. In *The freezing of mammalian embryos*: 97–108. (Ciba Found. Symp.) Amsterdam: Elsevier.

Whittingham, D. G. & Adams, C. E. (1976). Low temperature preservation of rabbit embryos. *J. Reprod. Fert.* **47**: 269–274.

Whittingham, D. G., Leibo, S. P. & Mazur, P. (1972). Survival of mouse embryos frozen to −196°C and −269°C. *Science, N.Y.* **178**: 411–414.

Willadsen, S. M. (1977). Factors affecting the survival of sheep embryos during deep-freezing and thawing. In *The freezing of mammalian embryos*: 175–194. (Ciba Found. Symp.) Amsterdam: Elsevier.

Willadsen, S. M., Polge, C. & Rowson, L. E. A. (1978). The viability of deep-frozen cow embryos. *J. Reprod. Fert.* **52**: 391–393.

Willadsen, S. M., Polge, C., Rowson, L. E. A. & Moor, R. M. (1976). Deep freezing of sheep embryos. *J. Reprod. Fert.* **46**: 151–154.

Wilmut, I. (1972). Effect of cooling rate, warming rate, cryoprotective agent and stage of development on survival of mouse embryos during cooling and thawing. *Life Sci.* **11**: 1071–1079.

Wilmut, I., Polge, C. & Rowson, L. E. A. (1975). The effect on cow embryos of cooling to 20, 0 and −196°C. *J. Reprod. Fert.* **45**: 409–411.

Wilmut, I. & Rowson, L. E. A. (1973). Experiments on the low temperature preservation of cow embryos. *Vet. Rec.* **92**: 686–690.

Symp. zool. Soc. Lond. (1978) No. 43, 317–328

Handling and Sedation for Procedures Associated with Artificial Breeding

D. M. JONES

Institute of Zoology, Zoological Society of London, Regent's Park, London, England

SYNOPSIS

The study of reproductive cycles and the manipulations associated with semen collection and insemination of non-domestic mammals invariably require a degree of sedation or light anaesthesia. As the effects of these drugs on the physiology of ejaculation and the resultant semen quality are not entirely understood, sedation is usually kept to a minimum providing the animal is relaxed and safe to handle. The same principle is applied for insemination. Suggestions are also given for sedation in birds.

Where surgical intervention is required for a study of changes in the reproductive system, full anaesthesia is indicated. Most of the techniques described have not been used specifically for reproductive studies, but are based on established methods of providing sedation and short periods of anaesthesia for routine veterinary management. Most workers have their own preferences for particular drugs and techniques and inevitably this contribution tends to be a summary of the methods used by the Zoological Society of London rather than a comprehensive review of the subject.

INTRODUCTION

Techniques applicable to one group of animals are not necessarily relevant to another, a point that has been emphasized by other authors in this Symposium. Most domestic animals can be handled without sedation, although they sometimes have to be trained to tolerate a particular manipulation. This is not generally the case with non-domestic species, although in some instances training is possible and in others physical restraint is feasible because of the animal's small size and unaggressive nature.

A successful artificial breeding programme involves several distinct stages of investigation, each of which may require different methods of handling. These stages may be defined as: monitoring and/or control of female reproductive cycles, semen collection, insemination and, possibly, pregnancy diagnosis. Ideally all these stages require the following conditions to be satisfied to some degree depending on the species, the individual and the nature of the manipulation.

  1. The animal should be safe to handle so preventing damage to itself and to the operator.

2. It should be possible to maintain the animal in a suitable position for the manipulation required.
3. A state of complete relaxation may be necessary.
4. The animal's neural reflexes must be intact.

Our present knowledge of the use of drugs for handling non-domestic animals enables almost any species to be "immobilized" with relative safety. Much of this information has been summarized by the author (D. M. Jones, 1973; 1977a) and by Harthoorn (1976). Unfortunately, the degree of analgesia and relaxation achieved with many of the drugs used may not be sufficient for finely controlled manipulations such as laparoscopy or insemination. Techniques for inducing sedation and anaesthesia in a wide range of species have been evolved over a 25-year period in the animal hospital of the Zoological Society of London and some of these are effective for the studies involved in a programme of artificial breeding. Seager (1974) and Seager & Platz (1976) have published the methods used for their studies on carnivores, and Ball (1976) has briefly reviewed methods of restraint on a range of species for electroejaculation. The information in this paper represents suggestions as to the best method of handling particular groups of animals, but may not necessarily have been used for reproductive studies in particular.

REPRODUCTIVE STUDIES IN MAMMALS

Monitoring Reproductive Cycles and Pregnancy Diagnosis

The recent introduction of fibre-optic laparoscopes into the veterinary field is facilitating the examination of the female reproductive tract. Full surgical anaesthesia is the ideal state for laparoscopy because of the degree of immobility and relaxation attained. As inflation of the abdominal cavity is needed in most cases for laparoscopy, it may be necessary to control breathing artificially and this can only be satisfactorily achieved after endotracheal intubation has been carried out. Correct intubation also eliminates the risk of the animal inhaling regurgitated fluids. The tendency for regurgitation to occur is increased if the animal is lying on its back and in a sloping position with the hind-quarters raised, but this is often a necessary manoeuvre to ensure that most of the abdominal contents fall forward leaving the reproductive tract visible. Laparoscopy is useful for pregnancy diagnosis and examination of the reproductive tract during the first half of gestation, but after this stage the procedure may be difficult to use without an increased risk of damaging abdominal viscera. There is no evidence that carefully controlled anaesthesia is harmful to the foetus during the first half of gestation.

Both Primates and Carnivora can be anaesthetized by the same general method. One of three cyclohexamines, phencyclidine, ketamine or tiletamine (Table I), is usually given first by intramuscular injection in

Table I

Drugs described in the text

Drug	Trade name	Manufacturer
Acepromazine	ACP injection	C-Vet
Alphaxalone and Alphadolone	Saffan or Althesin	Glaxo
Diazepam	Valium	Roche
Diprenorphine (M.5050)	—	Reckitt and Colman
Etorphine (M.99)	—	Reckitt and Colman
Etorphine with Acepromazine	Large animal "Immobilon"	Reckitt and Colman
Fentanyl	—	Janssen Pharmaceutica
Fentanyl with Fluanisone	Hypnorm	Crown Chemicals
Halothane	Fluothane	ICI
Ketamine	Vetalar	Parke Davis
Methoxyflurane	Penthrane V	Abbotts
Metomidate	Hypnodil	Crown Chemicals
Phencyclidine	Sernylan	Phillips Roxane (USA)
Sodium thiopentone	Pentothal	Abbotts
Tiletamine with Arylcyclo-alkylamine (CI 744)	—	Parke Davis (USA only)
Xylazine	Rompun	Bayer

combination with a promazine derivative or diazepam (Table I) at dose rates which enable the animal to be handled safely (see Table II). A short-acting barbiturate such as sodium thiopentone (Table I) is then given intravenously and, after intubation, anaesthesia is maintained with a mixture of halothane in oxygen (Table I). A similar method is applicable for almost any mammal where intravenous barbiturates can be given either after physical restraint or following the use of pre-anaesthetic agents such as those described.

A useful alternative to barbiturates in Primates, Felidae and macropod marsupials is a mixture of the steroids alphaxalone and alphadolone (Table I). A single dose of 4–5 mg kg^{-1} of the mixture given intravenously following the use of ketamine for initial restraint provides 20–30 min of surgical anaesthesia followed by a relatively rapid recovery. The endocrinological activity of this mixture in laboratory rodents is apparently negligible. Only a very slight anti-oestrogenic effect was detected but this had no influence on the reproductive cycle (Child *et al.*, 1972). In most Primates and Felidae it is also possible to use the steroid mixture intramuscularly as an alternative pre-anaesthetic to the cyclohexamine.

Initial trials indicate that the steroid mixture may be useful in rodent and lagomorph species. The small size of many of these animals usually necessitates giving the steroids intramuscularly, although the effect is less variable if it can be given intravenously. Many small mammals can be immobilized using ketamine but high dose rates are required and this drug tends to have a slight excitatory effect in some species which can usually be abolished by the administration of halothane through a face mask. A useful degree of narcosis in lagomorphs can also be produced by a mixture of fentanyl and fluanisone (Table I). However, many workers still prefer short-acting barbiturates given intraperitoneally as anaesthetic agents in small mammals despite the disadvantages of the lower safety margin and relatively slow recovery.

Anaesthesia of the pinnipeds and Cetacea is difficult and should not be attempted without facilities to control breathing mechanically on a closed circuit system (Ridgeway, 1965; Ridgeway & Simpson, 1969; Geraci, 1973). It is probable that the Sirenia would also require the same facilities.

For most management purposes, the majority of non-domestic ungulates are restrained with potent narcotics such as etorphine (Table I). Since narcotics tend to cause excitement in many of these species resulting in poor muscular relaxation after "immobilization", xylazine (Table I), with its powerful sedative action, is generally preferable as a premedicant for anaesthesia and surgery in ruminants. Nervous or aggressive species of the suborder Ruminantia and perissodactyl ungulates are not affected by xylazine alone except in very large doses which are often impractical to administer. The less tractable ruminants should be given a mixture of etorphine and xylazine while the perissodactyl ungulates may be handled with etorphine and acepromazine. The degree of immobility and relaxation achieved with these combinations would in most cases be sufficient to allow laparoscopy, but a local analgesic should be injected around the surgical site. If relaxation and analgesia are still insufficient for surgery, very small intravenous doses of a short-acting barbiturate can be given in addition, but great care must be taken to avoid respiratory depression.

Ideally, laparoscopy should be carried out on the standing ruminant, but the degree of immobilization achieved in a sedated, as opposed to an anaesthetized, animal is unlikely to be sufficient for detailed examination or satisfactory photography of the reproductive tract. After the animal is made tractable with xylazine, barbiturates are usually used to induce anaesthesia. Even in a fully anaesthetized ruminant intubation is not easy, and attempts to pass an endotracheal tube sometimes induce regurgitation of the ruminal contents. At present there is no information available on the effects of inflating the abdominal cavity of a ruminant lying on its back on a sloping table, but it is likely that without intubation the risk of dyspnoea and regurgitation would be considerable. An alternative method with apparently less risk of regurgitation has been used in sheep with some success (Taylor et al., 1972). Xylazine was given intramuscularly to achieve medium sedation. Ketamine was then given intravenously to induce a light plane of surgical anaesthesia and this state was maintained with further ketamine given in an intravenous drip or by intravenous injection at intervals. Preliminary trials of this method in two species of small non-domestic ruminants, fallow deer, *Dama dama*, and blackbuck, *Antilope cervicapra*, suggested that it may be useful in these species but that maintenance of a constant plane of anaesthesia for periods longer than half an hour may be difficult (D. M. Jones, unpublished).

Semen Collection and Insemination

As far as is known, the motor pathways involved in ejaculation are not markedly affected by anaesthesia (see I. C. A. Martin, this volume, p. 127). Manipulation, as opposed to electrical methods of semen collection, is not successful under anaesthesia because of the loss of central nervous involvement needed to induce ejaculation by this method of stimulation. It is not known whether successful conception may be prevented by a state of general anaesthesia and our knowledge of the physiology of coitus in most species is limited. Blake (1974) has reported that the release of luteinizing hormone in laboratory rats is inhibited by some anaesthetic agents.

Neither semen collection nor insemination strictly involves surgical invasion of the body, and anaesthesia is not usually necessary. In some larger species it may even be contraindicated on practical grounds. However, if complete immobilization, and therefore general anaesthesia, is necessary for these procedures, the suggestions given above for drugs and techniques apply.

In most instances it is possible to collect semen and inseminate the female by inducing in the animal a state of moderate to deep sedation using the methods described above. Under moderate sedation it is also possible to collect urine and blood, take swabs of vaginal and buccal mucosae, and carry out a thorough clinical examination of the animal, including abdominal palpation.

When semen collection and insemination is attempted in very large mammals that cannot be handled easily, a different approach has to be used. Although it is possible to electroejaculate and inseminate large ruminants when they are recumbent, it is not very practical and the risk of ruminal regurgitation appears to increase with the size of the animal and the pressure on its abdomen. Ideally, the initial dose of xylazine, which will differ with each individual, should be just sufficient to keep the animal standing but manageable. More aggressive or nervous animals which do not readily respond to xylazine alone are initially given the mixture of etorphine and xylazine to induce recumbency (Table II). The narcotic effect of the etorphine is then partially reversed with diprenorphine (Table I) given intramuscularly. As the animal begins to show signs of recovery a further dose of xylazine equal to the original quantity of that drug is also given intramuscularly. The animal usually rises with encouragement but remains moderately sedated. The mechanism of this effect is not known but it is probable that either diprenorphine or some residual or reabsorbed etorphine exerts a synergistic effect with the xylazine. The result is a much more marked effect than that of the same dose of xylazine if it had been given alone. At this stage the animal is walked to a point where it can be physically restrained and, if necessary, supported.

No reports exist of attempts to apply artificial breeding methods in non-domestic perissodactyl ungulates. This may in part be due to the fact that electroejaculation has not been altogether satisfactory in the horse. Captive examples of these species are not yet readily available for detailed reproductive studies. If attempts at insemination in these species were envisaged, it should be possible to carry out the procedure on the recumbent animal under narcosis.

In recent efforts to induce pregnancy in an African elephant at the Zoological Society of London (Regent's Park), frozen semen obtained from freshly culled wild individuals was used (R. C. Jones, 1973). Fortunately the female to be inseminated was very co-operative and no sedation was required. More fractious elephants can be sedated with xylazine alone but some form of physical support may be necessary until the optimal dose for an individual is known, since elephants occasionally collapse suddenly under the influence of the drug, with potentially disastrous results. The information given above is summarized in Table II.

REPRODUCTIVE STUDIES IN BIRDS

Other contributors to this Symposium have described some of the work which has been carried out on the artificial breeding of birds. It is probable that the practical application of these techniques will increase as the worldwide movement of animals is further restricted for reasons of animal health and conservation. Many workers prefer not to use a sedative or anaesthetic agent in birds, but excessive handling, especially of small species, can produce fatal shock. Because of this, the use of drugs ought to be considered in many cases. Moreover, if laparoscopic examination of testicular and ovarian function is to become part of any detailed study of avian reproductive cycles, deep sedation and analgesia will be necessary for the surgery and degree of immobilization required. A number of drugs which have recently become available are relatively safe to use for avian sedation and anaesthesia. The safest and most effective agents are ketamine, metomidate (Table I) and the steroid mixture of alphaxalone and alphadolone. Suggested dose rates for these drugs are summarized in Table III.

Of these drugs, ketamine has been the most widely used (Boever & Wright, 1975; D. M. Jones, 1977b) and appears to produce satisfactory sedation in individuals of most avian orders. It is the only one of the three

TABLE II

Suggested dose rates for the sedation and anaesthesia of mammals

Orders/families	Drugs for sedation	Dose mg kg^{-1} i.m.	Additional drugs for anaesthesia	Comments
Caprinae } Camelidae }	Xylazine	0·5–2·0	Sodium thiopentone 3–5 mg kg^{-1} i.v.	Xylazine-dose dependent more on temperament than species (see Jones & Manton, 1976)
Bovidae } Cervidae }	Xylazine or Etorphine plus Xylazine	0·5–6·0 0·01–0·02 0·2–0·5	or Ketamine 3–5 mg kg^{-1} i.v. plus Intubation—2% Halothane if necessary	
Suidae } Hippopotamidae }	Phencyclidine or Ketamine plus Acepromazine	1·0–1·5 12·0–15·0 0·1	Thiopentone 3–5 mg kg^{-1} i.v. or Halothane 3–4% by mask (Suidae)	Narcotics immobilize but tend to induce excitement and hyperthermia
Equidae } Tapiridae }	Etorphine plus Acepromazine	0·007–0·01 0·1	Thiopentone 3–4 mg kg^{-1} i.v. Intubate—Halothane 2%	Only the Przewalski's and domestic horse show significant excitement with etorphine
Rhinocerotidae } Proboscidea }	Etorphine plus Acepromazine	0·002 0·01–0·02	Unknown	Small elephants not difficult to intubate

TABLE II—*continued*

Orders/families	Drugs for sedation	Dose mg kg^{-1} i.m.	Additional drugs for anaesthesia	Comments
Pinnipedia	Ketamine	4·5–11·0	4–5% Halothane by mask Intubate—2% Halothane Halothane after intubation of conscious animal	Poor anaesthetic risk in debilitated animals
Cetacea	—			Specialized equipment needed
Carnivora over 1 kg } Primates }	Phencyclidine or	0·8–1·2	Thiopentone 4–6 mg kg^{-1} i.v. or Alphaxalone + Alphadolone 5·0 mg kg^{-1} i.v.	Steroid mixture sometimes not very effective when given i.m.
	Ketamine plus Acepromazine or Diazepam	10·0–15·0 0·1 0·5	Maintain 2% Halothane (intubate) 4% Halothane (mask)	Dose rates of cyclohexamines for primates generally lower than for Carnivora
	Alternative Alphaxalone + Alphadolone	10·0–15·0		
Macropodidae	Ketamine	15·0–20·0	Alphaxalone + Alphadolone 3–5 mg kg^{-1} i.v. or Thiopentone 7–10 mg kg^{-1} i.v.	This group more resistant to these drugs than most mammals

Group	Drug	Dose	Maintenance / Notes
Edentata, Tubulidentata			Unknown. Thiopentone i.v. probably most reliable
Lagomorpha	Ketamine or Phencyclidine	15·0 1·5	Maintain 2% Halothane (intubate) 4–5% Halothane (mask)
	Fentanyl citrate plus Fluanisone or	0·1 3·0	
	Alphaxalone + Alphadolone	{10·0 i.v. or 15·0–20·0 i.m.}	Limited trials only with the steroid anaesthetic
Rodentia, Carnivora under 1 kg, Primates	Ketamine or	20·0–30·0	Maintain 2% Halothane (intubate) 4–5% Halothane (mask)
	Alphaxalone + Alphadolone	15·0–20·0	
Dermoptera, Chiroptera, Insectivora			Limited information available suggests that ketamine and halothane at above dose rates for rodents are effective
Small Marsupialia, Hyracoidea, Sirenia, Monotremes			No information

TABLE III

Suggested drugs and dose rates for sedation and anaesthesia of birds

Order	Drugs for sedation/immobilization	Dose mg kg^{-1} i.m.	Additional drugs for anaesthesia	Comments
Galliformes ⎫ Anseriformes ⎭	Ketamine	11–15	2–3% Halothane or Methoxyflurane	Metomidate produces less excitement in these species than ketamine
	or Metomidate	10–15		
Ciconiiformes	Ketamine	12–15	2–3% Halothane or Methoxyflurane	Mild excitement occasionally seen with ketamine
Pelecaniformes Gruiformes ⎫ Sphenisciformes ⎬ Charadriiformes Columbiformes ⎭	Ketamine	15–20	No information 2–3% Halothane or Methoxyflurane	
	Ketamine	20	Methoxyflurane	
Strigiformes	Ketamine or Alphaxalone + Alphadolone	15–25	Alphaxalone + Alphadolone 3–4 mg i.v.	Steroid also given intravenously after ketamine but only 8 cases to date with no problems
Falconiformes ⎫ Psittaciformes ⎭	Alphaxalone + Alphadolone	8–14	or 2–3% Halothane or Methoxyflurane	
Passeriformes	Ketamine	20–35	2–3% Halothane or Methoxyflurane by mask	Sometimes more practical to use halothane alone
Struthioniformes	Ketamine	50–100	No information	Produces dissociation and some degree of immobilization but excitement occurs as well

drugs that has any significant immobilizing effect on the ostrich family, but even at dose rates up to 100 mg kg^{-1} complete immobilization was not achieved by the author in young emus, *Dromaius novaehollandiae*. In two orders, Anseriformes (ducks and geese) and Galliformes (turkeys, chickens and pheasants) some degree of excitement is often produced by ketamine. In these cases metomidate produces better relaxation, but the chief disadvantage of metomidate is that the degree of analgesia is poor in all the species in which it has been used. The most effective analgesic is the mixture of alphaxalone and alphadolone, and in those species in which it is known to have a clinical effect, the degree of relaxation is also marked. However, when injected intramuscularly in some birds, notably the Anseriformes and Galliformes, the steroid mixture produces very little effect even at high dose rates, but when injected intravenously it is more effective. Cooper & Frank (1973) used the steroid mixture in 15 raptorial species, and this is the group of birds in which it has been most extensively tested.

Without considerable expertise it is often difficult to administer intravenous injections in fully conscious birds unless they are sufficiently large and effectively restrained. A practical method of inducing and maintaining surgical anaesthesia for up to 30 min is carried out in two stages as follows: the steroid mixture or ketamine is given first by intramuscular injection to achieve immobilization, and approximately one-third of the usual intramuscular dose of the steroid is then given intravenously to induce surgical anaesthesia.

Two other methods can also be used to induce anaesthesia once immobilization is achieved. Short-acting barbiturates can be given intravenously or intraperitoneally, but the margin of safety is poor and this method is not recommended. Alternatively, halothane or methoxyflurane (Table I) in oxygen can be administered either through a face mask or, more preferably, through an endotracheal tube using an Ayres T-piece system. In the hands of an experienced person, the use of volatile anaesthetics in birds is a reasonably safe procedure.

CONCLUSIONS

The techniques for sedation and anaesthesia described in this paper are those which have been developed over the last 25 years for routine veterinary management in zoological collections. The time has been reached where almost any species of mammals and birds can be handled with relative safety to both animal and operator, but we still know virtually nothing of the physiological effects of the techniques and drugs used.

Because the use of such drugs is essential for any detailed physiological investigation of non-domestic animals, there is a need to discover whether these agents are likely to affect the results observed.

In the context of this Symposium, future work on artificial breeding in non-domestic species will have to consider the possibility that sedative drugs may affect the central nervous and endocrine control of reproduction. Investigation of this possibility should be an important priority in this field and the early dissemination of information on the subject should be encouraged.

REFERENCES

Ball, L. (1976). Electroejaculation. In *Applied electronics in veterinary medicine and physiology*: 394–441. Klemm, W. R. (ed.). Springfield, Illinois: Charles C. Thomas.

Blake, C. A. (1974). Differentiation between the "critical period", the "activation period" and the "potential activation period" for neurohumoral stimulation of LH release in proestrous rats. *Endocrinology* **95**: 572.

Boever, W. J. & Wright, W. (1975). The use of ketamine for restraint and anaesthesia of birds. *Vet. Med. small Anim. Clin.* **70**: 86–87.

Child, K. J., English, A. F., Gilbert, H. G. & Woollett, E. A. (1972). An endocrinological evaluation of Althesin (CT 1341) with special reference to reproduction. *Postgrad. Med. J.* June supplement **1972**: 51–55.

Cooper, J. E. & Frank, L. (1973). Use of the steroid anaesthetic CT 1341 in birds. *Vet. Rec.* **98**: 474.

Geraci, J. R. (1973). An appraisal of Ketamine as an immobilising agent in wild and captive pinnipeds. *J. Am. vet. med. Ass.* **163**: 574–576.

Harthoorn, A. M. (1976). *The chemical capture of animals*. London: Baillière Tindall.

Jones, D. M. (1973). The use of drugs for immobilization, capture and translocation of non-domestic animals. In *The veterinary annual* **13**: 320–354. Grunsell, C. S. G. & Hill, F. W. G. (eds). Bristol: John Wright.

Jones, D. M. (1977a). Recent advances in the use of drugs for immobilization capture and translocation of non-domestic animals. In *The veterinary annual* **17**: 280–285. Grunsell, C. S. G. & Hill, F. W. G. (eds). Bristol: John Wright.

Jones, D. M. (1977b). The sedation and anaesthesia of birds and reptiles. *Vet. Rec.* **101**: 340–342.

Jones, D. M. & Manton, V. J. A. (1976). Report for Whipsnade Park 1973 and 1974. *J. Zool. Lond.* **178**: 494–507.

Jones, R. C. (1973). Collection, motility and storage of spermatozoa from the African elephant, *Loxodonta africana*. *Nature, Lond.* **243**: 38–39.

Ridgeway, S. H. (1965). Medical care of marine mammals. *J. Am. vet. med. Ass.* **147**: 1077–1082.

Ridgeway, S. & Simpson, J. G. (1969). Anaesthesia and restraint for the Californian sealion (*Zalophus californianus*). *J. Am. vet. med. Ass.* **155**: 1059–1062.

Seager, S. W. J. (1974). Semen collection and artificial insemination in captive wild cats, wolves and bears. *Proc. Am. Ass. Zoo Vet.* **1974**: 29–33.

Seager, S. W. J. & Platz, C. C. (1977). Semen collection and freezing in captive wild mammals. *VIII Int. Congr. Reprod. Artif. Insem.* [Krakow] **4**: 1075–1077.

Taylor, P., Hopkins, L., Young, M. & McFadyen, I. R. (1972). Ketamine anaesthesia in the pregnant sheep. *Vet. Rec.* **90**: 35–36.

Symp. zool. Soc. Lond. (1978) No. 43, 329–338

Disease Control in Semen Transfer and Artificial Insemination

G. F. SMITH

Milk Marketing Board, Thames Ditton, Surrey, England

SYNOPSIS

A commercial Artificial Insemination (AI) service for cattle was established in Britain in 1940. At that time it had two main objectives; first, to make more use of superior sires of various dairy breeds selected on the basis of pedigree and performance and secondly to introduce a disease control programme using semen from highly qualified bull studs. This was in the hope that fertility levels might be improved by the limitation of the spread of disease through natural service—a common occurrence of the day when ample evidence existed to show that brucellosis, trichomoniasis and other diseases could be transmitted through natural service contacts.

Government control measures were instituted from the onset of the commercial service and operators required a licence to distribute semen from approved premises. Government and industry jointly built up over the years a code of operation setting out standards of bull health and hygiene, as well as some control over staff movements and training programmes. The use of semen from non-domestic species has been relatively uncommon over the years, therefore information on disease transmission has to be treated with some caution at the present time.

The paper reviews disease hazards as they relate to domestic bull health, semen production and distribution, and develops the practical experience and risks entailed in handling a large commercial network for artificial insemination in cattle. Attention is drawn to the lack of evidence to support the belief that disease transmission is a serious hazard when pre-entry health tests are applied to bulls entering the AI service, when bull studs are subjected to regular health tests, and well-trained and supervised technical services are available for the operation of the insemination service. The development of international exchange programmes for semen and ova is discussed and some of the difficulties encountered in the provision of the service are mentioned.

It is suggested that the development of a wholly pathogen-free method of semen production and handling must be fraught with difficulty. However, by the application of an assiduous testing programme on the donor stud, combined with the close supervision of semen production and handling, as well as careful field routines, it has been shown that disease hazards can be successfully reduced. This has provided the basis for dependable national and international livestock improvement programmes, based on the widespread use of artificial insemination in cattle. These experiences would appear to merit investigation when semen exchange from non-domestic animals is under consideration.

INTRODUCTION

The national and international growth of artificial insemination (AI) in cattle has contributed to greater and more efficient use of animal resources throughout the world.

In order to attempt to lay a foundation for semen control in non-domestic animals it would appear necessary to draw upon the experiences of AI in cattle as there is no body of data which provides information on disease transmission in the non-domestic animals. Assuming that it can be expected that similar disease and technical problems are likely to exist in these animals AI in cattle could usefully act as a model for the development of AI in other species.

In Great Britain a combined research and commercial service was inaugurated in 1940, and by 1943 the techniques evolved were accepted as a means of breeding what then seemed to be a large number of cows from the semen of a single bull (Edwards & Ritchie, 1944).

The objectives of an AI service were two-fold; first, to maximize the use of what, at that time, were considered top quality sires and secondly, to make some contribution to the control of disease which could be fairly easily spread by direct service. In some measure it might be said that the objectives in using AI in non-domestic animals are somewhat different; isolated specimens are widely distributed throughout the world, there is a demand for international breeding programmes for endangered species of wildlife at present in captivity and under certain circumstances a dependable AI service might be applied to improve levels of fertility where disease is a limiting factor.

After a brief interlude without any controls, AI in cattle became subject to legislation which set down a code of practice supervised by the Ministry of Agriculture relating to the establishment of the domestic AI service. The bulls to be used were subjected to pre-entry health and livestock assessments; the premises where the bulls were housed had to be approved, and personnel responsible for semen handling and distribution had to be licensed. While the regulations may have seemed unnecessarily restrictive at the time, they laid a foundation, from which was to emerge an efficient commercial AI service.

Specific areas to be reviewed are
1. pre-entry health control;
2. on-centre health control;
3. staff and semen control;
4. semen handling at field level;
5. The international trade in semen distribution.

PRE-ENTRY HEALTH CONTROL

The production of pathogen-free semen would seem to be the necessary aim for a commercial AI service. It may be an extremely difficult thing to achieve but with the use of bacteriostats contaminants can be reduced to a minimum. Røzsa (1950) reported that semen from fertile healthy bulls contained very large numbers of bacteria, but that the use of antibiotics reduced the harmful effect of any pathogens that were present.

Initially, bulls were obtained by purchase directly from farms; it was therefore essential that freedom from disease should as far as possible be established, for it became evident as the AI service developed that disease transmission by semen could take place.

Brucellosis

Seit (1944) reported the spread of brucellosis in Denmark when infected bulls were in service. This was later confirmed by Bendixen & Blom (1947). Manthei, DeTray & Goode (1950) reviewed the risks associated with the use of brucellosis-positive bulls through AI. Serum agglutination tests were therefore applied to all bulls before they entered AI studs and so far under conditions prevailing in the Milk Marketing Board studs no dissemination of the disease has occurred within studs or in the field, although some blood samples with transient low titres have been recorded in donor bulls. The risk of spread becomes less as the cattle population becomes increasingly free of brucellosis. D. Barclay (pers. comm.) recorded the risk of using brucellosis-negative bulls from herds which had been infected; in no case, when bulls were used again on brucellosis-free cattle, were the bulls incriminated in the spread of the disease under natural service conditions.

A sensible precaution adopted from the outset of commercial AI has been the application of a regular six-monthly blood test for brucellosis for all bulls in the AI service and, where necessary, subsequent biological semen tests.

Tuberculosis

Tuberculosis, now virtually eliminated from the cattle population of this country, has been under some form of control since commercial AI services expanded, and no cases of field spread have been recorded in this country. Roumy (1966) recorded the spread of tuberculosis in France by the use of a bull with genital tuberculosis and Bartlett (1967) described an outbreak of tuberculosis in an AI bull stud. Macpherson & Fish (1954) had earlier shown that *Mycobacterium tuberculosis* could survive freezing in semen.

Bulls and bull calves for rearing are now only purchased from herds with a history of long-standing freedom from tuberculosis.

Trichomoniasis

Numerous reports have shown that *Trichomonas foetus* infection, a venereal disease, has been associated with lowered fertility. Kerr (1943) and Bartlett (1949) found that the causal organism could be isolated from the genital tract of the bull, while much later Parsonson & Snowdon

(1974), in a detailed study of naturally and experimentally infected bulls, showed that infection was limited to the penis and prepuce.

Pre-entry collection of sheath washings and subsequent microscopical examination for the presence of trichomonads has proved to be a satisfactory control measure.

Vibriosis

An enzootic sterility attributable to the presence of *Vibrio foetus* was recorded by Sjollema, Stegenga & Terpstra (1952) in Holland, and Lawson & Mackinnon (1952) also recorded *Vibrio foetus* infection as a cause of reduced fertility in dairy and beef cattle. The presence of *Vibrio foetus* infection was said to be as high as 50% in some studs in the USA in the 1950s (Hughes, 1956).

Pre-entry checks on herd health including examination of sheath washings have been used for many years and transmission by AI has not been recorded in this country.

Johne's Disease

Mycobacterium paratuberculosis (Johne's disease) has been recorded in breeding herds, especially in beef cattle where contamination of the teats and subsequent infection of calves at an early age may occur. Pre-entry tests are not applied but herd histories are investigated and although the causal organism has been isolated from semen it has been considered as an external contaminant.

Leptospirosis

Leptospirosis is not considered to be a disease of any importance within the British context, although records overseas showed it to be a cause of abortion in New Zealand (Te Punga & Bishop, 1953). A disease with so many different serotypes of varying pathogenicity in different countries, leptospirosis is not considered to be of any great clinical significance justifying routine pre-entry tests for bulls coming into British AI Centres—although pre-entry screening tests are applied from time to time when bulls may be expected subsequently to provide semen to fulfil export commitments.

Infectious Bovine Rhinotracheitis/Infectious Pustular Vulvovaginitis

Infectious bovine rhinotracheitis and infectious pustular vulvovaginitis infection is said to give penoposthitis in bulls and vaginal discharge in females. Huck *et al.* (1971) described the condition in a bull stud, and numerous reports have been made on the findings of the virus in bulls and cows. Diagnosis on the basis of blood samples and semen examination can be adopted, and regular screening of bulls takes place.

Foot-and-Mouth Disease

The full scale development of frozen semen services throughout the world has tended to reduce the risk of transmitting foot-and-mouth virus in semen, since quarantining of processed samples for a 28-day period post-collection allows the disease to develop in the donor bull if in the incubative stage at the time of collection. Cottral, Gailiunas & Cox (1968) reviewed the risks of transmitting foot and mouth virus through semen and Sellers et al. (1968) showed that bulls could excrete virus prior to the appearance of lesions in the mouth and on the feet. As Great Britain is considered free of foot-and-mouth disease, no pre-entry tests need to be applied to bulls entering AI Centres, other than to those from overseas at the discretion of the Ministry of Agriculture.

Leucosis

Enzootic bovine leucosis is as well considered now to be present in Great Britain, as being recorded in a number of other countries. Olson & Baumgartener (1975) drew attention to varying degrees of infection in many countries and suggested that while infection by semen transmission cannot be ruled out, the dam appears to be the most important link. Recent studies at Ames, Iowa, USA on six leukaemic bulls failed to show transmission of the condition through the semen (US Dept Agric., 1976: 105). Several surveys of AI bull studs have been carried out in this country recently and a nil incidence has been recorded.

Bluetongue

Bluetongue, a condition carried by cattle and causing serious losses among sheep, has been reported over a very wide area of the world (Hourrigan & Klingsporn, 1975a). Bluetongue does, however, have some geographical restriction since it depends on the presence of the fly, *Culicoides variipennis*, to transmit the virus. A. J. Luedke (pers. comm.) isolated bluetongue virus from the semen of a bull which had been experimentally infected. Great Britain is outside the normal range of *C. variipennis*, and although very large numbers of tests have been applied to our sires involved in exports, no trace of bluetongue has been found.

Mycoplasmosis

Natural mycoplasmosis in the genital tract of bulls was first recorded by Blom & Ernø (1967), and Afshar (1975) reviewed diseases of bovine reproduction associated with the presence of mycoplasmas. The full

G. F. SMITH

significance of *Mycoplasma* is still to be ascertained. However, the use of bacteriostatic antibiotics added to the processed semen (Stalheim, in preparation) appears to be satisfactory as a method of control.

ON-CENTRE HEALTH CONTROLS

Bull Studs and Semen Control

All bulls are routinely under veterinary supervision and subjected to regular six-monthly tests for tuberculosis and brucellosis, and such other health tests as may be required for export certification. These can include leucosis, foot-and-mouth disease, leptospirosis and Q fever. All tests are applied by Ministry of Agriculture veterinarians or through the established Ministry of Agriculture laboratory services. A routine preputial sheath lavage using antibiotics is applied every six months to all bulls as a protection against the introduction and spread of *Vibrio foetus* infection. Semen controls are based on a clean collecting routine using male teasers and a separate artificial vagina for each bull. Semen is subjected to antibiotic protection by the use of penicillin and streptomycin and after collection, processing and freezing all semen is held at a quarantine unit situated a considerable distance from the animal quarters. All semen remains in quarantine for at least 28 days before being re-introduced into the AI network. An increasing trade in semen collected from privately owned on-farm bulls has developed since frozen semen became an effective means of preservation. All bulls in this category are subjected to Ministry of Agriculture health certification before semen can be collected. Semen processing from this category of bulls takes place in specially constructed laboratory facilities.

Staff and Semen Control

The majority of the stockmen employed in handling the AI stud at active collecting centres are resident on the AI Centres; if not, their terms of contract prevent them becoming involved with the handling of animals outside their normal work. This prevents any risk of transmitting disease from outside to the bull stud, for example foot-and-mouth disease, brucellosis and other conditions to which the bull studs might be susceptible. As an added precaution, all new staff are subjected to medical examination and routine chest X-ray examination for evidence of tuberculosis.

At field level, all staff undertake a training course of up to eight weeks before approval, and once licensed, are subject to regular field supervision. Their training emphasizes the correct procedures which prevent the transmission of disease in the course of routine visits between farms. Under field conditions probably the most serious trial of effective field routines was put to the test during a very serious outbreak of foot-and-mouth disease in 1967–68 during which epidemic there were many occasions when transmission of disease could have arisen. However, no outbreaks were ever proved to be attributed to the cattle AI service. This situation can partly be attributed to the field procedures, and partly to the restrictions subsequently placed upon the movement of inseminators once disease has been diagnosed in a particular area. A new challenge has arisen in the past six years, while brucellosis is being eradicated from the national herd. Here again no breakdowns of disease have been attributed to the work of the inseminators.

SEMEN HANDLING AT FIELD LEVEL

Routine deliveries of frozen semen and liquid nitrogen take place at regular intervals, usually on a 14-day cycle, thus avoiding over-stocking of flasks, and unnecessary long-term contamination of flasks and contents. It is realised that nitrogen is an excellent medium in which to preserve a very large number of commensals and organisms of unknown importance; however, regular de-sludging of flasks can reduce the risks. Micro-biological cultures of the liquid nitrogen content of field flasks (Jacquet & Steig, 1967; Schafer et al., 1976) have shown that a great variety of non-pathogenic organisms, as well as pathogens, can be isolated. While the risk of contamination of stored semen straws may thus exist, there is no evidence to suggest that transmission of disease has resulted under the British system of operation.

THE INTERNATIONAL TRADE IN SEMEN DISTRIBUTION

The widespread development of livestock improvement programmes and the demand for the replenishment of gene pools has made it necessary to develop systems of health testing, semen storage and packaging in order to supply a very large number of overseas orders. Levels of health certification are set by the importing country and will vary with the health and livestock priorities of the countries concerned. For example, Australia, with a very considerable sheep population, demands careful screening of all donor bulls for the absence of bluetongue, while other countries may ask for leucosis screening tests, when the disease is known to exist in the exporting country, as a safeguard in preventing the re-introduction of infection.

Certification for traffic in semen must understandably be strictly supervised in view of the possible risks and health hazards which may arise in the country of importation. However, some may suggest that restrictions placed on imports of semen are sometimes based on too little information as to how supervision of the cattle AI service really works in the exporting country. Bartlett *et al.* (1976) have attempted to set out the objectives of all sensible semen producing organizations. It will, however, be a very considerable time before a simple international certificate will be acceptable to all, or even regional, parts of the world. However, if we do ever reach this stage we shall still have to be aware of emerging new disease patterns which will need regular surveillance (Parsonson & Snowdon, 1974). Rabies in non-domesticated animals would appear to call for caution in the testing and distribution of semen from a wide range of animals. The smaller the population of bulls producing the semen, the easier to subject them to disease control measures and strict supervision.

It is impossible entirely to omit a consideration of bovine ova from a paper of this type. Hourrigan & Klingsporn (1975b) have drawn attention to the disease risks associated with the international shipment of ova. The implementing of new techniques associated with the harvesting and freezing of eggs should, however, not be retarded by unnecessarily strict health standards.

We have seen the development of a cattle AI programme with good health and technical standards over the past 35 years; we may now be looking forward to new and exciting developments in the next decade when with wholehearted government participation it will be possible to make use of those new techniques to the advantage of non-domestic animal stocks in the wild and captive state.

CONCLUSIONS

In evaluating the success of an AI programme, its growth and technical development must be taken as a realistic measure of its trouble-free service to the herds which use it. In Great Britain the base for the commercial AI service in cattle has been a solid one, with both government and industry working together and playing their part.

Progress in the development of technical skills has provided methods of reducing the risk of disease transmission through the use of frozen semen and although it would be very wrong to minimize the risks that could still arise, most disease does appear to be under effective control.

It will be a very long time before an international exchange of semen without additional health controls can be expected; however, as health standards improve throughout the world, it may well give room for areas to accept similar health standards, and so simplify the passage of semen and ova. Certainly the expense of transporting the live animal will be under serious consideration in the future, especially when costs of building and maintaining quarantine facilities for live animals are taken into consideration. Even now these escalating costs are already representing a serious restriction on the development of international co-operation in breeding non-domestic animals in captivity.

REFERENCES

Afshar, A. (1975). Diseases of bovine reproduction associated with mycoplasma infections. *Vet. Bull.* **45**: 211–216.

Bartlett, D. E. (1949). Procedures for diagnosing bovine venereal trichomoniasis and handling infected herds. *J. Am. vet. med. Assoc.* **14**: 293–305.

Bartlett, D. E. (1967). Comments on the episode of bovine tuberculosis in a population of bulls used for artificial insemination. *Proc. A. Mtg USLSA.* No. 71: 166–171.

Bartlett, D. E., Larson, L. L., Parker, W. G. & Howard, T. H. (1976). Specific pathogen free (SPF) frozen bovine semen; a goal? *Proc. Tech. Conf. A.I. Reprod. NAAB.* No. 6: 11–22 (20/21 February 1976).

Bendixen, H. O. & Blom, E. (1947). Investigations on brucellosis in the bovine male with special regard to the spread of the disease by artificial insemination. *Vet. J.* **103**: 337–345.

Blom, E. & Ernø, H. (1967). Mycoplasmosis: infections of the genital organs of bulls. *Acta vet. Scand.* **8**: 186–188.

Cottral, G. E., Gailiunas, P. & Cox, B. F. (1968). Foot-and-mouth disease virus in semen of bulls and its transmission by artificial insemination. *Arch. ges. Virusforsch.* **23**: 362–377.

Edwards, J. & Ritchie, J. N. (1944). *Report on artificial insemination in the United States of America, submitted to the Agricultural Improvement Council.* Cambridge: Cambridge University Press.

Hourrigan, J. L. & Klingsporn, A. L. (1975a). Bluetongue: The disease in cattle. *Aust. Vet. J.* **51**: 170–174.

Hourrigan, J. L. & Klingsporn, A. L. (1975b). Certification of ruminants, semen and ova for freedom from bluetongue virus. *Aust. vet. J.* **51**: 211–212.

Huck, R. A., Miller, P. G., Evans, D. H., Stables, J. W. & Ross, A. (1971). Penoposthitis associated with infectious bovine rhinotracheitis/infectious pustular vulvovaginitis IBR/IPV virus in a stud of bulls. *Vet. Rec.* **88**: 292–297.

Hughes, D. E. (1956). Notes on vibriosis, with special reference to the isolation of *Vibrio foetus* from semen and preputial fluids. *Cornell Vet.* **46**: 249–256.

Jacquet, J. & Steig, L. (1967). Preliminary remarks on the conservation of microbes in liquid nitrogen. *Bull. Acad. vet. Fr.* **40**: 275.

Kerr, W. R. (1943). Trichomoniasis in the bull. *Vet. J.* **99**: 4–8.

Lawson, J. R. & Mackinnon, D. J. (1952). *Vibrio foetus* infection in cattle. *Vet. Rec.* **64**: 763–773.

Macpherson, J. W. & Fish, N. A. (1954). The survival of pathogenic bacteria in bovine semen preserved by freezing. *Am. J. Vet. Res.* **15**: 548–550.

Manthei, C. A., DeTray, D. E. & Goode, E. R. (1950). Brucella infection in bulls and spread of brucellosis in cattle by artificial insemination. I. Intrauterine infection. *J. Am. vet. Med. Ass.* **1950**: 177–184.

Olson, C. & Baumgartener, L. E. (1975). Lymphosarcoma (leukemia) of cattle. *Bovine Pract.* No. 10: 15–22.

Parsonson, I. M. & Snowdon, W. A. (1974). Ephemeral fever virus excretion in the semen of infected bulls and attempts to infect female cattle by the intra-uterine inoculation of virus. *Aust. vet. J.* **50**: 329–334.

Roumy, B. (1966). An enzootic of bovine TB transmitted by artificial insemination. *Rec. Med. Vet.* **142**: 729–741.

Røzsa, J. (1950). A Bikaondo Bakterimtartalmara vonatkozo vizsgalatok. *Agrar-tudomany* **2**: 599–601.

Schafer, T. W., Everett, J., Silver, G. H. & Came, P. E. (1976). Biohazard: virus-contaminated liquid nitrogen. *Science, N.Y.* **191**: 24 & 26.

Seit, B. (1944). Kastningsmitte Ved kingstig Inseminering. *Maandsskr. Dyrlaeg* **56**: 12–37.

Sellers, R. F., Burrows, F., Mann, J. A. & Dawe, P. (1968). Recovery of virus from bulls affected with foot-and-mouth disease. *Vet. Rec.* **83**: 303.

Sjollema, P., Stegenga, I. & Terpstra, J. (1952). Infectious sterility in cattle caused by *Vibrio foetus*. *Rep. Int. Vet. Congr.* **14(3)**: 123–127.

Stalheim, O. H. V. (In preparation). *Antibiotics for the elimination of mycoplasmas from bull semen for artificial insemination.*

Te Punga, W. A. & Bishop, W. H. (1953). Bovine abortion caused by infection with *Leptospira pomona*. *N.Z. Vet. J.* **1**: 143–149.

United States Department of Agriculture, Agricultural Research Service, National Programme Staff (1976). Livestock and veterinary sciences. *Rep. natn. Res. Progr. U.S. Dept. Agric., Agric. Res. Serv.* **1976**.

Symp. zool. Soc. Lond. (1978) No. 43, 339–343

Concluding Remarks

J. D. SKINNER

Mammal Research Institute, University of Pretoria, South Africa

First allow me to congratulate Dr Paul Watson for arranging this Symposium between so many talented participants. It is, perhaps, a measure of progress made in this aspect of animal breeding, that only a few years ago the holding of a two-day symposium on such a theme would have been impossible for lack of information. However, the Zoological Society of London has long been aware of the need for research in the important and difficult field covered by this symposium and Dr Watson paid tribute last night to Dr Idwal Rowlands who, he said, had pioneered the concepts which gave rise to the Symposium. A glance at the literature will show (Rowlands, 1965) that one of the aims of the Wellcome Institute of Comparative Physiology was the investigation of problems relating to the breeding of animals in the collections of the Zoological Society of London. A decade later, Rowlands (1974) when reviewing progress, could point to several successes and some disappointments, notably that in the giant panda. It had become obvious that there were no short-cuts to success in the application of techniques used in combatting infertility in domestic species. Freezing of spermatozoa was proving very difficult but one major project, that on the African elephant by Jones (1973) which combined field studies with laboratory research, had provided some answers to the collection and storage of these cells. The research of Dr R. C. Jones was continued by Dr Watson and Dr Frances D'Souza under the able direction of Dr R. D. Martin who succeeded Dr Rowlands. They had probably the more difficult assignment, namely that of establishing the duration of the oestrous cycle and time of ovulation in the female elephant (Watson & D'Souza, 1975).

Before reviewing the papers presented I think we should reflect for a moment on the significance of Dr Martin's statement yesterday that it will probably be primarily our task to ensure the conservation of many species still extant today. Unpleasant as it may seem, the expanding human populations of this world are going to be responsible for the disappearance of many species of animals primarily by the destruction of their habitat, as was emphasized by Professor Austin yesterday. One may pose the question: what right has man to destroy what the Creator has placed on this earth? Are we under an obligation to preserve some of the original species present when we arrived on the scene or is this just part of the evolutionary process and will they disappear at our expense? The resources of this world are not inexhaustible and what surely affects some species today may well affect others, including man, tomorrow.

In a most stimulating introduction Professor Austin mentioned that this Symposium was complementary to two other meetings on the breeding of endangered species held during the last five years. He emphasized that successful breeding in captivity often requires the application of sophisticated procedures that have been developed for the artificial control of breeding of domestic and laboratory species. Professor Austin went on to pose the questions why breed non-domestic animals, why use artificial methods, what are these methods and what will be the consequences of using such methods? In answering these questions he stressed how very few orders of birds and mammals had responded to artificial breeding in captivity. One important consequence worthy of particular mention is the genetic change which may take place when using artificial methods. He emphasized that one of the processes of evolution resulted from breeding within small groups. The smaller the group the sooner would a subspecies develop; he suggested that in rescuing an endangered species it would inevitably be modified according to specific human whims.

There is an urgent need for basic studies on many species to establish their genetic make-up so that they can be maintained as such or at least we can measure any changes that occur. The freezing of embryos discussed by Dr Polge may offer some solutions in this regard. If such embryos could be stored indefinitely we should be able to retain what we already have at this time.

Drs Lake, I. C. A. Martin and White emphasized the importance of studying the morphology of the reproductive tract of the male before semen collection and this theme was developed by others in respect of studying the female reproductive organs of different species of birds and mammals before embarking on artificial insemination. Dr Murton emphasized the problems in captive breeding of selecting birds from particular species to breed at different periods from those in wild populations. Dr Gee's experiments in producing hybrid cranes were fascinating but the purpose of such a programme should be clear to the researcher. Breeding for hybridization should be very carefully controlled or we shall end up with a great variety of new species. This also applies to Dr Boyd's work on artificial insemination of falcons.

In both his papers Dr Gee emphasized the advantages of obtaining bird semen and storing it artificially. Co-operation from the bird was only possible for a short period of the time over which forced collection methods may be employed. He discussed the relationship between spermatozoal abnormalities and fertilizing capacity. One can only stress here the dangers of extrapolating the situation in *Bos taurus* to other species. For example, in contrast to the situation in the bull, both rabbit and man have a very high proportion of abnormal sperm and a very high fertility rate.

   Dr Watson gave a comprehensive account of the different techniques available for semen collection from mammals and stressed the importance of avoiding risk. I am reminded here of a black rhinoceros bull from which semen was successfully collected by manual stimulation (Young, 1967); but when it got excited it chased its handlers out of the collecting pen!

   Like speakers before him Dr Graham had some difficulty in deciding how to classify non-domestic species, eventually opting for those other than stock farm animals. Indeed, this has been a problem throughout the Symposium with perhaps too much emphasis on domestic animals. On the other hand, in applying techniques to non-domestic animals biologists have little option other than to draw on experience gained with tame and more plentiful species. Success in freezing spermatozoa has been achieved for a number of species although the fertility of such spermatozoa has yet to be established in most cases. Moreover, very little information is available on the semen of endangered species.

   Dr Seager outlined the potential advantages of using frozen semen: breeding from incompatible pairs (which is frequently also a problem in rodents); introducing new blood lines, particularly from immobilized wild individuals; and in some instances, obtaining semen where only a single opportunity exists. He mentioned problems connected with manual collection from large carnivores. It may be of interest that I know of a male cheetah imprinted on its owner, which regularly submits to manual collection. The semen volume for five consecutive ejaculates ranged from 1·5 to 4·0 ml with a density of $100–200 \times 10^6$ ml^{-1} and this has been used regularly for artificial insemination (S. J. van Heerden, pers. comm.). It took 15 min by this method to collect an ejaculate (E. Young, pers. comm.), the first of which consisted mainly of accessory gland secretions. On the other hand R. Coubrough (pers. comm.) using electroejaculation obtained much smaller ejaculates from a number of male cheetahs, varying between only 0·25 and 0·78 ml but having normal spermatozoal density of $60–125 \times 10^6$ ml^{-1}.

   Dr Hendrickx discussed details of procedures used in the collection and evaluation of semen from primates and pointed out how important it is to determine the time of ovulation in any artificial breeding programme. Dr D'Souza emphasized that information concerning the temporal relations between ovulation and oestrus was restricted almost exclusively to domestic and laboratory animals and man. Species differences were evident in studies attempting to relate changes in the reproductive tract (such as vaginal cytology) to hormone levels and visual cues. Studies on behaviour tended to provide information of an imprecise nature especially if the duration of the studies was restricted. The importance of precise information about the events in the female was emphasized by the detailed studies of Dr Dukelow on three species of *Macaca* showing that small but important differences exist even at this taxonomic level.

Dr Polge discussed the great advances which have been made in our understanding of reproductive processes in the female domestic mammal. Spectacular successes have recently been achieved in ovum transplantation, transportation and the freezing of embryos, all of which could be applied to considerable genetic advantage in wild mammals. However, ovum transfer has only been reported in five species of wild mammals if we include the rabbit. Adams (1975) demonstrated the advantage of transferring ova from non-breeding feral females to domestic females to enable reproduction in captivity. Adams has also done equally important research on the hare and mink. Dr Polge pointed out that rare breeds could theoretically reproduce much faster by the induction of superovulation by exogenous hormones followed by the transfer and distribution of the fertilized eggs to several other females of the same species, an idea originally propagated by Rowlands. Indeed, non-surgical procedures recently developed for ova retrieval and transplantation offer distinct possibilities in the non-domestic mammal.

Drs Aamdal, Boyd, Cain and Jaczewski discussed the breeding of different species of animals with respect to human requirements, such as falconry and hunting and the provision of pelts and meat; there seems little doubt that these requirements will become more important in the future. Measurement of parameters in these situations is becoming more precise and the relevant information obtained will be helpful to conservation. It was pleasing to note that Dr Cain is measuring genetic parameters, something which is probably well documented for the fox, but I hope that in selecting for antler size for trophy hunting Dr Jaczewski will also monitor genetic changes. The importance of behaviour in achieving success in artificial breeding was a feature of all these papers.

Dr R. C. Jones and Dr D. E. Martin are to be complimented on successes achieved with species such as the elephant and the chimpanzee. Their attention given to detail in planning and execution of these two separate projects is particularly noteworthy and serves to pin-point the need for accuracy and precision when dealing with the complex mechanisms governing reproduction.

Sedation, to facilitate the capture and restraint of wild animals for artificial insemination, has an important bearing on any project of this nature. Mr D. M. Jones' wide experience of this subject made his task appear very simple and showed quite clearly the wide differences that exist between and within species in their tolerance to drugs. This part of any project on artificial breeding is obviously crucial and expertise in the correct dose rates for many species according to their temperament only comes with experience.

Mr G. F. Smith discussed the history of veterinary regulations related to the artificial breeding of cattle in the United Kingdom as an example of the legal requirements for disease control both within and between countries. Restrictions were modest at the beginning in 1943 but with the advance of research and knowledge of the transmission of disease they have necessarily become more stringent. He predicted that a similar pattern of controls will be applied to feral mammals and that eventually statutory restrictions will be imposed.

It would seem that we have some of the answers to the questions posed by Professor Austin. Much more research is however required. In particular, we need more information on the assessment of spermatozoal quality, the site of deposition of spermatozoa during artificial insemination, the total number and density of spermatozoa for satisfactory fertilization, the precise time of ovulation in different species related to the oestrous cycle and a host of other aspects. The daunting prospect is that every species may have its own optimum for each of the above characteristics.

It only remains for me to thank Dr Watson on behalf of all the participants for inviting us here for the past two days. To Dr Vevers and his willing helpers I offer our sincere thanks for the excellent arrangements and hospitality offered at this meeting so appropriately held at the Zoological Society of London.

REFERENCES

Adams, C. E. (1975). Stimulation of reproduction in captivity of the wild rabbit, *Oryctolagus cuniculus. J. Reprod. Fert.* **43**: 97–102.

Jones, R. C. (1973). Collection, motility and storage of spermatozoa from the African elephant, *Loxodonta africana. Nature, Lond.* **293**: 38–39.

Rowlands, I. W. (1965). Artificial insemination of mammals in captivity. *Int. Zoo Yb.* **5**: 105–106.

Rowlands, I. W. (1974). Artificial insemination of mammals in captivity. *Int. Zoo Yb.* **14**: 230–233.

Watson, P. F. & D'Souza, F. (1975). Detection of oestrus in the African elephant (*Loxodonta africana*). *Theriogenology* **4**: 203–209.

Young, E. (1967). Semen extraction by manipulative technique in the black rhinoceros, *Diceros bicornis. Int. Zoo Yb.* **7**: 166–167.

Author Index

Numbers in italics refer to pages in the References at the end of each article

A

Aamdal, J., 107, *118*, 157, 158, *170, 171*, 211, *217*, 241, 242, 244, 245, *247, 248*

Aarts, M. H., 307, *313*

Abbott, M., 304, 309, *312, 313*

Abdel-Raouf, M., 105, *119*

Ackerman, D. R., 159, *170*, 220, *237*

Adams, A. W., 82, *88*

Adams, C. E., 220, *237*, 304, 306, 308, 309, 310, 311, *312, 313, 314, 316*, 342, *343*

Afshar, A., 333, *337*

Ahmad, M. S., 102, *119*, 157, 158, 159, *170*

Allen, T. B., 89, *95*

Allen, W. R., 307, 308, 309, 310, *313*

Almquist, J. O., 57, *70*, 105, 111, 117, *119, 121, 123*

Amann, R. P., 101, 115, 117, *119, 123*

Andersen, K., 155, 156, 157, 158, *170, 171*, 241, 242, *248*

Anderson, B., 197, *204*

Anderson, L. L., 3, *6*

Andrews, L. G., 111, *119*

Anon., 114, *119*

Archibald, G. S., 53, 54, 55, 57, 59, 61, 66, 68, *70*

Ardran, G. M., 109, *122*, 133, *150*

Ariga, S., 197, 198, *204*, 307, *314*

Arslan, M., 234, *237*

Ashmole, N. P., 16, *25*

Asmundson, V. S., 32, *48*, 82, *88*

Assenmacher, I., 16, *25, 26*

Astier, H., 16, *25*

Atkinson, L. E., 235, *237*

Aulerich, R. J., 114, 115, *119, 125*, 134, 137, 138, *149, 152*

Austin, C. R., 2, *6*, 99, 116, *121, 125*, 133, 137, *152*, 159, 160, *172*, 221, 222, 223, 224, 225, 231, 232, 233, *237, 238, 239*, 304, *313*

Austin, J. W., 111, 112, 113, *119*, 138, *150*

B

Bailey, D. W., 110, 114, *122*, 161, *171*, 261, 266, *269*

Baker, R. D., 42, *48*, 131, *150*, 310, *314*

Balin, H., 196, *206*

Ball, L., 108, 109, *119*, 128, 132, 133, 134, 135, 136, 137, 147, 149, *150*, 318, *328*

Bank, H., 311, *313*

Baranczuk, R., 179, *191*

Barkay, J., 257, *259*

Barker, D. L., 234, *238*

Barranco, S., 115, *123*

Bartholomew, G. A., 37, *47*

Bartlett, D. E., 331, 336, *337*

Barton, R., 3, *6*

Barwin, B. N., 257, *259*

Batelli, F., 127, *150*

Batra, S., 179, *191*

Baumgartener, L. E., 333, *337*

Bautina, E. P., 107, 113, *119, 124*

Baxter, W. L., 81, *88*

Bearden, H. J., 225, *240*

Beaford, B. M., 117, *119*

Behrman, S. J., 220, 230, *237*

Bell, C., 100, *119*

Bendixen, H. O., 331, *337*

Bendon, B., 12, *26*

Benham, T. A., 128, *150*

Bennett, J. P., 99, 113, 117, *119*, 196, *204*, 220, 221, 227, *237, 238*, 257, *259*, 308, *315*

Benoit, J., 16, *25*

Bensenhaver, J. C., 220, 227, 228, 229, *240*, 257, *260*

Subject Index

A

ABP *see* Androgen binding protein
AV *see* Artificial vagina
Accipiter   gentilis   see   Goshawk,
    American
Acepromazine   *see   also   Acetyl-*
    promazine, 114, 115, 133, 294,
    319, 320, 323, 324
Acetylpromazine *see also* Aceproma-
    zine, 133, 134, 294
Acinonyx jubatus   see Cheetah
Age
    birds in
        semen quality effect on, 61, 65
    mammals in, 280
Ailurus fulgens   see Panda, lesser
Alces alces   see Moose
Alpaca
    semen collection, 101, 102
        artificial vagina, 105
        electroejaculation, 114
Alphaxalone/alphadolone, 112, 319,
    322, 324, 325, 326, 327
Anaesthesia *see also* Specific drugs
    birds in, 322, 326, 327
    mammals in, 111–115, 116, 132,
        133–134, 293–295, 317–322,
        323–325, 327–328
Anas erythrorhyncha   see Pintail, red-
    billed
Anas platyrhynchos   see Mallard
Androgen *see also* Testosterone, 11,
    16
Androgen binding protein, 11
Anseriformes *see   also   individual*
    species
    methods of semen collection, 37
Antilope cervicapra   see Blackbuck
Aquila chrysaetos   see Eagle, golden
Arctictis binturong   see Binturong
Armadillo, nine-banded
    electroejaculation of, 3, 113
Artificial breeding *see also* individual
    species
    birds of, 21, 31–33

Artificial breeding—(*cont.*)
    mammals of, 1–5, 207, 303, 305,
        339–343
Artificial   insemination   *see   also*
    Insemination dose and under
    specific species, 2, 219–220, 330
    birds in, 52–53, 69–70
        egg production effect on, 69
        fertility effect on, 68, 69, 94
        frequency of, 46, 57, 68, 91
        insemination site, 43, 46, 55, 56–
            57, 67, 76, 83, 91
        timing, 67–68, 91
        techniques of, 46, 54–57, 67, 76
    disease control, 329–336
    mammals in
        anaesthesia use of, 253, 321–322
        frequency of, 155
        insemination site, 155, 156, 157,
            161, 167, 226, 227–230, 242,
            245, 257, 258, 285
        timing, 170, 183, 190, 202, 255
    uses of, 207, 233, 330
Artificial vagina, 102–106
Ateles fusciceps   see Monkey, brown-
    headed spider
Ateles geoffroyi   see Monkey, Geoffroy's
    spider
Atherurus   africanus   see   Porcupine,
    West African brush-tailed
Atilax paludinosus   see Mongoose, West
    African water
Atropine, 116, 131, 133
Axis axis   see Deer, axis

B

Baboon
    electroejaculation, 112, 137, 138,
        208, 221, 224
    embryo transfer, 304, 305
    semen preservation, 159, 160, 165,
        215
Baboon, gelada
    electroejaculation, 113, 224

Felis pardalis *see* Ocelot
Felis temmincki *see* Cat, (Temminck's) golden
Felis wiedi *see* Margay
Fentanyl, 319, 320, 325
Fern patterns in cervical mucus, 178, 185–186, 190
Ferret
electroejaculation, 114, 134, 138
embryo transfer, 304, 310
Fertile period in birds *see* Fertilizing ability
Fertility *see also* Fertilizing ability
birds in,
relation to semen quality, 62, 63
mammals in,
frozen semen with, 154, 155, 156, 157, 158, 159, 160, 161, 162, 163, 164, 166, 202, 232, 245– 246, 247, 281, 284
Fertility rate
birds in
factors affecting, 67, 68, 69, 82
Fertilization
in vitro, 2, 3, 202
Fertilizing ability of spermatozoa *see also* Fertility
duration following insemination in birds, 43, 44–46, 63
factors affecting, 44–46, 67–68, 86–87
hybrid falcons in, 76
Finch (general)
semen collection, 55
Finch, house
semen collection, 59
Fluanisone, 319, 320, 325
Follicle
birds, growth of, 11–12
mammals, 177, 200, 201, 211, 258
Follicle stimulating hormone
birds in, 11
cyclic variation in levels, 177–178
injection of, 196, 197, 200
Forced-collection *see* Semen collection, birds in, massage method
Fowl, domestic, 14, 32
artificial insemination, 91, 94–95
dilution effect, 93
fertilizing ability of preserved semen, 44, 91, 95

Fowl, domestic—*(cont.)*
semen collection, 35, 38, 59, 61, 62, 90
semen preservation, 42–43, 89–93, 94
Fox (general)
semen collection, 241–242
Fox, blue
artificial insemination, 157, 158, 242, 245, 246, 247
semen collection, 107, 242, 244–245
semen preservation, 157, 158, 242
Fox, crab-eating
electroejaculation, 210
Fox, red
semen collection
electroejaculation, 210
other methods, 107
semen preservation, 165, 213
Fox, silver
artificial insemination, 242, 245
semen collection
electroejaculation, 113
manual stimulation, 107, 242, 244
Freezing of semen *see* Freezing procedures, Semen preservation, and individual species
Freezing procedures
birds in, 43–44
equilibration, 91, 92–93
freezing rate, 91, 92–93, 94
mammals in, 154, 155, 156, 157, 158, 159, 160, 161, 162, 163, 164, 165, 166, 210, 242, 277
equilibration, 155, 231–232, 277
comparison of methods, 155, 161
freezing rate, 232, 312
Freezing rate *see* Freezing procedures
Frequency of semen collection *see* Semen collection

G

Galliformes, *see also* individual species
male reproductive tract, 34
semen collection
methods of, 35–36
restraint, 36
training, 36

Z